Complex Dynamics and Geometry

SMF/AMS TEXTS *and* MONOGRAPHS • *Volume* 10

Panoramas et Synthèses • *Numéro* 8 • 1999

Complex Dynamics and Geometry

Dominique Cerveau, Étienne Ghys,
Nessim Sibony, and Jean-Christophe Yoccoz
(with the collaboration of Marguerite Flexor)

Translated by
Leslie Kay

American Mathematical Society
Société Mathématique de France

Dynamique et Géométrie Complexes
(Complex Dynamics and Geometry)

Dominique Cerveau, Étienne Ghys, Nessim Sibony, and Jean-Christophe Yoccoz
(with the collaboration of Marguerite Flexor)

Translated from the French by Leslie Kay

2000 *Mathematics Subject Classification.* Primary 30D05, 57R30.

For additional information and updates on this book, visit
www.ams.org/bookpages/smfams-10

Library of Congress Cataloging-in-Publication Data

Dynamique et géométrie complexes. English.
Complex dynamics and geometry / Dominique Cerveau ... [et al.] ; translated by Leslie Kay.
p. cm. — (SMF/AMS texts and monographs, ISSN 1525-2302 ; v. 10)
Includes bibliographical references.
ISBN 0-8218-3228-X (alk. paper)
1. Differentiable dynamical systems. 2. Geometry, Algebraic. I. Cerveau, D. (Dominique) II. Title. III. Series.

QA614.8.D9613 2003
514′.74—dc22 2003052359

∞ The paper used in this book is acid-free and falls within the guidelines
established to ensure permanence and durability.
Visit the AMS home page at http://www.ams.org/

10 9 8 7 6 5 4 3 2 1 08 07 06 05 04 03

Contents

Abstract

In the last twenty years, the theory of holomorphic dynamical systems has had a resurgent activity, in particular concerning the fine analysis of Julia sets associated to polynomials and rational maps in one complex variable. Simultaneously, closely related theories have had a similar rapid development, for instance, the qualitative theory of differential equations in the complex domain.

The meeting "État de la recherche" held in the ENS Lyon in January 1997 aimed at a presentation of the present state of the art in this area, emphasizing the unity linkng the various sub-domains. This volume contains four survey articles, corresponding to the lectures presented at the meeting. The paper by D. Cerveau describes the structure of polynomial differential equations in the complex plane focusing on the local analysis around singular points. The second paper, by É. Ghys, is a survey of the theory of laminations by Riemann surfaces which occcur in many dynamical or geometrical situations. N. Sibony describes the present state of the generalization of the Fatou-Julia theory for polynomials or rational maps in complex dimension at least 2. Finally, the lecture of J.-C. Yoccoz, written by M. Flexor, considers polynomials of degree 2 in one complex variable and in particular deals with hyperbolic properties of these polynomials centered around the Jakobson theorem. A general introduction gives the rudiments of the history of holomorphic dynamical systems, the aim being to show the reader that these questions are intimately linked and that their interactions are numerous and fruitful.

In the spirit of the meetings "États de la recherche de la Société Mathématique de France", these articles are not written for experts but rather for students or mathematicians working in different fields.

HOLOMORPHIC DYNAMICAL SYSTEMS

Étienne Ghys

This volume contains four articles dealing with holomorphic dynamics. In this introduction we describe some historical roots of the theory. But the point is not to give a detailed historical analysis of the evolution of holomorphic dynamical systems. Our only goal is to present a point of view and help the reader understand the unity of the subject.

The contribution of the nineteenth century: monodromy.

> *After the invention of the integral calculus, the problem that confronted mathematicians was the integration of* differential equations *(of one or several independent variables). The goal of the first efforts was to represent the integral by means of known elementary functions and symbols. Once mathematicians realized that such a representation was impossible in general, they resigned themselves to studying the properties of the integral directly, from the differential equation itself.*
>
> *The natural development of this research soon led geometers to consider imaginary as well as real values of the variable. The theory of the Taylor series, elliptic function theory, and the vast doctrine of Cauchy made the fruitfulness of this generalization obvious. It became clear that between two truths in the real domain, the shortest and easiest path often leads through the complex domain.*

Painlevé wrote this at the end of the nineteenth century [**8**]. He was right! Though it is impossible to give a precise analysis here of the contribution of complex variables in mathematics, we can nonetheless emphasize a fundamental discovery that is "invisible" from a real point of view: the *phenomenon of monodromy*, which lies at the heart of holomorphic dynamics.

The study of algebraic functions, that is, "multivalued" functions $y(x)$ of one variable x satisfying a polynomial relation $P(x, y) = 0$, led Riemann to his brilliant insight of "Riemann surfaces lying over the x-plane", on which the algebraic functions become single-valued. If one considers a point (x_0, y_0), draws a loop γ in the x-plane starting at x_0 and avoiding the "singular points", and follows the value of $y(x)$ "by continuity", then the value attained by y when x "returns" to its starting point x_0 is in general not y_0. Thus, for each path, one obtains a permutation of the set of solutions of $P(x_0, y_0)$; this is a "monodromy group" (finite in this case). A great deal of time and work went into giving a precise meaning to all these terms in quotation marks...

From the point of view of differential equations, the most innocuous examples can lead to multivalued solutions: the solutions of the linear differential equation

$xdy = \lambda ydx$ are $y = \text{const} \cdot x^\lambda$, and are not single-valued functions of the variable x when λ is not an integer. When x makes "one circuit" around the "singular point" 0, a solution is multiplied by $\exp(2i\pi\lambda)$. In this case, the "monodromy group" acting on the solution space is the cyclic group generated by $\exp(2i\pi\lambda)$, which is infinite in general. Of course, this example is too elementary. Consider a Riccati equation:

$$P(x)\frac{dy}{dx} = A(x) + B(x)y + C(x)y^2.$$

Here P, A, B, and C are polynomials in one complex variable x. Let us start with an initial condition (x_0, y_0) for which x_0 is not a singular point (that is, $P(x_0) \neq 0$). There exists a local solution of the equation, defined in a neighborhood of x_0. Choosing a path starting at x_0 and not passing through any singular point, one can show that it is possible to extend this solution along the path as a *meromorphic* solution. If γ is a loop starting at x_0, then the value y_1 of the solution y when γ returns to x_0 is in general not y_0. The transformation assigning this new value y_1 to the initial value y_0 is a linear fractional transformation $y_1 = (ay_0+b)/(cy_0+d)$ that depends only on the homotopy class of the loop γ (which avoids the singular points). This defines a homomorphism from the fundamental group of the complex plane minus the singular points to the group of linear fractional transformations acting on the Riemann sphere $\mathbb{C} \cup \{\infty\}$. But this fundamental group is a nonabelian free group (if P has at least three roots), so the monodromy group, the image of this homomorphism, can be extremely complicated; for instance, it may be dense in the group of linear fractional transformations. The "solutions" of the Riccati equation are thus "very multivalued", which intimidated many of our predecessors...

If we consider an equation of the preceding type but replace the polynomial $A(x) + B(x)y + C(x)y^2$ by $A(x) + B(x)y + C(x)y^2 + D(x)y^3$, a new phenomenon appears: when we continue a solution analytically along a path γ, in general we run into (branch-type) singularities at points that are not singular for the equation. For example, the equation $-2dy/dx = y^3$ has solutions $y = 1/\sqrt{x - \text{const}}$, which exhibit a "movable singularity" at the point $x = \text{const}$. In other words, the monodromy group cannot even be defined because the values of a solution can no longer be followed along a path that avoids the singular points of the equation. How can we analyze the "solutions" of such an equation if we cannot measure their failure to be single-valued by a group? When we consider equations in which dy/dx appears implicitly, or equations of higher order, other (even worse) phenomena appear, such as the existence of movable transcendental singular points: for example, the equation $y^2y'' + 2yy'^2 + 1 = 0$ has solutions $y = 1/(\text{const} + \ln(x - \text{const}'))$.

The reaction of nineteenth-century mathematicians to this "excess of monodromy" of the solutions of algebraic differential equations was to try to classify differential equations that have little monodromy, in the hope that such equations would be sufficiently rich to produce "new transcendentals that would enrich analysis". The choice was clear: to neglect the study of "generic" differential equations, which were too complicated, in order to concentrate on the rare examples where the monodromy is controlled. Here are some examples. Briot-Bouquet, and later Fuchs and Poincaré, looked among polynomial equations of the form $P(y', y, x)$ for those whose solutions are single-valued or, at worst, assume only finitely many values. They showed that aside from the algebraic functions, only the elliptic functions can satisfy this criterion. Schwarz described the hypergeometric equations whose monodromy group is *finite*, that is, whose solutions are algebraic functions. In the same spirit, Painlevé studied second-order equations with single-valued solutions: he discovered the famous "Painlevé transcendents". For a historical study of these questions, the reader will do well to consult the remarkable book by Gray [**3**].

Thus, at the turn of the century, the vast majority of differential equations were still unexplored.

Concurrently with this rich theory of differential equations in the complex domain, a more modest theory of iteration of holomorphic functions—especially of rational functions—emerged toward the end of the nineteenth century. A fascinating book by Alexander tells this story [**1**]. Here one learns, among other things, that the motivations of the forerunners (such as Schröder, Koenigs, and Leau) were far removed from differential equations; the analogy between iterations in discrete and continuous time (in a differential equation, for instance) was not clear to them. Instead, the object of study was the "theory of functional equations". One example is Abel's equation: given an analytic function $\phi(z)$, can one find another analytic function f such that $f(\phi(z)) = f(z) + 1$? This was an opportunity not only to develop ideas such as conjugacy between two dynamical systems, but also to sketch a qualitative description of iteration in a neighborhood of a fixed point. One common feature with differential equations should, however, be pointed out. The basic goal was not to study dynamics (of a differential equation or an analytic function) but rather to search for new and interesting transcendentals that could be defined by the process. As was true for differential equations at the beginning of the century, no general study was attempted: at best, a few nice examples were analyzed. In any case, this primitive theory of iteration of rational functions cannot be compared *in volume* with that of differential equations, which already had a long tradition: there were as yet few important results.

The beginning of the twentieth century: Poincaré, Painlevé, Fatou, and Julia. It's time to move on to the twentieth century! Let's cheat a bit and consider Poincaré's work on Fuchsian and Kleinian groups as dating from the twentieth century (they're so revolutionary!). Here are at least three new fundamental ideas.

Poincaré plunged into a systematic study of (discrete but infinite) groups of linear fractional transformations of the Riemann sphere: Kleinian groups. He recognized their dynamical richness and the complicated limit sets they generate, and showed their fundamental advantage: Fuchsian and Kleinian functions can be used to "resolve" and "uniformize" the solutions[1] of linear algebraic differential equations (or, which is almost the same thing, Riccati equations). The important thing is that here one is studying general equations. Fuchsian groups proved to be remarkably powerful: the Koebe-Poincaré uniformization theorem shows, for example, that *every* connected Riemann surface is isomorphic to either the Riemann sphere, a quotient of $\mathbb{C}$ by a discrete group of translations, or a quotient of the unit disk by a Fuchsian group (the "generic" case).

For nonlinear differential equations, Poincaré understood the importance of local analysis in a neighborhood of a singular point. His theorem on linearization of the equation $(ax+by+\cdots)dx = (cx+dy+\cdots)dy$ in a neighborhood of the origin,

[1]That is, make them explicit and single-valued; the French words for multivalued and single-valued are *multiforme* and *uniforme*. [Translator]

under a *generic* condition on the linear part, is a good example of the importance he assigned to the most general equations.

In the field of real dynamics—in his work on celestial mechanics, for instance—he revealed the phenomena that are now called *chaotic*. For a generic system, there is no longer any point in trying to "integrate" the equation, which would usually be impossible; instead, one must try to describe the behavior qualitatively.

Still in the field of real dynamics, he defined the concept of *first return map* on a Poincaré section. This trick can (sometimes) be used to replace the study of a differential equation by that of the iteration of a transformation. But it does not seem to have given him an inkling of any connection between the dynamics of complex differential equations and the theory of iteration of rational maps, which was still in its infancy (and seems not to have interested him). Of course, the first return maps introduced by Poincaré are injective by definition, while a rational map is not!

Painlevé, whom we quoted above, began research into the nature of the singularities of solutions of general differential equations. He had the intuition of the idea of *foliation structures* and recognized the advantages of considering the solutions as nonparametrized Riemann surfaces, which are nonsingular in general but whose projections on the coordinate axes exhibit singularities. The door was opened for a global study of their qualitative behavior, independent of the coordinates used.

As far as the iteration of polynomials and rational maps is concerned, between 1917 and 1923 Fatou and Julia developed a remarkably interesting theory. They analyzed the dynamics both locally, in a neighborhood of a periodic point, for example, and globally, by bringing out the dichotomy between the ("chaotic") *Julia set* and its complement. They also recognized the necessity of understanding the dynamics of a generic rational map and discovered the unbelievable wealth of possibilities. We can imagine this dynamics becoming the primary object of study and pushing functional equations into the background in the minds of Fatou and Julia.

After Fatou and Julia, the theory of iteration of rational maps seems to have dozed off... There is, however, one notable exception: in 1942, Siegel proved his famous theorem on the linearization of a germ of a holomorphic transformation in the neighborhood of a fixed point, when the derivative at the origin has modulus 1 and satisfies a diophantine equation. The existence of *Siegel disks* in the dynamics of polynomials had been discussed by Fatou and Julia; the positive answer supplied by Siegel added to the richness of the theory. The difficulty presented by *small divisors* had been revealed by Poincaré in his work on celestial mechanics.

Similarly, Poincaré's work on Kleinian groups and the qualitative study of complex differential equations had few repercussions among his immediate successors. The same is true to a lesser degree of his work in real dynamics, with significant exceptions such as Birkhoff and Denjoy.

Thus the field of holomorphic dynamical systems, which was not yet unified, entered a rather long period without significant activity. As we know, mathematicians were not idle for all that. This was the period of development of topology, algebraic geometry, the theory of functions of one and several complex variables, and ergodic theory, which are quite useful for the holomorphic dynamical systems of the present...

The years 1960–1980. The situation would change in the sixties. Reeb, in the line of descent of Painlevé, began the systematic study of *foliations*. He was dealing with dynamical objects where the leaves, replacing the trajectories, are not assumed a priori to be parametrized by (real or complex) time. The basic example could be the integral curves of a differential form $P(x, y)dx + Q(x, y)dy = 0$, where one gives up thinking of the solutions as functions $y(x)$ and thinks of them instead as Riemann surfaces in some space (projective space, for instance). But the theory of foliations is not intended only for the complex case. Among the fundamental contributions of this theory, we must emphasize the notion of *holonomy*, a more highly developed version of monodromy. Given a path in the leaves, one tries to follow it in nearby leaves. At first, still because of the "fear of monodromy", mathematicians tried to understand the cases where there is little holonomy: during this period they studied foliations without holonomy, or almost without holonomy, or where all the leaves are compact, etc.

We owe to Haefliger the idea that studying a foliation reduces in large part to studying its *holonomy pseudogroup*. This is a collection of homeomorphisms between open sets in $\mathbb{R}^n$; compositions are considered *where they are defined*. This is a natural setting that is ideally suited to the problems of monodromy that we encountered above. The tools were finally available for a general study of holomorphic, or even simply differentiable, foliations. A whole school began to develop around this body of ideas. But we should point out that this extremely active and productive group of mathematicians had no (or few) connections with Poincaré's work on Kleinian groups. The history of *exceptional minimal sets* illustrates this fact. For over ten years this group tried to understand whether a (sufficiently differentiable) foliation of real codimension 1 on a compact manifold can be of only one of two types: either all the leaves are dense, or there exists a compact leaf. It would have sufficed to observe, as Raymond did in 1972, that Poincaré's papers on Fuchsian groups are overflowing with counterexamples!

The local study of singularities was not forgotten. The advance of algebraic geometry made possible a successful approach to the *desingularization* of singular points of holomorphic differential equations in dimension 2 (Seidenberg's theorem). This was also the beginning of a rich theory of *local* dynamics. After desingularization the singular point becomes a divisor, in a neighborhood of which one tries to understand the germ of the dynamics, in particular via the holonomy of the divisor—which may be quite complicated.

During this period and more or less independently, real dynamics made fantastic progress, especially through the impetus of Smale, who proposed the ambitious program of understanding the dynamics of a generic diffeomorphism (or, since the analogy is now familiar to everyone, of a generic vector field). This was the glorious period that saw the emergence of such important concepts as *structural stability* and the *hyperbolicity* of diffeomorphisms. A diffeomorphism is structurally stable if every sufficiently close diffeomorphism is topologically conjugate to it (i.e. has the same topological dynamics). To simplify, a diffeomorphism is hyperbolic if the tangent bundle of the ambient manifold splits, at least over the nonwandering part, as a direct sum of a subbundle on which the differential of the diffeomorphism is expanding and a subbundle on which it is contracting. Such a hyperbolicity property (almost) implies structural stability. A great deal of activity developed around this type of diffeomorphism: topological analysis, ergodic analysis, etc. The hope that a generic diffeomorphism would be hyperbolic and structurally stable was unfortunately frustrated... It is nonetheless true that these hyperbolic diffeomorphisms occur frequently, and that their study was a preliminary to the study of *partially hyperbolic* diffeomorphisms, which is now underway.

Among the examples of diffeomorphisms that are not hyperbolic but whose dynamics deserve thorough study, we must mention Hénon's example: a polynomial bijection of the real plane, of the form $(x, y) \in \mathbb{R}^2 \mapsto (y + x^2 + c, ax)$. Restricted to certain values of the real numbers a, c, these examples exhibit a *non-hyperbolic attractor*, that is, an invariant compact set Λ (which does not reduce to a periodic orbit) such that every trajectory starting at a point near Λ accumulates on Λ. Moreover, even if these dynamics are unstable, the existence of the attractor is persistent: it remains in "many" small perturbations, even if one leaves the family of Hénon maps.

As far as the iteration of rational maps is concerned, this period was very quiet... We must, however, point out the work of Brolin, which established a connection with potential theory that would be very fruitful after 1980.

The theory of Kleinian groups developed in the setting of complex analysis and its connections with Riemann surfaces. In particular, the deep study of their deformation spaces was undertaken. We must mention the remarkable *Ahlfors finiteness theorem*, which dates from this period: if a Kleinian group has a finite generating set, and if we consider the quotient of the domain of discontinuity in the Riemann sphere by the action of the group, we obtain a Riemann surface *of finite type*: one that is isomorphic to a compact Riemann surface with finitely many points removed.

Since 1980, unification? The principal architect of this unification is Sullivan. As early as the beginning of the 1980s, he proposed a "dictionary" relating several theories: Kleinian groups, the iteration of rational maps, algebraic differential equations, and, more generally, transversely holomorphic foliations. The common point is a pseudogroup of holomorphic transformations. For a rational map, a non-invertible transformation of the Riemann sphere, one considers the pseudogroup generated by all the branches of the "inverse". For Kleinian groups, this pseudogroup is a global group: there is no longer a critical point, hence no branching for the inverse; but in place of a single transformation, one has to consider the orbits of a group that is very "large" in general. For algebraic foliations, the issue is the Reeb-Haefliger holonomy pseudogroup; here the difficulties are twofold because not only is one dealing with compositions of several transformations, but the transformations in question are not globally defined. The dictionary contains, for instance, the analogy (clearly established at last) between the limit set of a Kleinian group and the Julia set of a rational map. These analogies would prove to be fruitful. The most famous example is Sullivan's theorem on the nonwandering components of the complement of the Julia set of a rational map, which is analogous to the Ahlfors finiteness theorem. The comparative study of rational maps/Kleinian groups is now natural for the experts. The contribution was immense; each theory had something to offer the other. The use of quasiconformal maps passed from Kleinian groups to rational maps, where it was remarkably successful with holomorphic surgery on rational maps. Conversely, the problems surrounding structural stability/hyperbolicity passed from rational maps to Kleinian groups. Mañé, Sad, and Sullivan proved that structural stability is generic among rational maps, but the problem of the genericity of hyperbolicity is still central to current research.

Unfortunately, Sullivan's dictionary seems to have had little influence on the theory of differential equations. This is regrettable, but we are convinced that the techniques of Kleinian groups/rational maps will soon become useful in the context of algebraic differential equations...

Actually, the dictionary goes further and proposes an analogy with *real* dynamics in dimension 1. The keyword is *conformal pseudogroup*: a local diffeomorphism of $\mathbb{R}$ is clearly conformal because its differential at each point is a similarity (in real dimension 1!). Koebe's distortion lemmas, which are so useful in the study of rational maps, find their real analogues here, and one obtains a series of remarkable results on the dynamics of differentiable maps from an interval to itself. For example, one obtains precise analogues of the Ahlfors finiteness theorem for real dynamics in dimension 1. Similar techniques had in fact already been developed in the 1970s by specialists in foliations of real codimension 1, such as Sacksteder, in the context of exceptional minimal sets referred to above, for instance. One may regret that these new points of view coming from Kleinian groups and rational maps have not yet led to a rethinking of this theory of exceptional minimal sets; the "theory of levels" is well developed, but many essential problems in the real dynamics of codimension-one foliations are still unexplored (perhaps simply because of a lack of communication between the specialists in these subjects...).

This comparative study of real/complex dynamics in dimension 1 is responsible to a large extent for the successes obtained in understanding "phenomena of renormalization and universality" that were first discovered experimentally and whose final justification is obtained by a clever mix of dynamics, Teichmüller theory, and holomorphic surgery...

Another recent development should be mentioned. It has long been known that the study of Hénon attractors is more complicated than but related to that of real quadratic polynomials. It was tempting to complexify Hénon's polynomial diffeomorphism and study the dynamics of the resulting polynomial automorphism of $\mathbb{C}^2$. In doing this, one tries as far as possible to mimic the results of Fatou and Julia, but a number of surprises appear—which will not surprise the reader familiar with the theory of several complex variables. In particular, one can use the beautiful theory of currents on complex manifolds, as well as pluripotential theory, to obtain analogues of the ergodic theory of polynomials. This recent theory of holomorphic dynamics in several complex variables seems quite promising and is likely to have repercussions in real dynamics.

Thus there are now, at the end of the twentieth century, grounds for optimism about the future of holomorphic dynamical systems. On the one hand, the field has been considerably broadened because our present ambition is to understand generic dynamical systems, not just a few very special examples. On the other hand, the subjects studied and the techniques used have been generalized, consolidating the "classical" study of differential equations, holomorphic foliations, and Kleinian groups with that of iterations of polynomials or rational maps in one or several complex—or even real!—variables.

We hope we have convinced the reader of the coherence of the four articles that comprise this volume.

A few useful references on the history of holomorphic dynamics are suggested below.

Bibliography

[1] D. S. Alexander, *A history of complex dynamics, from Schröder to Fatou and Julia*, Aspects of Mathematics, E24, Vieweg & Sohn, Braunschweig, 1994.

[2] C. Godbillon, *Feuilletages: Études géométriques*, Progress in Mathematics, vol. 98, Birkhäuser Verlag, Basel, 1991. With a preface by G. Reeb.

[3] J. Gray, *Linear differential equations and group theory from Riemann to Poincaré*, Birkhäuser Boston Inc., Boston, MA, 1986.

[4] E. Hille, *Ordinary differential equations in the complex domain*, Wiley-Interscience [John Wiley & Sons], New York, 1976.

[5] K. Iwasaki, H. Kimura, S. Shimomura, and M. Yoshida, *From Gauss to Painlevé. A modern theory of special functions*, Aspects of Mathematics, E16, Vieweg & Sohn, Braunschweig, 1991.

[6] A. Kolmogorov and A. Yushkevich, *Mathematics of the 19th century: Geometry, analytic function theory*, Birkhäuser Verlag, Basel, 1996. Translated from the Russian by R. Cooke.

[7] J. Milnor, *Dynamics in one complex variable: Introductory lectures*, Vieweg & Sohn, Braunschweig, 1999.

[8] P. Painlevé, *Œuvres, tome I*, Éditions du Centre National de la Recherche Scientifique, Paris, 1973; see in particular the "Leçons de Stockholm".

[9] H. Poincaré, *Œuvres, tomes I, II, III*, Gauthier-Villars, 1951, 1952, and 1956; see in particular "Analyse des travaux scientifiques; équations différentielles" in volume I.

[10] G. H. Reeb, *Feuilletages: résultats anciens et nouveaux (Painlevé, Hector et Martinet)*, vol. 55, Les Presses de l'Université de Montréal, Montreal, Que., 1974.

CODIMENSION-ONE HOLOMORPHIC FOLIATIONS, REDUCTION OF SINGULARITIES IN LOW DIMENSIONS, AND APPLICATIONS

Dominique Cerveau

Abstract: This article is an introduction to the resolution of singularities of codimension-one holomorphic foliations. After some discussion and examples of the desingularization of curves and hypersurfaces, we state the theorem on reduction of singularities for foliations in dimensions 2 and 3, with a precise description of the terminal singularities. We then apply this tool to such classical problems as the singular Frobenius theorem and the construction of invariant hypersurfaces.

Introduction

Let X be a smooth complex analytic variety of dimension n. As usual, we let Ω^p_X denote the sheaf of holomorphic p-forms on X, and $\mathcal{O}_X = \Omega^0_X$. A codimension-one holomorphic foliation $\mathcal{F}$ on X is defined by giving a covering $\mathcal{U} = (U_i)_{i\in I}$ by open sets U_i, and holomorphic 1-forms $\omega_i \in \Omega^1(U_i)$ satisfying the following conditions:

(i) $\omega_i \wedge d\omega_i = 0$ (integrability).

(ii) $\operatorname{Sing}\omega_i := \{m \in U_i : \omega_i(m) = 0\}$ has codimension ≥ 2.

(iii) On $U_i \cap U_j$, $\omega_i = g_{ij}\omega_j$, where $g_{ij} \in \mathcal{O}^*(U_i \cap U_j)$ = holomorphic units on $U_i \cap U_j$.

This definition can clearly be synthesized in a sheaf-theoretic way. $\operatorname{Sing}\mathcal{F}$, the singular locus of $\mathcal{F}$, is defined by $(\operatorname{Sing}\mathcal{F}) \cap U_i = \operatorname{Sing}\omega_i$; it is an analytic set of codimension ≥ 2. A nonsingular point is also called regular. The local structure of $\mathcal{F}$ near a regular point $m \in U_i$ is described by the classical Frobenius theorem, which ensures the existence of a local submersion x_i and a unit g_i such that $\omega_i = g_i dx_i$. We then have the notion of local leaves (the level sets of x_i) and, by maximal gluing, of global leaves. On some varieties, every codimension-one foliation $\mathcal{F}$ will have a nonempty singular locus; on others, all foliations will be nonsingular. As we will see in the examples that follow, not only the topology but also the complex structure come into the picture. It therefore seems useful to have a reasonable description of the singularities of foliations, and we will endeavor to show how statements about the reduction of singularities can be exploited for this. Since this discussion is directed particularly toward nonspecialists, we have recalled some classical statements, a number of which are old but still indispensable in practice. We barely skim the surface of some important themes. This is a pretext for mentioning the bibliography—in particular, everything that touches on global problems, whose introduction would require much heavier machinery.

1. Examples of Foliations (with and without Singularities)

1.1. On tori. Consider a complex torus $\mathbb{T}^n(\Lambda) = \mathbb{C}^n/\Lambda$, where Λ is a lattice in $\mathbb{C}^n$: $\Lambda = \oplus\mathbb{Z}e_i$, (e_i) an $\mathbb{R}$-basis of $\mathbb{C}^n$. If L is a linear form on $\mathbb{C}^n$, then the foliation given by the global form $\omega = dL$ passes to the quotient and defines a nonsingular foliation $\mathcal{F}(L,\Lambda)$ on $\mathbb{T}^n(\Lambda)$. Its leaves are the projections of the hyperplanes $L =$ constant. If the lattice Λ is sufficiently general, there are no nonconstant meromorphic functions on $\mathbb{T}^n(\Lambda)$. A foliation $\mathcal{F}$ on $\mathbb{T}^n(\Lambda)$ (with or without singularities) lifts to a foliation $\widetilde{\mathcal{F}}$ on $\mathbb{C}^n$, which can be defined by a 1-form $\omega = dx_1 + \sum_{i\geq 2} a_i dx_i$, where the a_i are meromorphic and Λ-periodic, hence constant. Hence all foliations on such tori are of type $\mathcal{F}(L,\Lambda)$, and in particular have no singularities. This argument breaks down on algebraic tori (products of elliptic curves, for instance). In [**22**], E. Ghys gives, in particular, the classification of foliations on tori.

1.2. On projective spaces. Through the projection $p : \mathbb{C}^{n+1}\setminus\{0\} \to \mathbb{CP}(n)$, a foliation $\mathcal{F}$ on the projective space $\mathbb{CP}(n)$ induces a foliation $p^*\mathcal{F}$ on $\mathbb{C}^{n+1}\setminus\{0\}$. H. Cartan proved in 1938, in [**14**], that $H^1(\mathbb{C}^{n+1}\setminus\{0\},\mathcal{O}^*) = 0$ for $n \geq 2$. This implies, in particular, that $p^*\mathcal{F}$ can be defined by a global 1-form $\underline{\omega} \in \Omega^1(\mathbb{C}^{n+1}\setminus\{0\})$ that extends, by the theorem of Hartogs (here, Cauchy's formula), to an element $\omega \in \Omega^1(\mathbb{C}^{n+1})$. Invoking the fact that the kernel of ω at the point m must contain the line $[Om]$, one shows easily that $\omega = \sum_{i\geq 0} a_i dx_i$ can be chosen to be projective, homogeneous, and integrable (i.e. the a_i are homogeneous polynomials of the same degree, and $\sum x_i a_i = 0$). This is the analogue for foliations of Chow's theorem, which states that an analytic subset of $\mathbb{CP}(n)$ is actually algebraic. Bezout's theorem or the de Rham–Saito division theorem (which we will mention again) shows that such a foliation has a nonempty singular locus of codimension ≥ 2. There is currently a great deal of activity relating to foliations on projective spaces, which are far from being understood—in particular on $\mathbb{CP}(2)$, which is ultimately the most difficult case. The reader will surely recall that the study of foliations on $\mathbb{CP}(2)$ can be identified with the study of differential equations $y' = R(x,y)$, where $R \in \mathbb{C}(x,y)$; indeed, if $R = A/B$, $A, B \in \mathbb{C}[x,y]$, we see that the solutions $x \mapsto y(x)$ of our differential equation produce (local) parametrizations of the leaves of the foliation defined in affine coordinates by $\omega = Adx - Bdy$.

Here are three classical problems about foliations of $\mathbb{CP}(n)$.

1.2.1. *The Poincaré problem (*[**38**]*, vol. III, pp. 35 and 59).* Let $\mathcal{F}$ be a foliation of $\mathbb{CP}(2)$, and let Γ be an algebraic curve. Γ is called $\mathcal{F}$-invariant if $\Gamma \setminus \operatorname{Sing}\mathcal{F}$ is a leaf of $\mathcal{F}$. If $\mathcal{F}$ has infinitely many invariant algebraic curves, then $\mathcal{F}$ has a rational first integral: there exist homogeneous polynomials P and Q such that the leaves of $\mathcal{F}$ are precisely the curves $\lambda P + \mu Q = 0$, where $\lambda, \mu \in \mathbb{C}$. This is a known result of Darboux; a proof can be found in [**25**]. If $\mathcal{F}$ is defined by the homogeneous form $\omega = \sum a_i dx_i$, the degree of $\mathcal{F}$ is by definition $\deg\mathcal{F} = \deg a_i - 1$. Geometrically, $\deg\mathcal{F}$ is the number of tangencies of $\mathcal{F}$ with a generic line L, i.e. the number of points of L where $\mathcal{F}$ is not transverse to L, in an obvious sense. In the case where $\mathcal{F}$ does not have a rational first integral, the Poincaré problem consists of estimating the degree of an invariant algebraic curve Γ as a function of the degree of $\mathcal{F}$. This problem has been solved in special cases including, for example, smooth or, more generally, nodal curves ([**13**] and [**15**]): $\deg \Gamma \leq \deg \mathcal{F} + 2$, and it is conjectured that this estimate is general. But the reader should not be led to think that any foliation on $\mathbb{CP}(2)$ has an invariant algebraic curve; for any $n \geq 2$, Jouanolou constructs in [**25**] a foliation of degree n on $\mathbb{CP}(2)$ that has no invariant algebraic curve (see §5.1). Note that the set $\mathcal{F}(2,n)$ of foliations of degree n on $\mathbb{CP}(2)$ is a Zariski open set in a projective space because each $\mathcal{F} \in \mathcal{F}(2,n)$ has an associated 1-form ω of degree $n+1$ (defined up to a multiplicative scalar), $\omega = \sum a_i dx_i$, where $\gcd(a_0, a_1, a_2) = 1$. Jouanolou's result implies [**13**] that the subset of $\mathcal{F}(2,n)$ consisting of those foliations that have no invariant algebraic curves contains a dense open set (in the ordinary topology, that of the coefficients of the polynomials a_i).

1.2.2. *The "minimal set problem"* [**7**]. This problem revolves around the following question: if $\mathcal{L}$ is a leaf of a foliation $\mathcal{F}$ on $\mathbb{CP}(2)$, does the closure $\overline{\mathcal{L}}$ of $\mathcal{L}$ contain a singular point of $\mathcal{F}$? For instance, if Γ is an invariant algebraic curve of $\mathcal{F}$, then Γ contains at least one singular point of $\mathcal{F}$ [**25**]. At this date the problem above is still open; it leads to another, which is also open: does there exist a smooth (real) hypersurface $\Sigma \subset \mathbb{CP}(2)$ that is Levi flat? (If $m \in \Sigma$, the tangent space $T_m\Sigma \subset T_m\mathbb{CP}(2)$ contains a unique complex line E_m; to say that Σ is Levi flat means that the plane field $m \mapsto \Sigma_m$ is Frobenius integrable.) The Levi-flat hypersurfaces of $\mathbb{CP}(2)$—if they exist—are, in the view of specialists, natural candidates for the closures of leaves.

1.2.3. *The problem of components.* We denote by $\Omega^1(n,d)$ the vector space of 1-forms $\omega = \sum_{i=0}^n a_i dx_i$ on $\mathbb{C}^{n+1}$ that are homogeneous of degree $d+1$ and satisfy the condition $\sum x_i a_i = 0$. The integrability condition $\omega \wedge d\omega = 0$ defines an algebraic subset

$$\Sigma(n,d) := \{\omega \in \Omega^1(n,d) : \omega \wedge d\omega = 0\}$$

of $\Omega^1(n,d)$; it is an intersection of quadrics. The space $\mathcal{F}(n,d)$ of foliations of degree d on $\mathbb{CP}(n)$ can be naturally identified with a subspace of the projectivization $\mathbb{P}\Sigma(n,d) \subset \mathbb{P}\Omega^1(n,d)$:

$$\mathcal{F}(n,d) \cong \mathbb{P}\{\omega \in \Sigma(n,d) : \omega = \sum a_i dx_i,\ \gcd(a_0,\ldots,a_n) = 1\}.$$

To obtain a reasonable description of the space $\mathcal{F}(n,d)$ (in order to talk about deformations, for instance), it would be reasonable to know the decomposition of $\Sigma(n,d)$ into irreducible factors; for $n=2$, $\Sigma(2,d) = \Omega^1(2,d)$, and the question is nontrivial for $n \geq 3$. This decomposition ($n \geq 3$) is known only for $d = 0,1,2$. For $d=0$ there is a single component: every foliation of degree 0 is linearly conjugate to the foliation given by $x_0 dx_1 - x_1 dx_0$ (open book or pencil of hyperplanes). In degree 1 there are two components, Σ_0 and Σ_1; a generic point of Σ_0 produces, up to conjugation, a foliation given in affine coordinates by $x_0 dx_1 - \lambda x_1 dx_0$, $\lambda \neq 1$. Up to linear conjugation, a generic point of Σ_1 is always given in affine coordinates by dQ, where Q is the standard quadratic form $Q = \sum_0^b x_i^2$. In degree 2 there are six components, and we refer the reader to [**16**].

One can exhibit some kind of list of components in every degree, but this list is incomplete; all the known components are unirational, i.e. parametrized by a dominant morphism $\mathbb{C}^N \to \Omega^1(n,d)$ [**16**].

1.3. On the affine space $\mathbb{C}^n$. Consider a rational *closed* 1-form α on the affine space $\mathbb{C}^n$. The set $\mathrm{Pol}(\alpha)$ of poles of α is a hypersurface with equation

$P_1^{n_1+1}\dots P_s^{n_s+1}=0$, where $P_i \in \mathbb{C}[x_1,\dots,x_n]$ and the n_i+1 are the multiplicities of the poles $P_i=0$. The (rather delicate) analogue of the theorem on decomposition of rational functions into simple factors ensures the existence of complex numbers λ_i (the residues of α_i) and a polynomial H such that

$$\alpha = \sum \lambda_i \frac{dP_i}{P_i} + d\frac{H}{P_1^{n_1}\cdots P_s^{n_s}} = d\left[\sum \lambda_i \log P_i + \frac{H}{P_1^{n_1}\cdots P_s^{n_s}}\right].$$

The form α clearly has an associated foliation $\mathcal{F}(\alpha)$ (chase the denominators of α), which in general has singularities (the crossings $P_i = P_j = 0$, in particular). The leaves of $\mathcal{F}(\alpha)$ are the connected components of the level sets of the multivalued function $\sum \lambda_i \log P_i + H/P_1^{n_1}\cdots P_s^{n_s}$. Even when the P_i are simple—of degree 1, for instance—the qualitative description of these leaves turns out to be relatively complicated. This kind of example, which can be seen in $\mathbb{CP}(n)$, will play an important role in what follows. Let us keep in mind that a theory of reduction of singularities for foliations will have to contain a theorem on reduction of singularities for meromorphic closed forms.

1.4. On Hopf manifolds. We will consider only the simplest kind of Hopf manifolds: the quotient $H(n,\lambda)$ of $\mathbb{C}^n \setminus \{0\}$ by a homothety $x \mapsto \lambda x$ of ratio λ, with $|\lambda| \neq 1$.

A codimension-one foliation $\mathcal{F}$ on $H(n,\lambda)$ can be lifted to a foliation $\widetilde{\mathcal{F}}$ on $\mathbb{C}^n \setminus \{0\}$ that is equivariant under the action of λ. As in 1.2, $\widetilde{\mathcal{F}}$ will be defined by a global 1-form $\underline{\omega} \in \Omega^1(\mathbb{C}^n \setminus \{0\})$ that extends to $\omega \in \Omega^1(\mathbb{C}^n)$. The equivariance of $\widetilde{\mathcal{F}}$ under λ allows us to choose ω to be homogeneous, with an isolated singularity at 0, if $\mathcal{F}$ has no singularities. The description we will give later of homogeneous forms allows us to classify the foliations on $H(n,\lambda)$. It is amusing to observe here that certain singularities of a foliation permit the construction of foliations without singularities: for $n=2$, the only singularity of the homogeneous form ω is the origin, which disappears under projection to the Hopf surface.

1.5. On surfaces. The nonsingular foliations of complex surfaces have recently been classified by Brunella in [**3**]. This classification uses the work of Kodaira in a deep way; tori and Hopf manifolds appear here, of course, but not the projective plane; the projective plane blown up at a point, however (cf. 3.1), carries a foliation without singularities: one considers the radial foliation at a point $0 \in \mathbb{CP}(2)$, whose leaves are the lines through 0, and blows up the point 0.

2. Germs of Foliations

This is the local version of the initial definition. We work at the origin in $\mathbb{C}^n$ and denote by $\mathcal{O}(\mathbb{C}^n,0)$ (resp. $\Omega^p(\mathbb{C}^n,0)$) the ring of germs of holomorphic functions (resp. the $\mathcal{O}(\mathbb{C}^n,0)$-module of germs of holomorphic p-forms) at the origin in $\mathbb{C}^n$. A germ of a foliation $\mathcal{F}$ at 0 is defined by giving $\omega \in \Omega^1(\mathbb{C}^n,0)$, $\omega = \sum a_i dx_i$, $a_i \in \mathcal{O}(\mathbb{C}^n,0)$, satisfying

(i) $\omega \wedge d\omega = 0$,
(ii) $\operatorname{Sing}\omega = \{a_1 = \cdots = a_n = 0\}$ is of codimension ≥ 2.

Two such forms ω and ω' define the same $\mathcal{F}$ if and only if $\omega = g\omega'$ for some unit $g \in \mathcal{O}^*(\mathbb{C}^n,0)$. This is equivalent to the condition $\omega \wedge \omega' = 0$ by the de Rham–Saito lemma [**40**]:

LEMMA. *Let* $\alpha \in \Omega(\mathbb{C}^n, 0)$ *such that* $\operatorname{codim}\operatorname{Sing}\alpha \geq p+1$. *The following assertions are equivalent for* $\beta \in \Omega^q(\mathbb{C}^n, 0)$, *with* $q \leq p$:

(i) $\alpha \wedge \beta = 0$.
(ii) *There exists* $\gamma \in \Omega^{q-1}(\mathbb{C}^n, 0)$ *such that* $\beta = \alpha \wedge \gamma$.

The set $\operatorname{Sing}\mathcal{F}$ is $\operatorname{Sing}\omega$ by definition. The notion of leaves can also be localized.

2.1. The Frobenius theorem in the singular case. Proved by Malgrange in 1976 in [**28**], this theorem says that if $\operatorname{Sing}\mathcal{F}$ has codimension ≥ 3, then $\mathcal{F}$ has a (necessarily irreducible) holomorphic first integral f. This means that one can define $\mathcal{F}$ by df, and the leaves of $\mathcal{F}$ are the fibers $f^{-1}(c)$ of f. Note that Malgrange's theorem is not valid in dimension 2 ...except in the regular case. Later we will give a proof based on the reduction of singularities in dimension 2.

2.2. Homogeneous foliations of $\mathbb{C}^n$. These are the foliations that are invariant under homothety; such a foliation $\mathcal{F}$ is given by an integrable homogeneous 1-form

$$\omega_\nu = \sum a_i dx_i,$$

where the a_i are homogeneous polynomials of degree ν and $\gcd(a_i) = 1$. Let $R = \sum x_i \partial/\partial x_i$ be the radial field, and let $P_{\nu+1}$ be the polynomial of degree $\nu+1$ defined by

$$P_{\nu+1}(x) = \sum x_i a_i = i_R \omega_\nu.$$

We say that $\mathcal{F}$ is *dicritical* ($\mathcal{F}$ passes to $\mathbb{CP}(n-1)$) if $P_{\nu+1} \equiv 0$, and *non-dicritical* otherwise. Suppose $\mathcal{F}$ is non-dicritical. Then the rational 1-form $\omega/P_{\nu+1}$ is closed, and as in 1.3 we have

$$\frac{\omega_\nu}{P_{\nu+1}} = \sum \lambda_i \frac{dP_i}{P_i} + d\left(\frac{H}{P_1^{n_1} \cdots P_p^{n_p}}\right),$$

where the P_i and H are homogeneous, $\lambda_i \in \mathbb{C}$, $\sum \nu_i \lambda_i = 1$, $\nu_i = \deg P_i$, and $P_{\nu+1} = P_1^{n_1+1} \cdots P_p^{n_p+1}$. Note that if the ambient dimension is greater than 3 and the singularities of the homogeneous foliation $\mathcal{F}$ are of codimension ≥ 3, then $P_{\nu+1}$ is necessarily irreducible; indeed, the intersections $P_i = P_j = 0$, $i \neq j$, are singularities of $\mathcal{F}$. Hence

$$\frac{\omega_\nu}{P_{\nu+1}} = \lambda \frac{dP_{\nu+1}}{P_{\nu+1}},$$

and $P_{\nu+1}$ is a first integral of $\mathcal{F}$: this is the homogeneous version of Malgrange's singular Frobenius theorem. In particular, we obtain the description of the foliations of the Hopf manifolds $H(n, \lambda)$. For $n \geq 3$, they are the projections of homogeneous foliations with a first integral—which doesn't keep the leaves from being rather complicated!

One of the advantages of homogeneous foliations is the following. Consider a germ of a foliation $\mathcal{F}$ given by

$$\omega = \omega_\nu + \omega_{\nu+1} + \cdots,$$

where the ω_i are homogeneous of degree i. Clearly ω_ν is integrable. We introduce the homothety $\varepsilon : x \mapsto \varepsilon \cdot x$; we have $\omega_\varepsilon = \dfrac{\varepsilon^* \omega}{\varepsilon^{\nu+1}} = \omega_\nu + \varepsilon \cdot \omega_{\nu+1} + \cdots$, and the path $\varepsilon \mapsto \omega_\varepsilon$ connects the forms ω_ν and ω through integrable forms. It turns out that ω_ν determines $\mathcal{F}$ for sufficiently general ω_ν. We give two examples.

At the origin in $\mathbb{C}^3$, consider a dicritical ω_ν such that $\operatorname{codim}\operatorname{Sing}\omega_\nu \geq 2$ and $\operatorname{Sing} d\omega_\nu \subset \{0\}$. The last condition means that at each singular point of the foliation of $\mathbb{CP}(2)$ induced by ω_ν, the Baum-Bott residue [**1**] is nonzero (this is a Zariski-dense condition in the set of ω_ν). Let $\mathcal{F}_\omega$ be defined at the origin of $\mathbb{C}^3$ by $\omega = \omega_\nu + \cdots$. Then $\mathcal{F}_\omega$ and $\mathcal{F}_{\omega_\nu}$ are interchanged by a holomorphic diffeomorphism [**5**]. The proof is relatively simple. If $\nu = 1$, then ω_1 is of type $xdy - ydx$, and one considers the field Z such that $d\omega = i_Z \operatorname{vol}$; Z is nonzero at 0 and tangent to $\mathcal{F}$. Straightening Z along $\partial/\partial z$, we can define $\mathcal{F}$ by

$$\omega = a(x,y)dx + b(x,y)dy = xdy - ydx + \text{h.o.t.}$$

We then invoke Poincaré's linearization theorem ([**38**], thèse, vol. I), applied to

$$X = b\frac{\partial}{\partial x} - a\frac{\partial}{\partial y} = x\frac{\partial}{\partial x} + y\frac{\partial}{\partial y} + \cdots.$$

If $\nu > 1$, then by the de Rham–Saito theorem there exists a vector field X such that $\omega = i_X d\omega$; clearly $X(0) = 0$ and $\omega_\nu = i_{X_1} d\omega_\nu$, where X_1 is the 1-jet of X. Since $\omega_\nu = i_{R/(\nu+1)} d\omega_\nu$ (Euler's identity), we have $i_{R-(\nu+1)X_1} d\omega_\nu = 0$. If $\nu \geq 3$, we have $R - (\nu+1)X_1 = 0$, and to conclude the argument we apply Poincaré's linearization theorem to X. If $\nu = 2$, we find a field Y "by hand" such that Y is in the kernel of ω and has 1-jet R; we conclude in the same way. Note that the foliation associated with ω_ν is also given by the pair of vector fields R and $Z_{\nu-1}$, with $i_{Z_{\nu-1}}\operatorname{vol} = d\omega_\nu$; since the Lie bracket $[R, Z_{\nu-1}]$ equals $(\nu-2)Z_\nu$, there is a germ of a group action behind this ($\mathbb{C}^2$ if $\nu = 2$, an affine group if $\nu \neq 2$), and we can interpret the above as a result on finite determination for this group action.

Now consider $\mathcal{F}$ given at the origin of $\mathbb{C}^3$ (or more generally $\mathbb{C}^n$, with $n \geq 3$) by $\omega = \omega_\nu + \cdots$, and suppose that ω_ν is logarithmic and non-dicritical, i.e.

$$\frac{\omega_\nu}{P_{\nu+1}} = \sum \lambda_i \frac{dP_i}{P_i}, \quad \sum \lambda_i \nu_i = 1.$$

If the $P_i = 0$ have normal crossings (in $\mathbb{CP}(n)$) and some λ_i is not real (or is irrational, badly approximated by rationals), then there are $f_i = P_i + \cdots \in \mathcal{O}(\mathbb{C}^n, 0)$ such that $\dfrac{\omega}{f_1 \cdots f_s} = \sum \lambda_i \dfrac{df_i}{f_i}$; i.e., $\mathcal{F}$ is also given by a logarithmic closed form. The initial proof in [**17**] uses the Deligne-Fulton theorem, which ensures the commutativity of the fundamental group of the complement of a nodal curve in $\mathbb{CP}(2)$. At the end we will give a different version, at least in a generic case.

3. Reduction of Singularities

We begin by introducing the blow-up of a point.

3.1. The blow-up of a point in $\mathbb{C}^n$. We denote by $\widetilde{\mathbb{C}^n}$ the set

$$\widetilde{\mathbb{C}^n} := \{(z, D) \in \mathbb{C}^n \times \mathbb{CP}(n-1) : z \in D\}.$$

This is a smooth analytic variety of dimension n; its construction produces two projections, $E : \widetilde{\mathbb{C}^n} \to \mathbb{C}^n$ and $\pi : \widetilde{\mathbb{C}^n} \to \mathbb{CP}(n-1)$. The second projection turns $\widetilde{\mathbb{C}^n}$ into a line bundle over $\mathbb{CP}(n-1)$ (the tautological, or Hopf, bundle); the first is called the blow-up of the origin. Note that $E^{-1}(0)$ can be identified with the projective space $\mathbb{CP}(n-1)$; $E^{-1}(0)$ is called the exceptional divisor. Since an element (z, D) is uniquely determined by z when $z \neq 0$, the map E is an isomorphism between $\widetilde{\mathbb{C}^n} \setminus E^{-1}(0)$ and $\mathbb{C}^n \setminus \{0\}$. In practice, one uses the system of charts on $\widetilde{\mathbb{C}^n}$ induced by the standard charts on $\mathbb{CP}(n-1)$. More precisely, one considers $U_i := \{(y_1, \ldots, y_i, \ldots, y_n)\} \cong \mathbb{C}^n$ and $\delta_i : U_i \to \widetilde{\mathbb{C}^n}$ defined by

$$(y_1, \ldots, y_i, \ldots, y_n) \mapsto (y_i(y_1, \ldots, y_{i-1}, 1, y_{i+1}, \ldots, y_n), (y_1 : \ldots : y_{i-1} : 1 : y_{i+1} : y_n)).$$

In the chart U_i, the blow-up E is described by $E_i = E \circ \delta_i$, where

$$E_i(y_1, \ldots, y_i, \ldots, y_n) = (y_1 y_i, \ldots, y_i, \ldots, y_n y_i).$$

Thus blowing up the origin in dimension 2 consists of setting $y = tx$; the homogeneous differential equations $y' = f(y/x)$ studied in a first-year university course are treated by blowing up...

The preceding discussion allows us to blow up a point in any complex manifold.

3.2. Reduction of singularities of plane curves. A germ of an irreducible curve $\gamma \subset (\mathbb{C}^2, 0)$ is given by the vanishing of a germ of an irreducible holomorphic function $f \in \mathcal{O}(\mathbb{C}^2, 0)$:

$$\gamma := \{f(x, y) = 0\}.$$

The curve γ is nonsingular if and only if f is a submersion; γ is irreducible if and only if $\gamma \setminus \{0\}$ is connected. By the preparation theorem, we may suppose f is a Weierstrass polynomial:

$$f(x, y) = y^\nu + a_1(x) y^{\nu-1} + \cdots + a_0(x), \quad a_i \in \mathbb{C}\{x\},$$

where $\nu = \nu(f) =$ algebraic multiplicity of f. The projection $(x, y) \mapsto x$ restricted to $\gamma \setminus \{0\}$ is then a ν-fold covering over a punctured disk D^*. It follows easily that there exists a holomorphic function of one variable $s \mapsto \phi_2(s)$ such that the image of $s \to (s^\nu, \phi_2(s)) = \phi(s)$ is exactly the curve γ. This is Puiseux's theorem, which identifies the irreducible curves and the parametrized curves. It is already a desingularization theorem because it puts a smooth curve (a disk) into one-to-one correspondence with γ. We will see that it produces an "ambient desingularization" by a finite sequence of blow-ups. We expand ϕ_2 in a power series:

$$\phi_2(s) = a_q s^q + \cdots, \quad \text{where } a_q \neq 0.$$

After possibly applying a diffeomorphism of the type $(x, y) \mapsto (x, y - \varepsilon(x))$, we may assume that the a_i have multiplicity greater than $\nu - i + 1$, and hence that $q > \nu$. Blowing up the origin, we see that in the first chart, where $E_1(y_1, y_2) = (y_1, y_1 y_2)$, the map $\phi : (\mathbb{C}, 0) \to (\mathbb{C}^2, 0)$ lifts to a map $\widetilde{\phi} : (\mathbb{C}, 0) \to (\widetilde{\mathbb{C}}^2, 0)$ such that $E \circ \widetilde{\phi} = \phi$. To be precise, $\widetilde{\phi}$ can be written as

$$\widetilde{\phi}(s) = (s^\nu, a_q s^{q-\nu} + a_{q+1} s^{q-\nu+1} + \cdots),$$

and the multiplicity of the second component has been reduced. Iterating this procedure, we will eventually make the multiplicity of the "second component" less than ν.

We then continue if necessary, switching the roles of the components until one of them has multiplicity one. Finally, we construct a smooth surface $M(\gamma)$, a morphism $\underline{E}$ obtained by composition of a finite sequence of blow-ups, and an immersion $\psi : (\mathbb{C}, 0) \to M(\gamma)$ such that $\underline{E} \circ \psi = \phi$. We have then desingularized the curve γ in an ambient way. The exceptional divisor $\underline{E}^{-1}(0)$ is a union of "projective lines" D_i with normal crossings, and $\underline{E}$ realizes an isomorphism of $M \setminus \underline{E}^{-1}(0)$ onto $\mathbb{C}^2 \setminus \{0\}$. The smooth curve $\widetilde{\gamma}$ parametrized by ψ passes through a unique point $m \in \underline{E}^{-1}(0)$. One is more demanding in general, and requires m to be a smooth point of $\underline{E}^{-1}(0)$ and $\widetilde{\gamma}$ to be transverse to $\underline{E}^{-1}(0)$ at m. This can easily be obtained by taking finitely many point blow-ups, if necessary. The curve $\widetilde{\gamma}$ is called the strict transform of γ by $\underline{E}$, and $\underline{E}^{-1}(\gamma)$ is called the total transform of γ; the latter is a curve with normal crossings.

Consider the classical example of the cuspidal cubic $f(x,y) = y^2 - x^3 = 0$, which has Puiseux parametrization $s \to (s^2, s^3)$. The first blow-up $E_1(x,t) = (x, tx)$ gives $f \circ E_1(x,t) = x^2(t^2 - x)$, and the strict transform of the cuspidal cubic is a parabola $x = t^2$. It is smooth, but tangent to the divisor $x = 0$.

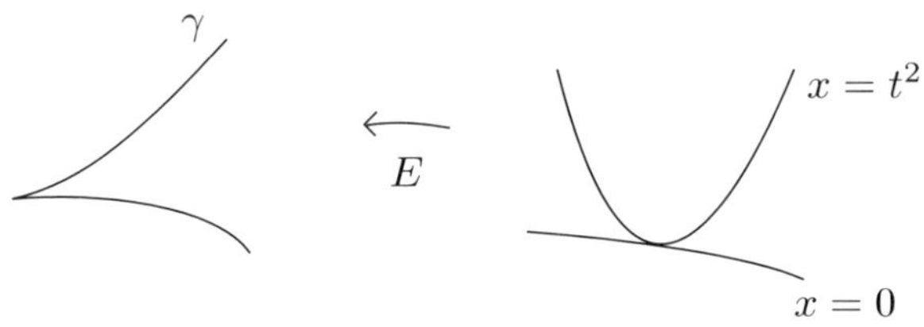

FIGURE 1

We must take two more blow-ups to obtain transversality at a smooth point of the exceptional divisor.

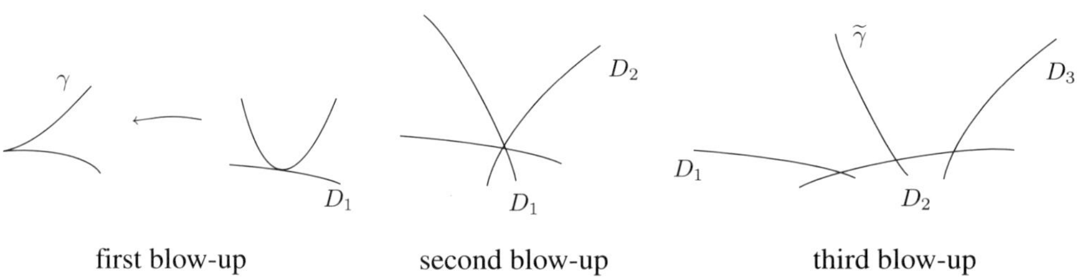

FIGURE 2

3.3. Arbitrary curves. It is easy to obtain the reduction of an arbitrary curve $\gamma = \gamma_1 \cup \cdots \cup \gamma_p$, where each γ_i is irreducible: indeed, it suffices by 3.2 to know how to desingularize a union of smooth curves $\prod(y - \varepsilon_i(x)) = 0$. This presents no difficulty: we encountered this problem in the example of the cuspidal cubic, where, after a blow-up, one is faced with the curve $x(x - t^2) = 0$. The final statement is the following: *There exist a surface $M(\gamma)$ and $\underline{E} : M(\gamma) \to (\mathbb{C}^2, 0)$, obtained by composing finitely many blow-ups, such that the total transform $\underline{E}^{-1}(\gamma)$ is a curve with normal crossings.* We will call $\underline{E} : M(\gamma) \to (\mathbb{C}^2, 0)$ a *reduction of singularities* of γ. There is a minimal reduction: the one that "uses the fewest blow-ups". Two minimal reductions are isomorphic. Note that the curves for which $M(\gamma) = \mathbb{C}^2$ (i.e. those for which a blow-up serves no purpose) are smooth curves and normal crossings.

3.4. Functions of two variables. Let $f \in \mathcal{O}(\mathbb{C}^2, 0)$, let $f = f_1^{n_1} \cdots f_p^{n_p}$ be the decomposition of f into irreducible factors, and let $\gamma = \cup \gamma_i$, where γ_i are the curves $f_i = 0$. Let $\underline{E} : M(\gamma) \to (\mathbb{C}^2, 0)$ be the minimal reduction of γ, as in 3.3. If $m \in \underline{E}^{-1}(0)$, then there exist a coordinate system (u, v) at m and integers p and q such that $f \circ \underline{E}(u, v) = u^p v^q$. Since the monomials $u^p v^q$ cannot be simplified by blowing up (try it), the reduction of γ leads to that of the function and thus even to that of the foliation associated with f, the foliation given by $f_1 \cdots f_p \sum n_i df_i/f_i$. Thus we have to expect the 1-forms $uv(p\frac{du}{u} + q\frac{dv}{v})$ to be part of the terminal models for the reduction of singularities of foliations in dimension 2.

3.5. Arbitrary dimension: hypersurfaces and functions.

Let $f \in \mathcal{O}(\mathbb{C}^n, 0)$ and $X = f^{-1}(0) \subset (\mathbb{C}^n, 0)$. We denote the singular locus of X by $\operatorname{Sing} X \subset X$. The theorem on resolution of singularities, proved in complete generality by H. Hironaka, can be stated in this context as follows:

There exist a smooth analytic variety M and a proper morphism $\underline{E} : M \to (\mathbb{C}^n, 0)$ that realizes an isomorphism from $M \setminus \underline{E}^{-1}(\operatorname{Sing} X)$ into $\mathbb{C}^n \setminus \operatorname{Sing} X$ such that the total transform $\underline{E}^{-1}(X)$ is a hypersurface with normal crossings. If $m \in \underline{E}^{-1}(\operatorname{Sing} X)$, then there exist a local coordinate system $(u_1, \ldots, u_n)$ at m and integers $p_1, \ldots, p_n$ such that $f \circ \underline{E}(u_1, \ldots, u_n) = u_1^{p_1} \cdots u_n^{p_n}$.

This statement gives the resolution of singularities of the foliation by the level sets of f in this setting as well. There is a more costly result—which, at least as far as the theory of foliations is concerned, leads to pointless and delicate combinatorial complications—stating that if $X = \cup X_i$ is the decomposition of X into irreducible components, then one may assume the strict transforms $\overline{\underline{E}^{-1}(X_i) \setminus \underline{E}^{-1}(\operatorname{Sing} X)}$ to be smooth and disjoint. We will not use it.

3.6. Surfaces. Although the ingredients involved are relatively simple for curves, we can already see that the degree of complexity in the case of surfaces is another story altogether. One reason is the lack of a Puiseux-type theorem. To desingularize curves (in an ambient way), all we did was blow up points in the ambient space. This is no longer sufficient for surfaces, whether they are irreducible or not. The (nonirreducible) example of a saddle surface and its tangent plane, $0 = z(z - xy) = f(x, y, z)$, is edifying. In the first chart, the blow-up $E_1(y_1, y_2, y_3) = (y_1, y_1y_2, y_1y_3)$ transforms f into $f \circ E_1(y_1, y_2, y_3) = y_1^2 f(y_1, y_2, y_3)$. Thus, at the point $(0, 0, 0)$, we recover a surface isomorphic to the initial surface together with the divisor $y_1 = 0$! So we haven't simplified anything at all! On the other hand, if we blow up the x-axis (that is, if we think of $\mathbb{C}^3$ as $\mathbb{C} \times \mathbb{C}^2$ and blow up 0 in $\mathbb{C}^2$), we obtain the desired reduction.

Strictly speaking, in the case of foliations, we cannot in general (in high dimensions) extract consistent information from a reduction theorem unless we have a relatively simple blow-up procedure, i.e. unless we can effectively control the topology of the various divisors. Here is an example.

3.7. An easy case: reduction of singularities of homogeneous surfaces. We start with the case of curves. Let $X \subset (\mathbb{C}^3, 0)$ be a homogeneous cone; i.e., $X = (P_\nu = 0)$, where P_ν is a homogeneous polynomial of degree ν. Consider the total transform of X under the blow-up $E : \widetilde{\mathbb{C}^3} \to \mathbb{C}^3$. It is given by $P_\nu \circ E = 0$; in the first chart, we obtain $P_\nu \circ E_1(y_1, y_2, y_3) = y_1^\nu (P_\nu(1, y_1, y_2))$. The strict transform of X is the inverse image under $\pi : \widetilde{\mathbb{C}^3} \to \mathbb{CP}(2)$ of the algebraic curve $C(X)$ associated with P_ν. It is smooth and transverse to the exceptional divisor at every point m such that $\pi(m)$ is a regular point of $C(X)$ (Figure 3).

Clearly the points that are not reduced are the inverse images $\pi^{-1}(m)$ of those points $m \in C(X)$ that are not reduced for the curve $C(X)$. It is easy to imagine

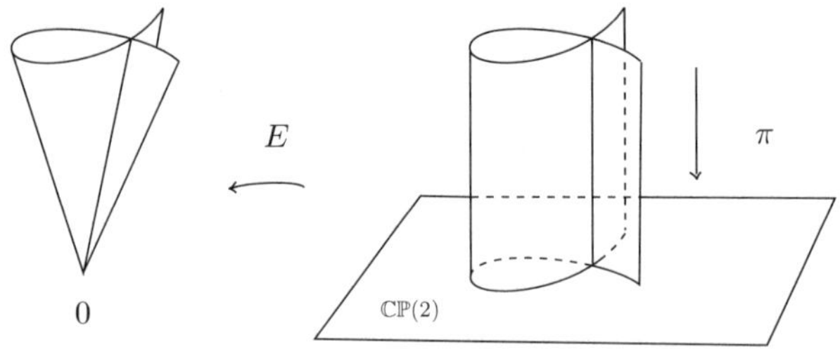

FIGURE 3

how to construct a reduction of X or P_ν. We just observe that the process of reducing a germ of a plane curve $\gamma \subset \mathbb{C}^2$ leads naturally to a reduction of the surface $\gamma \times \mathbb{C} \subset \mathbb{C}^3$; blow-ups of points of the type $E : \widetilde{\mathbb{C}^2} \to (\mathbb{C}^2, 0)$ are replaced by blow-ups of lines $E \times \mathrm{Id} : \widetilde{\mathbb{C}^2} \times \mathbb{C} \to \mathbb{C}^2 \times \mathbb{C}$. We note in passing that a homogeneous surface $X \subset \mathbb{C}^3$ as above, such that $C(X) \subset \mathbb{CP}(2)$ is smooth or has normal crossings, can be reduced by blowing up the origin.

4. Reduction of Singularities of Foliations in Dimension 2

In every result on reduction of singularities, one introduces terminal objects that will model the terminal singularities. For example, for hypersurfaces these are the normal crossings $x_1 \cdots x_p = 0$ (where $x_1^{s_1} \cdots x_p^{s_p} = 0$, counting multiplicity). We introduce the class of simple singularities first in dimension 2. This class is invariant under blow-ups; i.e., the blow-up of a simple singularity produces singularities that are all simple. The theorem on reduction of singularities will say that every holomorphic foliation on a surface can be transformed after finitely many blow-ups into a foliation that has only simple singularities, with, moreover, a particular transversality condition that we will make precise. One of the major difficulties in passing from dimension 2 to dimension 3 was finding a suitable definition of the notion of simple singularities in dimension 3; it was necessary to forget the standard definition in dimension 2 presented in 4.1, for which no reasonable generalization could be found. The important fact, explained in 4.7, is that a simple singularity has a "formal integrating factor" that turns out to be maximally desingularized, i.e. conjugate to a monomial $x^p y^q$.

4.1. Simple singularities. Let $\mathcal{F}$ be a germ of a foliation singular at the origin in $\mathbb{C}^2$, given by the 1-form $\omega = adx + bdy$. We say that $\mathcal{F}$ is *simple* if, in a well-chosen coordinate system (x, y), the linear part ω_1 of ω is of one of the types

(∗) $\omega_1 = xdy - \lambda ydx$, $\lambda \notin \mathbb{Q}_{\geq 0}$,
(∗∗) $\omega_1 = xdy$ (saddle node).

In particular, if $f \in \mathcal{O}(\mathbb{C}^2, 0)$, then the foliation associated with f is simple if and only if $f = 0$ has normal crossings; that is, if and only if f is conjugate to a monomial $x^p y^q$ (cf. 3.4).

4.2. Separatrices; the Briot-Bouquet theorem. Let $\mathcal{F}$ be a germ of a foliation, and X a hypersurface at the origin of $\mathbb{C}^n$; X is a *separatrix* (or integral hypersurface) of $\mathcal{F}$ if the smooth part $X_{\mathrm{smooth}} = X \setminus \mathrm{Sing}\, X$ of X is a union of leaves of $\mathcal{F}$. If $\mathcal{F}$ is defined by $\omega \in \Omega^1(\mathbb{C}^n, 0)$ and $X = \{f = 0\}$, where $f \in \mathcal{O}(\mathbb{C}^n, 0)$, then X is a separatrix of $\mathcal{F}$ if and only if f divides $\omega \wedge df$. There is also a notion of formal separatrix: $\widehat{X}$, defined by $\{\widehat{f}=0\}$, where $\widehat{f} \in \widehat{\mathcal{O}}(\mathbb{C}^n,0)$ = formal completion of $\mathcal{O}(\mathbb{C}^n,0)$, is a formal separatrix of $\mathcal{F}$ if $\widehat{f}$ divides $\omega \wedge d\widehat{f}$ as above.

The Briot-Bouquet theorem will allow us to construct separatrices for simple singularities by a blow-up technique. It says that the differential equation

$$xy' - \lambda y = ax + \phi(x,y),$$

where $\lambda \in \mathbb{C} \setminus \mathbb{N}$, $a \in \mathbb{C}$, and $\phi \in \mathcal{O}(\mathbb{C}^2,0)$ of multiplicity ≥ 2, has a unique holomorphic solution $x \to y_0(x)$ such that $y_0(0)=0$. This is a classical exercise (Ince [**24**] or Valiron [**42**]) in majorant series. In other words, the Briot-Bouquet theorem ensures that the foliation $\mathcal{F}$ defined by $xdy - (\lambda y + ax + \phi(x,y))dx$ has a unique smooth separatrix $\{y = y_0(x)\}$, which is tangent to the axis $y=0$. According to Briot and Bouquet, the existence of separatrices for simple singularities will result from an effective blow-up, which we present in the general setting.

4.3. Blow-up of $\mathcal{F}$. Let $\mathcal{F}$ be defined by

$$\omega = \sum a_i dz_i, \quad a_i \in \mathcal{O}(\mathbb{C}^n,0), \ \operatorname{codim}(\operatorname{Sing}\omega) \geq 2.$$

We denote by ν the algebraic multiplicity of ω: $\nu = \inf \nu(a_i)$, where $\nu(a_i)$ = multiplicity of a_i, and by $\mathrm{In}_\nu a_i$ the initial segments (jets of order ν) of the a_i. We introduce the homogeneous polynomial $P_{\nu+1} = \sum z_i \cdot \mathrm{In}_\nu a_i$; as in 2.2 we say that ω is dicritical if $P_{\nu+1} \equiv 0$, and non-dicritical otherwise. Thus ω, or $\mathcal{F}$, is dicritical if and only if the initial segment ω_ν of ω is dicritical in the sense of 2.2.

4.3.1. *The non-dicritical case.* Carrying out the computation in the chart U_1, where the blow-up is written $E_1(y_1,\dots,y_n) = (y_1, y_1y_2, \dots, y_1y_n)$, we have

$$\begin{aligned} E_1^*\omega &= \Big[a_1(y_1,y_1y_2,\dots,y_1y_n) + y_1\sum_{i\geq 2} a_i(y_1,y_1y_2,\dots,y_1y_n)\Big]dy_1 \\ &\quad + y_1\sum_{i\geq 2} a_i(y_1,y_1y_2,\dots,y_1y_n)dy_i \\ &= y_1^\nu\Big[P_{\nu+1}(1,y_2,\dots,y_n)dy_1 + y_1\sum_{i\geq 2}\mathrm{In}_\nu a_i(1,y_2,\dots,y_n)dy_i\Big] + y_1^{\nu+1}\omega_1' \\ &= y_1^\nu \widetilde{\omega}_1, \end{aligned}$$

where ω_1' and $\widetilde{\omega}_1$ are holomorphic 1-forms in a neighborhood of $y_1 = 0$. The 1-form $\widetilde{\omega}_1$ is called the *strict transform* of ω and defines $E^{-1}\mathcal{F}$, the *strict transform* of $\mathcal{F}$, in the chart U_1. The singularities of $E^{-1}\mathcal{F}$ contained in the exceptional divisor $E^{-1}(0) = \{y_1 = 0\}$ coincide with the projective hypersurface of the equation $(P_{\nu+1} = 0)$. This is the *tangent cone* of ω. Note that the exceptional divisor is a separatrix of $E^{-1}\mathcal{F}$.

The reader should check that the blow-up of a simple singularity in dimension 2 produces a foliation with two simple singularities.

4.3.2. *The non-dicritical case.* This is the same computation, but $E_1^*\omega$ is divisible by $y_1^{\nu+1}$ since $P_{\nu+1} = 0$. The exceptional divisor here is not a separatrix; the singularities of $E^{-1}(\mathcal{F})$ on $E^{-1}(0)$ are given by $\{\mathrm{In}_\nu(a_i) = 0,\ \mathrm{In}_{\nu+1}(\sum x_i a_i) = 0\}$. For example, the blow-up of the radial foliation $xdy - ydx$ produces a nonsingular foliation whose leaves are the fibers of the Hopf fibration: y/x = constant. We observe immediately that a dicritical foliation in dimension two has infinitely many separatrices. Indeed, if $m \in E^{-1}(0)$ is nonsingular for $E^{-1}(\mathcal{F})$, then the leaf through m is an analytic curve (distinct from $E^{-1}(0)$), and its projection under E gives a separatrix of $\mathcal{F}$.

4.4. Briot-Bouquet for separatrices. Let $\mathcal{F}$ be a simple singularity at the origin of $\mathbb{C}^2$, given by $\omega = \omega_1 + \cdots$, with ω of type $(*)$ or $(**)$. We use the traditional notation in dimension 2, $x = y_1$ and $t = y_2$. Then, up to a multiplicative *unit*,

$$\widetilde{\omega} = \frac{E_1^*(\omega)}{x} = ((1-\lambda)t + xA)dx + xdt,$$

where A is holomorphic in a neighborhood of $x = 0$. We associate with $\widetilde{\omega}$ the differential equation

$$\text{(B.B.)} \qquad xt' - (\lambda - 1)t = xA(x,t),$$

which is of Briot-Bouquet type because $\lambda \notin \mathbb{Q}_{>0}$. Let $t_0(x)$ be the holomorphic solution of (B.B.) such that $t_0(0) = 0$; then $x \mapsto E_1(x, t_0(x)) = (x, xt_0(x))$ parametrizes a smooth separatrix of $\mathcal{F}$ tangent to the x-axis. Furthermore, in both $(*)$ and $(**)$ it is the unique one with this property. In case $(*)$, another separatrix, tangent to the y-axis, can be constructed in the same way.

4.5. Divergence. In the saddle-node case, a direct formal calculation gives a smooth formal separatrix transverse to the convergent separatrix. It is divergent in general, as is shown by the famous example of Euler's equation

$$x^2dy - (y - x)dx,$$

which has the formal curve $y = \sum n!x^{n+1}$ as a separatrix.

4.6. Exercise. Simple singularities have only two separatrices (smooth or not, formal or not).

4.7. Integrating factors and formal normal forms [29]. Let $\mathcal{F}$ be defined by $\omega = adx + bdy$, with a simple singularity at the origin in $\mathbb{C}^2$. By Briot and Bouquet, $\mathcal{F}$ has two smooth, transverse separatrices—at least formal ones. Thus, up to formal diffeomorphism and a formal multiplicative unit, we will assume that ω is of type

$$xdy - y(\lambda + B)dx, \quad \lambda \notin \mathbb{Q}_{>0}, \quad B \in \mathfrak{m} \cdot \widehat{\mathcal{O}}(\mathbb{C}^2, 0),$$

where $\mathfrak{m}$ is the maximal ideal of $\mathcal{O}(\mathbb{C}^2, 0)$.

The vector field $X = x\dfrac{\partial}{\partial x} + (\lambda + B)y\dfrac{\partial}{\partial y}$ has linear part $x\dfrac{\partial}{\partial x} + \lambda y\dfrac{\partial}{\partial y}$, and generates the kernel of ω; this is the condition $\gcd(a,b) = 1$. Considering X as a derivation of the ring $\mathfrak{m} \cdot \widehat{\mathcal{O}}(\mathbb{C}^2, 0) = E$ and noting that $X(\mathfrak{m}^k) \subset \mathfrak{m}^k$, we recover, for every integer k, a factorization

$$\begin{array}{ccc} E & \xrightarrow{X} & E \\ \downarrow & & \downarrow \\ E_k & \xrightarrow{X_k} & E_k \end{array}$$

where $E_k = E/\mathfrak{m}^{k+1}$; clearly $X = \lim_k X_k$. Since $\dim E_k < \infty$, we can talk about the Jordan decomposition $X_k = X_{kS} + X_{kN}$ into commuting semisimple and nilpotent operators. Passing to the limit [29], we obtain

$$X = X_S + X_N,$$

where X_S and X_N are commuting formal vector fields (derivations of $\widehat{\mathcal{O}}(\mathbb{C}^2,0)$); by construction, X_S is semisimple, i.e. formally linearizable (more precisely, formally conjugate to the linear vector field $x\partial/\partial x + \lambda y\partial/\partial y$), and X_N is nilpotent [**29**]. Obviously we choose coordinates in which $X_S = x\partial/\partial x + \lambda y\partial/\partial y$, and try to exploit the commutation relation $[X_S, X_N] = 0$. When $\lambda \notin \mathbb{Q}$, the direct formal calculation shows that $X_N \equiv 0$; in this case, of course, the foliation $\mathcal{F}$ is (formally) conjugate to the linear foliation given by $\omega_1 = xdy - \lambda ydx = xy(\frac{dy}{y} - \lambda\frac{dx}{x})$. Here the monomial xy is an integrating factor for ω_1, i.e. the meromorphic form ω_1/xy is closed. Going back through our calculations, we observe that the form ω that we started with has a formal integrating factor $\widehat{f} \in \mathfrak{m}\cdot\widehat{\mathcal{O}}(\mathbb{C}^2,0)$, i.e. $\omega/\widehat{f}$ is closed. As we will see, this is a general fact. Consider the "resonant case", where $\lambda = -r/s$, $r, s \in \mathbb{N}$, $\gcd(r,s) = 1$. Note that the monomial x^ry^s is annihilated by X_S, and in fact generates the kernel of X_S: $\operatorname{Ker} X_S = \mathbb{C}[\![x^ry^s]\!]$. The commutation relation $[X_S, X_N] = 0$ immediately gives

$$X_N = x\alpha(x^ry^s)\frac{\partial}{\partial x} + y\beta(x^ry^s)\frac{\partial}{\partial y},$$

where α and β are formal series in the "resonant variable" x^ry^s, $\alpha(0) = \beta(0) = 0$. The foliation $\mathcal{F}$ is then (formally) defined by

$$xdy + y\frac{\left(\frac{r}{s} + \beta(x^ry^s)\right)}{1+\alpha(x^ry^s)}dx = xdy + y\left(\frac{r}{s} + h(x^ry^s)\right)dx,$$

which we write as follows when h is nonzero:

$$xyh(x^ry^s)\left[\frac{dy/y + (r/s)dx/x}{h(x^ry^s)} + \frac{dx}{x}\right].$$

Clearly we recover a linearizable case when $h \equiv 0$. Once again, noting that the expression in brackets defines a closed meromorphic form, we find an integrating factor. Using only the implicit function theorem, we can improve the reasoning above and unify all the formal models of simple singularities in the same expression. To be precise, if $\mathcal{F}$ is simple, then $\mathcal{F}$ is formally conjugate to a foliation given by

$$x^{p+1}y^{q+1}\left[\frac{dx}{x} - \lambda\frac{dy}{y} + d\frac{1}{x^py^q}\right],$$

where p and q are positive integers and λ is a complex number (which will be positive and irrational when $p = q = 0$).

4.8. Linearization. When $\mathcal{F}$ is formally linearizable, we actually have holomorphic linearization in the following instances of case $(*)$:

(i) $\lambda \in \mathbb{C}\setminus\mathbb{R}^-$: Poincaré's theorem applies to the field X in 4.7 ([**38**], thèse, vol. I);

(ii) λ is a negative irrational that is badly approximated by rationals (Siegel-Brjuno) [**2**];

(iii) $\lambda \in \mathbb{Q}_{<0}$ (this follows from a result of Malgrange-Mattei-Moussu but also from a linearization of the action of a compact group near a fixed point, which we will discuss in §8).

4.9. Classification of simple singularities. The holomorphic classification of simple resonant singularities and saddle-node singularities is due to Jean Ecalle [**21**] and Martinet-Ramis [**30, 31**]. Ecalle is more interested in the classification of vector fields, and Martinet-Ramis in foliations. Both approaches are difficult. When the invariant λ is a negative irrational that is well approximated by rationals, there are still plenty of mysteries; but the concept of holonomy, which will be introduced later and creates a correspondence between the study of simple singularities and that of germs of diffeomorphisms in one variable (cf. §8), allows us to derive important results via the work of Pérez Marco and Yoccoz [**37**].

4.10. Noether's formulas. Let $\alpha \in \Omega^1(M,m)$ be a germ of a 1-form at the point m of the smooth surface M. If $\mathcal{F}$ is a germ of a foliation defined by α, we write $\nu_m(\mathcal{F})$ for the algebraic multiplicity of $\mathcal{F}$ and $\mu_m(\mathcal{F})$ for the Milnor number of $\mathcal{F}$ at m. If (x,y) is a local coordinate system at m, where we write $\alpha = Adx + Bdy$, then

$$\nu_m(\mathcal{F}) := \inf(\nu(A), \nu(B)) \quad \text{and} \quad \mu_m(\mathcal{F}) = \dim_{\mathbb{C}} \frac{\mathbb{C}\{x,y\}}{(A,B)}.$$

The Milnor number can be interpreted as the number of singular points near m in a generic deformation of $\mathcal{F}$. Let $E : \widetilde{M} \to M$ be the blow-up with center m and $\widetilde{\mathcal{F}}$ the strict transform of $\mathcal{F}$. Setting $\nu = \nu_m(\mathcal{F})$ gives *Noether's formulas*:

$$\mu_m(\mathcal{F}) = \nu^2 - (\nu+1) + \sum_{p \in E^{-1}(m)} \mu_p(\widetilde{\mathcal{F}})$$

in the non-dicritical case and

$$\mu_m(\mathcal{F}) = (\nu+1)^2 - (\nu+2) + \sum_{p \in E^{-1}(m)} \mu_p(\widetilde{\mathcal{F}})$$

in the dicritical case.

For instance, if $\mathcal{F}$ is non-dicritical and can be desingularized by one blow-up, with singularities that are all of type $(*)$, then $\widetilde{\mathcal{F}}$ has $(\nu+1)$ singular points p_i, with $\mu_{p_i}(\widetilde{\mathcal{F}}) = 1$; we find that $\mu_m(\mathcal{F}) = \nu^2$, which can also be checked by using Bezout's theorem.

We see (whether we are in the dicritical cases or not) that if $\nu_m(\mathcal{F}) > 1$, then *all* the Milnor numbers $\mu_p(\widetilde{\mathcal{F}})$ are less than $\mu_m(\mathcal{F})$. Hence, after finitely many blow-ups, we reduce to singularities that each have algebraic multiplicity one.

4.11. Theorem on reduction of singularities. *Let M be a complex analytic manifold of dimension 2, $\mathcal{F}$ a foliation on M, and 0 a singular point of $\mathcal{F}$. Then there exists a morphism $\underline{E} : M' \to M$ consisting of a finite sequence of point blow-ups such that*

(1) *all the singularities of $\underline{E}^{-1}(\mathcal{F})$ are simple;*
(2) *if $m \in \underline{E}^{-1}(0)$, then the set $(\underline{E}^{-1}(0), m) \cup \Gamma_m$, where Γ_m denotes the union of the formal separatrices of $\underline{E}^{-1}$ at m, is a divisor with normal crossings.*

For example, (2) indicates that if m is nonsingular for $\underline{E}^{-1}(\mathcal{F})$, then the leaf $\mathcal{L}_m$ through m is either contained in or transverse to $(\underline{E}^{-1}(0), m)$.

Here is a diagram of the reduction of singularities of the foliation given by the level sets of $(y^2+x^3)/x^2$, showing all the possible configurations. The strategy is the same as in the resolution of the cuspidal cubic γ (3.2).

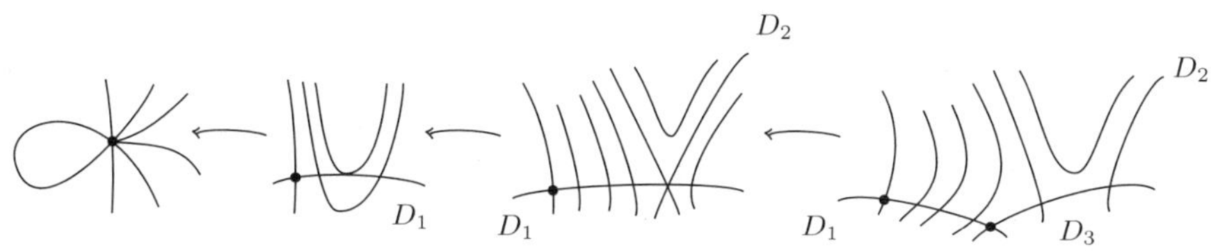

FIGURE 4

To pass from the discussion of 4.10 to the proof of 4.11, it suffices to treat the case of foliations $\mathcal{F}$ such that $\nu_m(\mathcal{F}) = 1$. When the linear part α_1 of the 1-form α defining $\mathcal{F}$ is diagonizable or of maximal rank, this is easy. The hard but edifying case is the one where α_1 is nilpotent, i.e. of type ydy. We refer the reader to [**32**].

4.12. The Camacho-Sad theorem [8]**.** This states that *every holomorphic foliation at the origin in* $\mathbb{C}^2$ *has a convergent separatrix.*

When the singularity is simple, this is the Briot-Bouquet theorem. To fix our ideas, we examine the foliations $\mathcal{F}$ that are reduced after one blow-up. It suffices to study the non-dicritical case because the appearance of a dicritical case in a reduction process guarantees the existence of infinitely many separatrices. Let ω define $\mathcal{F}$, $\nu = \nu_0(\mathcal{F})$, $\mathrm{In}_\nu\omega = a_\nu dx + b_\nu dy$, $P_{\nu+1} = xa_\nu + yb_\nu$. Choose coordinates x and y such that $P_{\nu+1}(0,1) \neq 0$. The strict transform $\widetilde{\omega}$ of ω is given (in the chart $y = tx$) by

$$\widetilde{\omega} = P_{\nu+1}(1,t)dx + xb_\nu(1,t)dt + x \cdot \omega',$$

and the singularities of $\widetilde{\omega}$ are $m_j = (0, t_j)$, where the t_j are the roots of $P_{\nu+1}$. At a point m_j, the linear part of $\widetilde{\omega}$, which is given by

$$[P'_{\nu+1}(1,t_j)(t - t_j) + x?]dx + xb_\nu(1,t_j)dt,$$

must be of type $(*)$ or $(**)$. Suppose $\mathcal{F}$ does not have a convergent separatrix. Then each singular point must be of type $(**)$ and, more precisely (Briot-Bouquet), we must have $b_\nu(1,t_j) = 0$ and $P'_{\nu+1}(1,t_j) \neq 0$. Thus $P_{\nu+1}(1,t)$ has $\nu + 1$ distinct roots. This implies that $b_\nu \equiv 0$ and $a_\nu \equiv 0$, which is stupid.

The preceding discussion conceals a deep global argument. For each irreducible component C of the divisor $\underline{E}^{-1}(0)$ associated with a resolution of a foliation $\mathcal{F}$, Camacho and Sad prove an index formula linking the Chern class of C (a negative integer that describes how $C \cong \mathbb{CP}(1)$ is embedded in the manifold M') with a sum of numerical invariants (essentially the λ) at the singular points of $\underline{E}^{-1}(\mathcal{F})$ on C, invariants whose nonvanishing marks, in particular, the existence of a convergent separatrix. A subtle combinatorial argument shows that there is a smooth point of $\underline{E}^{-1}(0)$ at which this invariant is nonzero.

4.12.1. *Application.* Consider a differential equation of type $y'' = f(y,y')$, where f is holomorphic or, more generally, meromorphic at $(0,0)$; this type of object is traditionally studied by introducing the variable $z = y'$. We obtain the system

$$y' = z$$
$$z' = f(y,z)$$

describing the flow of the vector field $z\dfrac{\partial}{\partial y} + f(y,z)\dfrac{\partial}{\partial z}$, with which we associate the foliation with equation $zdz - f(y,z)dy$. If $\gamma = \{P(y,z) = 0\}$ is a separatrix given by the Camacho-Sad theorem at the point under consideration, then the solutions $x \mapsto y(x)$ with initial condition $(y(x_0), z(x_0))$ on γ satisfy $P(y(x), y'(x)) = 0$. This type of problem was a major concern in the work at the beginning of the twentieth century on second-order equations (Painlevé, Gambier, Boutroux, et. al. ...): to determine whether certain solutions are also solutions of first-order equations. For example, Painlevé showed that no solution of the equation (I of Painlevé) $y'' = 6y^2 + x$ satisfies an equation $P(y, y') = 0$, where P is any polynomial in (y, z) (with coefficients that are analytic in the variable x). The study of equations $zdz - f(y,z)dy$ associated with second-order equations, where f is holomorphic, was outlined in [**19**] and recently completed by R. Méziani in his thesis (Kenitra 1998).

4.12.2. *Application to Hopf surfaces.* A holomorphic foliation $\mathcal{F}$ of a Hopf surface M always has a compact leaf. This can be done by hand, by listing the Hopf surfaces and their foliations, but can also be obtained by a direct application of Camacho-Sad to the foliation $\overline{\pi^{-1}\mathcal{F}}$ at $0 \in \mathbb{C}^2$, where $\pi : \mathbb{C}^2 \setminus \{0\} \to M$ is the canonical projection and $\overline{\pi^{-1}\mathcal{F}}$ is the extension of $\pi^{-1}\mathcal{F}$ to $\mathbb{C}^2$.

4.13. Dicritical foliations in the general sense (dimension 2). These are the foliations that have infinitely many separatrices, a condition that is equivalent, in the notation of 4.11, to having at least one branch of $\underline{E}^{-1}(0)$ transverse to $\underline{E}^{-1}(\mathcal{F})$. The typical example of a dicritical foliation in the general sense is given by the foliations associated with meromorphic functions f/g = constant, $\gcd(f, g) = 1$. Dicritical foliations have hardly been studied; the reader may consult M. Klughertz's nice work [**26**], which describes foliations whose "reduction trees" have the same properties as those of foliations given by the level sets of meromorphic functions. Although they do not always have such a first integral, they are topologically conjugate to foliations that do.

4.14. Generalized curves [6]. By definition, these are the foliations $\mathcal{F}$ at the origin in $\mathbb{C}^2$ that are non-dicritical in the sense of 4.13 and such that all their singularities, after reduction, are of type $(*)$. These foliations are a priori very reassuring; if $f = f_1 \cdots f_p$ is a reduced equation of the set of separatrices of $\mathcal{F}$ (all separatrices are convergent in this case), then the reduction of singularities of $\{f = 0\}$ is a reduction for $\mathcal{F}$. Moreover, the numerical invariants of f and $\mathcal{F}$ coincide: $\mu_0(\mathcal{F}) = \mu_0(f)$ and $\nu_0(df) = \nu_0(f) - 1 = \nu_0(\mathcal{F})$. But one shouldn't think that the study of generalized curves is limited to a description of their separatrices. We will return to this.

4.15. A typical approach. Henri Dulac proves in [**20**] that a foliation $\mathcal{F}$ at the origin in $\mathbb{C}^2$ of multiplicity $\nu(\mathcal{F})$ that has at least $\nu(\mathcal{F}) + 2$ separatrices has infinitely many, i.e. is dicritical in the sense of 4.13. Following [**17**], we prove this result by induction on the integer $N(\mathcal{F})$ = minimal number of blow-ups necessary to reduce $\mathcal{F}$. If $N(\mathcal{F}) = 0$, there is nothing to prove because a simple foliation is of order 1 and has at most two separatrices. Suppose the result has been proved for those $\mathcal{F}'$ such that $N(\mathcal{F}') < N$, and let $\mathcal{F}$ be such that $N(\mathcal{F}) = N$; blow up $\mathcal{F}$ by $E : \widetilde{\mathbb{C}^2} \to \mathbb{C}^2$. If $\mathcal{F}$ is dicritical, i.e. $E^{-1}(0)$ is generically transverse to $E^{-1}(\mathcal{F})$, we immediately obtain infinitely many separatrices. Otherwise, let t_i, $i = 1, \dots, p$, denote the singular points of $E^{-1}(\mathcal{F})$ on $E^{-1}(0)$, and let $n_i + 1$, $n_i \in \mathbb{N} \cup \{\infty\}$, denote the "number" of separatrices of $E^{-1}(\mathcal{F})$ at the point t_i (there is at least one: $(E^{-1}(0), t_i)$).

The formulas 4.3.1 give the estimate

$$\nu_{t_1}(E^{-1}(\mathcal{F})) + \cdots + \nu_{t_p}(E^{-1}(\mathcal{F})) \leq \nu(\mathcal{F}) + 1.$$

On the other hand,

$$\sum n_i \geq \nu(\mathcal{F}) + 2$$

by hypothesis. Hence there is a point t_i for which

$$\#\{\text{separatrices of}\,(E^{-1}(\mathcal{F}, t_i)\} \geq \nu_{t_i}(E^{-1}(\mathcal{F})) + 2.$$

But clearly $N(E^{-1}(\mathcal{F}, t_i)) < N$ at t_i, and we argue by induction.

5. Reduction of Singularities in Dimension 3

We would like to explain (without proof) the following statement [**9**]:

Let $\mathcal{F}$ be a germ of a foliation that is non-dicritical at the origin in $\mathbb{C}^3$. There exist a smooth manifold $M(\mathcal{F})$ and a proper morphism

$$\underline{E} : M(\mathcal{F}) \to (\mathbb{C}^3, 0)$$

that induces an isomorphism from $M(\mathcal{F}) \setminus \underline{E}^{-1}(\operatorname{Sing}\mathcal{F})$ onto $\mathbb{C}^3 \setminus \operatorname{Sing}\mathcal{F}$ such that the singularities of the strict transform $\underline{E}^{-1}\mathcal{F}$ are all simple. If $m \in \underline{E}^{-1}(\operatorname{Sing}\mathcal{F})$, then the set $(\underline{E}^{-1}(\operatorname{Sing}\mathcal{F}), m) \cup \Gamma_m$, where Γ_m denotes the set of (formal) separatrices of $\underline{E}^{-1}(\mathcal{F})$ at m, is a divisor with normal crossings.

In particular, we have to make precise what we mean by dicritical and by simple singularities.

5.1. Dicriticality in dimension 3. There are several equivalent definitions. It may be a good idea to return to dimension 2; a foliation $\mathcal{F}$ on $(\mathbb{C}^2, 0)$ is dicritical in the general sense if, for every sufficiently small polydisk B in which one represents $\mathcal{F}$, the set of leaves whose closure is an analytic curve passing through 0 has nonempty interior. We adapt this definition to arbitrary dimension. Let the foliation $\mathcal{F}$ be defined by ω. An integral curve of $\mathcal{F}$ is a holomorphic curve Γ such that Γ is not contained in $\operatorname{Sing}\mathcal{F}$; $\Gamma \cap \operatorname{Sing}\mathcal{F} \neq \emptyset$; and $\Gamma \subset \mathcal{F}$. If $\gamma : (\mathbb{C}, 0) \to \Gamma$ parametrizes Γ (a curve in $(\mathbb{C}^n, 0)$ can in fact be expanded in a Puiseux series), this means that $\gamma^*\omega = 0$ and $\gamma(0) \in \operatorname{Sing}\mathcal{F}$. We will say that $\mathcal{F}$ is dicritical if the set covered by the integral curves in a small polydisk in which our objects are represented has nonempty interior.

For example, consider the foliation $\mathcal{F}$ given by

$$\omega_\lambda = xyz\left(\lambda\frac{dx}{x} + \frac{dy}{y} - \frac{dz}{z}\right), \quad \text{where } \lambda \notin \mathbb{Q}.$$

The singular locus $\operatorname{Sing}\mathcal{F}$ is the union of the three coordinate axes. The lines $t \mapsto (x_0, t, at)$, $a \in \mathbb{C}$, $x_0 \in \mathbb{C}$, are integral curves and fill an open set. In contrast, the integral curves that pass only through 0 must lie in $xyz = 0$; thus, by themselves, they do not fill an open set.

Dicriticality can be properly interpreted in terms of blow-ups; in dimension 3, this says that $\mathcal{F}$ is dicritical if there exists a finite sequence

$$X_0 = (\mathbb{C}^3, 0) \xleftarrow{\pi_1} X_1 \longleftarrow \cdots \xleftarrow{\pi_N} X_N,$$

where each π_i is either a point blow-up or the blow-up of a smooth curve contained in the singular locus of $\mathcal{F}_{i-1} = (\pi_{i-1} \circ \cdots \circ \pi_1)^{-1}\mathcal{F}$ and such that the exceptional divisor of the last morphism is not a separatrix of $\mathcal{F}_N$.

In the example of ω_λ above, we blow up the x-axis; the exceptional divisor is isomorphic to $\mathbb{CP}(1) \times \mathbb{C}$ and the blown-up foliation is transverse to it.

Although the Camacho-Sad theorem guarantees the existence of separatrices in dimension 2, we know that they need not exist in dimension 3 or higher. It is customary to cite the following example, due to Jouanolou [**25**], which has no separatrix:

$$\omega = (y^m x - z^{m+1})dy + (z^m y - x^{m+1})dz + (x^m z - y^{m+1})dx.$$

This example is "projective", i.e. it is a cone over a foliation of $\mathbb{CP}(2)$, and the fact that it has no separatrix implies that "many" dicritical foliations in dimension 3 will not have separatrices.

5.2. Study of an example, or how to predict the terminal models in a reduction of singularities. Let $\mathcal{F}$ be a foliation at the origin in $\mathbb{C}^3$, and suppose we know that $\mathcal{F}$ has two smooth, transverse separatrices. One can come across this situation in a reduction process as (for instance) the result of two successive blow-ups—where one comes up with two components of divisors—or as a consequence of a Briot-Bouquet-type result. Choose coordinates x, y, z such that the separatrices are given by $x = 0$ and $y = 0$. Let $\mathcal{F}$ be given by $\omega \in \Omega^1(\mathbb{C}^3, 0)$, of type

$$\omega = ya\, dx + xb\, dy + xyc\, dz, \quad \text{where } a, b, c \in \mathcal{O}(\mathbb{C}^3, 0).$$

An important invariant (the *adapted order* of $\mathcal{F}$ to the divisor $xy = 0$) is

$$r = \inf(\nu(a), \nu(b), \nu(c)).$$

We assume it is zero and examine the different cases that can occur. If $a(0)$ or $b(0)$ is nonzero, we can find α and β in $\mathcal{O}(\mathbb{C}^3, 0)$ such that $xyc = \alpha ya + \beta xb$; the vector field

$$Z = \frac{\partial}{\partial z} - \alpha\frac{\partial}{\partial x} - \beta\frac{\partial}{\partial y}$$

is tangent to $\mathcal{F}$ (it annihilates ω) and trivializes it. We say that $\mathcal{F}$ is of 2-*dimensional type*; we view the reduction of singularities of such a foliation of 2-dimensional type as being modeled on that of the foliation of $\mathbb{C}^2$ given by $\mathcal{F}_0 = \mathcal{F}_{|z=0}$. The interesting case occurs when $a(0) = b(0) = 0$ and $c(0) \neq 0$; we may assume that $c = 1$. Consider the vector fields

$$X = x\frac{\partial}{\partial x} - a\frac{\partial}{\partial z},$$
$$Y = y\frac{\partial}{\partial y} - b\frac{\partial}{\partial z}.$$

Since they are tangent to $\mathcal{F}$, so is their Lie bracket. But $[X, Y]$, which is clearly collinear with $\partial/\partial z$, must annihilate ω; thus $[X, Y] = 0$. Hence, returning to the idea of 4.7 and considering X and Y as derivations of the ring $\widehat{\mathcal{O}}(\mathbb{C}^3)$, we can put the vector fields X and Y simultaneously into Jordan form. Roughly speaking, this means that we can write $X = X_S + X_N$, $Y = Y_S + Y_N$, where X_S and Y_S are formal vector fields that can be chosen to be linear in an ad hoc formal coordinate system, X_N and Y_N are nilpotent, and all these vector fields commute pairwise. Suppose, for instance, that the nilpotent fields X_N, Y_N are zero (which happens when the 1-jets of X and Y have no common "resonances"); we can then find formal coordinates x_1, x_2, x_3 such that

$$X = X_S = \sum \lambda_i x_i \frac{\partial}{\partial x_i} \quad \text{and} \quad Y = Y_S = \sum \mu_i x_i \frac{\partial}{\partial x_i}.$$

Note that $\mathcal{F}$ is given in these coordinates by $x_1x_2x_3\sum_{i=1}^{3}\alpha_i dx_i/x_i$, a logarithmic 1-form.

Now consider the situation where the eigenvalues λ_i and μ_i of the X_S and Y_S satisfy a common, and essentially a single, resonance. What we mean by this is that the monoid $\{(m_1, m_2, m_3) \in \mathbb{N}^3 : \sum m_i\lambda_i = \sum m_i\mu_i = 0\}$ is generated by an element (p'_1, p'_2, p'_3). Now we can write

$$X_S = \sum \lambda_i x_i \frac{\partial}{\partial x_i}, \qquad Y_S = \sum \mu_i x_i \frac{\partial}{\partial x_i},$$

$$X_N = \sum A_i(x_1^{p'_1}x_2^{p'_2}x_3^{p'_3})x_i\frac{\partial}{\partial x_i}, \quad Y_N = \sum B_i(x_1^{p'_1}x_2^{p'_2}x_3^{p'_3})x_i\frac{\partial}{\partial x_i},$$

where the A_i and B_i are in $\widehat{\mathcal{O}}(\mathbb{C}, 0)$.

After manipulations as in 4.7, we can find a formal coordinate system in which $\mathcal{F}$ is defined by a 1-form of type

$$x_1^{p_1+1}x_2^{p_2+1}x_3^{p_3+1}\left(\sum \alpha_i \frac{dx_i}{x_i} + d\left(\frac{1}{x_1^{p_1}x_2^{p_2}x_3^{p_3}}\right)\right), \quad \alpha_i \in \mathbb{C},$$

where (p_1, p_2, p_3) is an integer multiple of (p'_1, p'_2, p'_3).

The complete treatment of this type of examples leads one to think that simple singularities in dimension 3 (in fact, arbitrary dimension) are natural generalizations of the normal forms given in 4.7. Here is the description.

5.3. Simple singularities in dimension 3 and higher. We say that $\mathcal{F}$ has a *simple singularity* at the origin in $\mathbb{C}^n$ if, up to formal diffeomorphism, $\mathcal{F}$ is defined by a 1-form of type

$$\prod_{i=1}^{r} x_i^{p_i+1}\left(\sum \lambda_i \frac{dx_i}{x_i} + d\left(\frac{1}{x_1^{p_1}\cdots x_r^{p_r}}\right)\right),$$

where $(x_1, \dots, x_n)$ is a coordinate system for $\mathbb{C}^n$ and $p_i \in \mathbb{N}$, $\lambda_i \in \mathbb{C}$, $i = 1, \dots, r$ satisfy the following relations:

(1) $|\lambda_i| + p_i \neq 0$ for $i = 1, \dots, r$;
(2) if $p_i = p_j = 0$ and $i \neq j$, then $\lambda_i/\lambda_j \notin \mathbb{Q}_{<0}$.

As in dimension 2, the problem of analytic classification of simple singularities arises naturally. It is essentially a question of understanding whether the conjugating diffeomorphism above can be taken to be convergent and, if it cannot, of understanding the nature of the divergence. This involves small divisors and large resurgences. Although, to my knowledge, the details are not written down anywhere, one knows how to go about this classification in almost all cases (cf. some examples later on).

5.4. Dimension 3; simple singularities of 2-dimensional type. This is the case where the integer r is 2. We recover (formal) models of simple singularities in dimension 2. The singular locus is smooth, and the foliation has a product structure in the sense that a nonsingular vector field is tangent to it (by Artin or by flatness, we may assume that this field is convergent). Thus we get the properties of simple singularities in dimension 2; in particular, there are one (case of a saddle node $\times\,\mathbb{C}$) or two smooth, transverse separatrices.

5.5. Dimension 3; simple singularities of 3-dimensional type. This is the case where $r = 3$; the singular locus of a formal model is the three coordinate axes, and the separatrices are the three coordinate planes. Thus, for a simple singularity $\mathcal{F}$ of this type ($r = 3$), $\operatorname{Sing}\mathcal{F}$ is holomorphically diffeomorphic to the three coordinate axes, and we will in fact assume that it is equal to these three axes. At a smooth point of $\operatorname{Sing}\mathcal{F}$, i.e. away from 0, we see that $\mathcal{F}$ is of 2-dimensional type. We need not get three holomorphic separatrices for $\mathcal{F}$; we will study the zoology of the different cases.

5.5.1. *Logarithmic simple singularities of formal type.*

$$x_1x_2x_3\left(\sum \lambda_i \frac{dx_i}{x_i}\right), \quad \frac{\lambda_i}{\lambda_j} \notin \mathbb{Q}_{<0}.$$

At each smooth point of $\operatorname{Sing}\mathcal{F}$ there are two transverse, smooth separatrices (Briot-Bouquet + product structure). These separatrices (defined away from the big singular point 0) merge in a reasonable way at 0, and at the origin $\mathcal{F}$ has three smooth, transverse separatrices, which are holomorphically diffeomorphic to the planes $xyz = 0$.

As in 5.2, the foliation is given by a local holomorphic action of $\mathbb{C}^2$ on $\mathbb{C}^3$ that is formally linearizable (often—depending on the values of λ—holomorphically).

5.5.2. *(Resonant) simple singularities of formal type.*

$$x_1^{p_1+1}x_2^{p_2+1}x_3^{p_3+1}\left(\sum \lambda_i \frac{dx_i}{x_i} + d\left(\frac{1}{x_1^{p_1}x_2^{p_2}x_3^{p_3}}\right)\right),$$

where $p_1 \cdot p_2 \cdot p_3 \neq 0$. Here, at each smooth point of the singular locus, the singularity is resonant and of 2-dimensional type, and again there are two separatrices; as in 5.5.1, they merge to give rise to three separatrices diffeomorphic to $xyz = 0$.

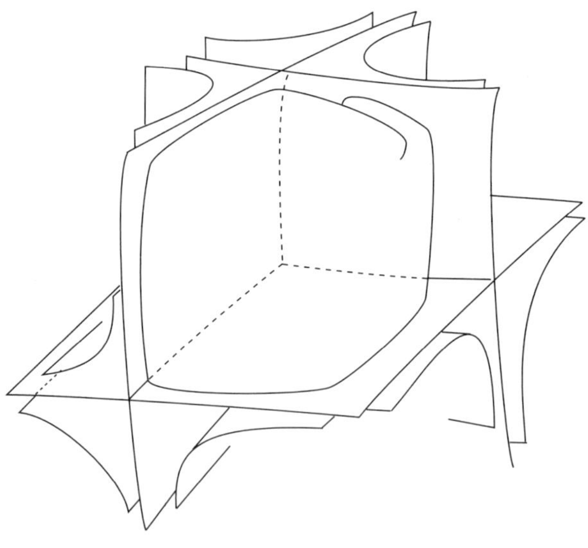

FIGURE 5

5.5.3. *(Resonant saddle-node) simple singularities of formal type.*

$$x_1^{p_1+1}x_2^{p_2+1}x_3\left(\sum \lambda_i \frac{dx_i}{x_i} + d\left(\frac{1}{x_1^{p_1}x_2^{p_2}}\right)\right), \quad \lambda_3 \cdot p_1 \cdot p_2 \neq 0.$$

Here the three branches of the singular locus do not play the same role. Along the x_3-axis, away from 0, we have a resonant singularity of 2-dimensional type; along the other two axes, we have a saddle-node singularity of 2-dimensional type. The two holomorphic local separatrices along the x_3-axis "pass through" the origin to give rise to two smooth separatrices diffeomorphic to $x_1 \cdot x_2 = 0$. In contrast, along the other two axes we can have a transverse divergent formal separatrix that merges with the formal separatrix given by the normal form.

5.5.4. *(Logarithmic saddle-node) simple singularities of formal type.*

$$x_1^{p_1+1}x_2x_3\left(\sum \lambda_i \frac{dx_i}{x_i} + d\left(\frac{1}{x_1^{p_1}}\right)\right), \quad p \neq 0,\ \lambda_2 \cdot \lambda_3 \neq 0.$$

As always, the singular locus is holomorphically diffeomorphic to the three axes; at a point $\neq 0$ on the x_1-axis, we have a logarithmic singularity of 2-dimensional type with, in particular, two convergent separatrices. Along the other two axes, we again have a saddle-node singularity of 2-dimensional type, with the same conclusion as in 5.5.3.

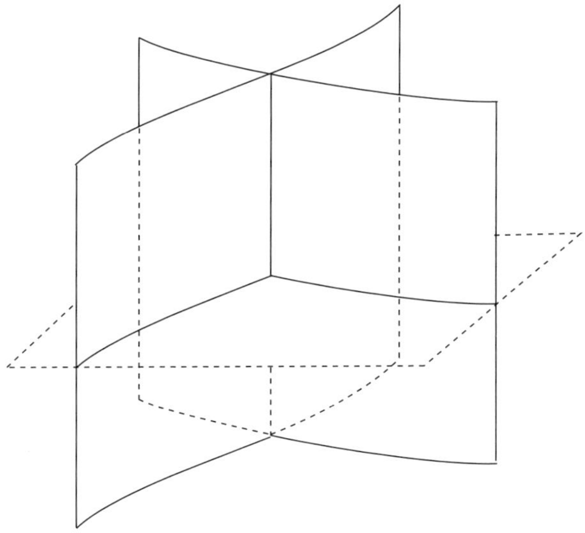

FIGURE 6

5.6. What to remember. Every simple foliation is defined by a holomorphic action of $\mathbb{C}^2$; indeed, the reader can check that we are in the situation studied in 5.2. Moreover, we note the following phenomenon. Let m be a singular point near the singularity 0; let γ be an integral curve of $\mathcal{F}$ passing through m and not contained in Sing $\mathcal{F}$. Then γ is in exactly one smooth, convergent separatrix of $\mathcal{F}$ at 0.

6. Application to the Existence of Separatrices

Let $\mathcal{F}$ be a non-dicritical foliation at the origin in $\mathbb{C}^3$. Then there exists a surface $S \subset (\mathbb{C}^3, 0)$ that is a separatrix.

The idea of the proof is as follows (cf. [**9**]). We consider a reduction of singularities $\underline{E} : M(\mathcal{F}) \to (\mathbb{C}^3, 0)$ of $\mathcal{F}$, then a general plane section $i : (\mathbb{C}^2, 0) \to (\mathbb{C}^3, 0)$. The Camacho-Sad theorem guarantees the existence of at least one separatrix for $i^{-1}\mathcal{F}$. This produces an integral curve γ for $\mathcal{F}$; the strict transform $\widetilde{\gamma}$ of γ by $\underline{E}$ cuts the exceptional divisor associated with $\underline{E}$ in a singular point of $\widetilde{\mathcal{F}} = \underline{E}^{-1}(\mathcal{F})$. All the singularities are simple, so according to 5.6 the "saturation" of $\widetilde{\mathcal{F}}$ by $\widetilde{\gamma}$ produces a hypersurface $\widetilde{S} \subset M(\mathcal{F})$ that is a global separatrix of $\mathcal{F}$. Since $\underline{E}$ is proper, $S = \underline{E}(\widetilde{S})$ is a suitable holomorphic surface.

In fact, it has been proved that a foliation $\mathcal{F}$ that is non-dicritical at the origin in $\mathbb{C}^3$ has the following property: every integral curve γ of $\mathcal{F}$ is contained in a separatrix (and there are finitely many separatrices).

REMARK 1. One can also "extend" the formal integral curves to formal separatrices by proceeding more or less as above (but with unwieldy formalism). This is also true for a formal foliation.

REMARK 2. It is not easy to predict whether a foliation is dicritical or not, and the results above should be understood as a dichotomy: the presence of separatrices, or an abundance of holomorphic integral curves that go to the singular locus.

7. Dimension > 3

The statement of §6 remains true in dimension > 3: *a foliation $\mathcal{F}$ that is non-dicritical at the origin in $\mathbb{C}^n$ has a hypersurface that is a separatrix.* But the strategy of the proof in [**12**] is conceptually different from that above: the theorem in dimension 3 is actually used in a deep way. One proceeds by first cutting $\mathbb{C}^n$ by a generic $\mathbb{C}^3$, in which the statement of §6 produces a separatrix S, then studying $\mathcal{F}$ in a neighborhood of an annulus $\Delta \times \{0\}$, $\Delta \subset \mathbb{C}^3$, and showing that there is a generic equireduction property over Δ. This makes it possible to "push" S onto the thickening $\Delta \times B$, where $B \subset \mathbb{C}^{n-3}$ is a small polydisk, to obtain a separatrix $X_{|\Delta\times B}$ of $\mathcal{F}_{|\Delta\times B}$. Then $H^1(\Delta \times B, \mathcal{O}^*) = 0$ because $n \geq 3$; hence $X_{|\Delta\times B}$ is defined by an equation $F = 0$, $F \in \mathcal{O}(\Delta \times B)$. One extends F to a neighborhood of $0 \in \mathbb{C}^n$ by Cauchy's formula, which finally produces a separatrix X for $\mathcal{F}$. Note that this argument cannot be used to pass directly from dimension 2 to dimension 3 because if $n = 3$, Δ will be an annulus in $\mathbb{C}^2$, and here $H^1(\Delta \times B, \mathcal{O}^*)$ will be nontrivial!

8. What Can We Do with the Reduction of Singularities?

Although the reduction of singularities contains practically all the interesting information about a function $f \in \mathcal{O}(\mathbb{C}^n, 0)$, at least the topological information (for instance, in dimension 2, for a family f_t we have the equivalences $\mu(f_t)$ constant $\Leftrightarrow$ equireducibility $\Leftrightarrow$ constant topological type), nothing of the kind is true for general foliations. Here, reduction of singularities is only the skeleton around which the leaves will be organized. In what follows, we introduce some classical notions that will let us describe part of this organization.

8.1. Holonomy of a foliation in a neighborhood of an invariant hypersurface: case of a component of the exceptional divisor. Consider the following situation: $\mathcal{F}$ is a nonsingular foliation on the complex manifold V, and $X \subset V$ is a closed, smooth hypersurface that is invariant under $\mathcal{F}$. Suppose V is a locally trivial fibration (with disk fibers) over X. Since X is a leaf of $\mathcal{F}$, the other leaves of $\mathcal{F}$ are transverse to the fibers, at least near X. In particular, if Σ is a fiber, $t_0 = X \cap \Sigma$, α a loop in X with base point t_0, and z a point in Σ sufficiently near t_0, we can lift the loop α into the leaf through z, to a path with endpoints z and $h_\alpha(z)$ that lies in Σ. The transformation $h_\alpha : \Sigma_{t_0} \to \Sigma_{t_0}$, $z \mapsto h_\alpha(z)$, defines a germ of a diffeomorphism $\Sigma_{t_0} \cong (\mathbb{C}, 0)$ that depends only on the homotopy class of α; thus we have a representation (called the holonomy representation)

$$\mathrm{Hol} : \pi_1(X, t_0) \to \mathrm{Diff}(\Sigma, t_0) \cong \mathrm{Diff}(\mathbb{C}, 0)$$

that contains all "the structure of $\mathcal{F}$ in a neighborhood of X".

8.2. A word on the classification of simple singularities. Let $\mathcal{F}$ be a singularity at the origin in $\mathbb{C}^2$, which we will assume to be of type $(*)$. By Briot and Bouquet, $\mathcal{F}$ has two separatrices, which we take to be the coordinate axes. We can define $\mathcal{F}$ by a vector field

$$X = X(\mathcal{F}) = x\frac{\partial}{\partial x} + y(\lambda + B(x,y))\frac{\partial}{\partial y},$$

where

$$B \in \mathcal{O}(\mathbb{C}^2, 0), \quad B(0) = 0,$$

and

$$\lambda = \lambda(\mathcal{F}) \notin \mathbb{Q}_{\geq 0}.$$

After fiddling with a homothety, we may assume that $B \in \mathcal{O}(U)$, where U is a neighborhood of a polydisk $D(0, r_1) \times D(0, r_2)$, with $r_1 > 1$. The flow ϕ_t of X can be written as

$$\phi_t(x, y) = (e^t x, e^{\lambda t} y W(x, y, t)), \text{ with } W(x, y, 0) = 1.$$

The holonomy of the separatrix $y = 0$, i.e. the generator of $\mathrm{Hol}(\pi_1(\{y = 0\}\backslash\{0\}, *))$, computed relative to the vertical fibration on the transversal $x = 1$, is precisely the second component of

$$\phi_{2i\pi}(x, y) = (1, e^{2i\pi\lambda} y W(1, y, 2i\pi)) = (1, h_{\mathcal{F}}(y)).$$

The diffeomorphism $h_{\mathcal{F}}(y) = e^{2i\pi\lambda} y + \cdots$ is well defined for sufficiently small y. Mattei-Moussu [**32**] and Martinet-Ramis [**31**] have proved that if $\mathcal{F}$ and $\mathcal{F}'$, as above, have holonomies $h_{\mathcal{F}}$ and $h_{\mathcal{F}'}$ that are conjugate, then $\mathcal{F}$ and $\mathcal{F}'$ are holomorphically conjugate if the linear invariants $\lambda(\mathcal{F})$ and $\lambda(\mathcal{F}')$ coincide. The idea is to push a conjugating map defined along the transversal $x = 1$ along the leaves while preserving the vertical fibration. From another point of view, the work of Ecalle, Martinet, Ramis, Pérez Marco, and Yoccoz shows that every diffeomorphism $h \in \mathrm{Diff}(\mathbb{C}, 0)$ can be realized as a holonomy diffeomorphism $h = h(\mathcal{F})$ of a foliation $\mathcal{F}$ as above [**37**]. Thus the classification of foliations $\mathcal{F}$ of type $(*)$ reduces to that of diffeomorphisms of $(\mathbb{C}, 0)$. The classification of foliations of type $(**)$ is more complex.

A special case is the one where $\mathcal{F}$ has a formal first integral $\widehat{f} \in \widehat{\mathcal{O}}(\mathbb{C}^2, 0)$; this cannot occur unless $\lambda = -p/q \in \mathbb{Q}_{<0}$. From $\widehat{f} \circ \phi_t = \widehat{f}$, it follows in particular that $\widehat{f}(\phi_{2i\pi q}) = \widehat{f}$. An elementary calculation shows that $\phi_{2i\pi q} = \mathrm{Id}_{\mathbb{C}^2}$; hence the real flow $s \to \phi_{2i\pi qs}$ factors through an action of S^1, which can be linearized by averaging (Bochner-Cartan theorem). In particular, the foliation $\mathcal{F}$ can be linearized; i.e., in ad hoc coordinates one can choose $X(\mathcal{F})$ to be of type $x\frac{\partial}{\partial x} - \frac{p}{q}y\frac{\partial}{\partial y}$. But this linear vector field has first integral x^py^q, and thus $\mathcal{F}$ has a convergent first integral. Note that the periodicity of the flow, $\phi_{2i\pi q} = \mathrm{Id}_{\mathbb{C}^2}$, is equivalent to that of the holonomy, $h(\mathcal{F})^q = \mathrm{Id}_{\mathbb{C}}$. Note also that every formal first integral $\widehat{f}$ of the linear foliation associated with $x\frac{\partial}{\partial x} - \frac{p}{q}y\frac{\partial}{\partial y}$ factors through x^py^q: $\widehat{f} = \widehat{\ell}(x^py^q)$, $\widehat{\ell} \in \widehat{\mathcal{O}}(\mathbb{C}, 0)$, provided that p and q have been chosen to be relatively prime.

8.3. Simple singularities in dimension 3 (or higher). Simple singularities of 3-dimensional type are confluences of singularities of 2-dimensional type along the axes of singularities. The confluence occurs with constraints (relations among the three types of singularity, due to integrability, which rigidifies the situation). For instance, the classification of resonant singularities of type 5.5.2 is often more elementary than in dimension 2 [**31**] and, when it is not, it is a byproduct of that in dimension 2. More precisely, let $\mathcal{F}$ be a foliation at the origin in $(\mathbb{C}^3, 0)$ that is formally conjugate to the foliation $\mathcal{F}_{p,\lambda}$ given by

$$\Omega_{p,\lambda} = x_1^{p_1+1}x_2^{p_2+1}x_3^{p_3+1}\left[\sum \lambda_i \frac{dx_i}{x_i} + d\left(\frac{1}{x_1^{p_1}x_2^{p_2}x_3^{p_3}}\right)\right], \quad \lambda_i \in \mathbb{C},\ P_i \in \mathbb{N}^*.$$

We may assume that the planes $x_i = 0$ are separatrices of the foliation $\mathcal{F}$ and that its singularities are exactly the coordinate axes. We can compute the holonomy of $\{x_3 = 0\} \setminus \operatorname{Sing}\mathcal{F}$, whose Poincaré group is $\mathbb{Z} \times \mathbb{Z}$. The group $G \subset \operatorname{Diff}(\mathbb{C}, 0)$, the image of this holonomy representation, is an abelian group $\langle f, g\rangle$ with two generators, and it is not too hard to determine that it is formally conjugate to the holonomy group $G_{p,\lambda}$ of $\{x_3 = 0\}$ for $\mathcal{F}_{p,\lambda}$; since $G_{p,\lambda}$ can be explicitly computed, we note that if the λ_i/λ_j are not all rational, then $G_{p,\lambda}$ is isomorphic to $\mathbb{Z} \times \mathbb{Z}$. It turns out (Ecalle-Valiron, cf. [**19**]) that a subgroup of $\operatorname{Diff}(\mathbb{C}, 0)$ containing a noncyclic subgroup of elements tangent to the identity ($h'(0) = 1$) is rigid [**19**]; this is the case here, and it means that G is in fact holomorphically conjugate to $G_{p,\lambda}$. As in dimension 2, conjugation at the level of holonomy implies that $\mathcal{F}$ is conjugate to $\mathcal{F}_{p,\lambda}$. This generic rigidity phenomenon does not appear in dimension 2; it can be interpreted in terms of "confluence". When the λ_i/λ_j are rational (the phenomenon of double resonance), one can show that the moduli space of

$$\mathcal{F}_{p,\lambda} := \{\mathcal{F} \text{ formally conjugate to } \mathcal{F}_{p,\lambda}, \text{ modulo holomorphic conjugacy}\}$$

factors in a canonical way through a moduli space of resonant foliations in dimension 2. These results, worked out by Ecalle and myself in a correspondence that is already old, have now been written up by J. Mozo of the University of Valladolid. In principle, the classification of all simple singularities (at least in dimension 3) can be found there.

8.4. Frobenius; the formal step. At present, all proofs of the theorem of Malgrange that was stated in 2.1 ([**28**], [**32**]) pass through a formal step (the Godbillon-Vey algorithm), which we briefly describe. The starting point is a foliation $\mathcal{F}$ defined by ω_0 such that $\operatorname{codim}\operatorname{Sing}\mathcal{F} \geq 3$. Integrability ($\omega_0 \wedge d\omega_0 = 0$) and the de Rham–Saito lemma produce a 1-form $\omega_1 \in \Omega^1(\mathbb{C}^n, 0)$ such that $d\omega_0 = \omega_0 \wedge \omega_1$.

Differentiating this gives $\omega_0 \wedge d\omega_1 = 0$. Repeating the process, we construct ω_2 such that $d\omega_1 = \omega_0 \wedge \omega_2$; by induction, we find $\omega_j \in \Omega^1(\mathbb{C}^n, 0)$ such that

$$d\omega_j = \omega_0 \wedge \omega_{j+1} + \sum_{1\leq k\leq j} \binom{k}{j} \omega_k \wedge \omega_{j-k+1}.$$

Following an idea of J. Martinet, we add a variable t and consider the 1-form $\overline{\omega} = dt + \sum_{i=0}^{\infty} t^i\omega_i/i!$ on $\mathbb{C}^n \times \mathbb{C}$. By construction, $d\overline{\omega} = \overline{\omega} \wedge \partial\overline{\omega}/\partial t$; hence $\overline{\omega}$ is integrable and nonsingular at the origin. By the ordinary Frobenius theorem, we can find a submersion $F(x,t)$ and a unit $G(x,t)$ such that $\overline{\omega} = GdF$; setting $G_0(x) = G(x,0)$ and $F_0(x) = F(x,0)$, we obtain $\omega_0 = G_0dF_0$ by restriction to $t = 0$. The catch is that there is no guarantee that $\overline{\omega}$ is convergent (and one can produce some really divergent $\overline{\omega}$'s), so F and G, hence F_0 and G_0 are only formal! Malgrange's approach [**28**] consists of showing via analysis the existence of an $\overline{\omega}$ as above that converges.

8.5. The singular Frobenius theorem by restriction, reduction of singularities, and extension. With the notation of 8.4, note that the hypothesis $\operatorname{codim}\operatorname{Sing}\mathcal{F} \geq 3$ implies that the series F_0 is irreducible; if it were not, say $F_0 = F_1 \cdot F_2$, then the intersection $\{F_1 = F_2 = 0\}$ would be a branch of $\operatorname{Sing}\mathcal{F}$ of codimension exactly 2. Consider a general embedding $\tau : (\mathbb{C}^2, 0) \to (\mathbb{C}^n, 0)$, and let $\underline{\omega} = \tau^*\omega_0$, $\underline{F} = F_0 \circ \tau$. This time, $\underline{F}$ may not be irreducible, but it is certainly reduced: $\underline{F} = f_1 \cdots f_p$, with $\gcd(f_1, \ldots, f_p) = 1$. We proceed to the reduction of singularities of $\underline{\mathcal{F}}$ (where $\underline{\mathcal{F}}$ is defined by $\underline{\omega}$); it is also that of $\underline{F}$. We will find a point m in the exceptional divisor $\underline{E}^{-1}(0)$ (notation of 4.11) such that $\underline{E}^{-1}(0)$ is smooth and $\underline{E}^{-1}(\mathcal{F})$ is singular at m: m is the point through which there is a branch of the separatrix $\{\underline{F} = 0\}$ (note that it is necessarily convergent). At m we can apply the argument of 8.2 to $\underline{E}^{-1}(\underline{\mathcal{F}})$: because there is a formal first integral, we can linearize $\underline{E}^{-1}(\underline{\mathcal{F}})$ at m and find a first integral of monomial type x^py^q, with $\{x = 0\} = (\underline{E}^{-1}(0), m)$.

By the remark at the end of §8.2, $(\underline{F} \circ \underline{E}, m)$ factors through the monomial x^py^q: $(\underline{F} \circ \underline{E}, m) = \widehat{\ell}(x^py^q)$, where $\widehat{\ell} \in \widehat{\mathcal{O}}(\mathbb{C}, 0)$; since F is reduced, it must be true that $q = 1$ and $\widehat{\ell}$ is a formal diffeomorphism. The formal series $\widehat{\ell}^{-1}(\underline{F})$ is a first integral of $\underline{\mathcal{F}}$ and coincides somewhere in the reduction of singularities with the convergent series x^py. An innocuous upper bound argument shows that $\widehat{\ell}^{-1}(F)$ converges [**32**]; the same argument actually shows that $\widehat{\ell}^{-1}(F_0)$ converges. This is certainly a convergent first integral of the initial foliation $\mathcal{F}$ in $\mathbb{C}^n$. Note particularly that the approach above does not recover the statements of Mattei-Moussu, which are more general than those of Malgrange.

We can see in this sketch of a proof that an argument using general plane sections allows us to reduce to dimension 2; it turns out that all we need is the reduction of singularities in dimension 2. This is true in most problems about codimension-one foliations, at least in essentially algebraic or analytic problems. In contrast, as has already been pointed out, the construction of separatrices in dimensions greater than 3 requires a way of resolving singularities that does not reduce to dimension 2.

8.6. General philosophy. Let $\mathcal{F}$ be a foliation at the origin in $\mathbb{C}^n$, $n \leq 3$, that has a reduction of singularities $\underline{E} : M(\mathcal{F}) \to (\mathbb{C}^n, 0)$. Let D_i be a non-dicritical component of the exceptional divisor associated with $\underline{E}$. When the divisor D_i appears in the reduction process, it is accompanied by the construction of a fibration (one blows up a point or a smooth curve). This can be modified because one blows up points or curves in D_i but will keep a fibered neighborhood U_i of D_i, minus a neighborhood of $\operatorname{Sing} \underline{E}^{-1}(\mathcal{F}) \cap D_i$, and can apply the preceding construction. Thus we construct the holonomy of the component D_i, which describes the local organization of $\underline{E}^{-1}(\mathcal{F})$ near the smooth part of D_i. Clearly, we must glue this information together. This is done via the singular locus. The local description of the singularities is not always sufficient. In dimension 3, we can take advantage of the fact that, after reduction along the smooth part of the singular locus, the foliation is locally a product. If we draw a loop α with base point m in the smooth part of $\operatorname{Sing} \underline{E}^{-1}(\mathcal{F})$, we can follow the local trivializations over α to construct a singular holonomy: this time we are dealing with a diffeomorphism $\phi_\alpha \in \operatorname{Diff}(T, m)$, where T is a smooth surface transverse to $\operatorname{Sing} \underline{E}^{-1}(\mathcal{F})$ at m that preserves the restriction $\underline{E}^{-1}(\mathcal{F})_{|T}$. This leads naturally to studying the group of diffeomorphisms of $(\mathbb{C}^2, 0)$ that preserve a reduced foliation of $(\mathbb{C}^2, 0)$; this work is in progress (Berthier, Cerveau, Méziani, preprint, Rennes 1998).

8.7. The case of dimension 2. We will (as always) be satisfied with studying those generalized curves that are reduced after one blow-up $E : \widetilde{\mathbb{C}^2} \to \mathbb{C}^2$. Each singular point $m_i \in E^{-1}(0)$ contributes a transverse separatrix to the divisor. Consider small disks $\Delta_i \subset E^{-1}(0)$ centered at the m_i's, and let V be the open set $V = \pi^{-1}(E^{-1}(0) \setminus \cup \Delta_i)$, where π is the projection defined in 3.1. Working with V as in 8.1, we define the projective holonomy $\operatorname{Hol} : \pi_1(E^{-1}(0) \setminus V(m_i), m_0) \to \operatorname{Diff}(\mathbb{C}, 0)$. Let ν be the algebraic multiplicity of $\mathcal{F}$; we assume implicitly that $\nu \geq 2$ because a generalized curve of multiplicity 1 is desingularized either in zero blow-ups or after at least two blow-ups (exercise). We have the following results:

- If $\nu = 2$, then $\mathcal{F}$ is completely determined, up to diffeomorphism, by the representation Hol and the data of the 1-jet ω_2 of the defining form ω for $\mathcal{F}$.
- If $\nu = 3$, the same statement as above, adding "and the cross ratios of the m_i".
- If $\nu \geq 4$, then the projective holonomy gives the topological type of $\mathcal{F}$.

In the first two cases, one exploits the fact that the separatrices, which are smooth and transverse to the divisor $E^{-1}(0)$ after blow-up, can be "put" in the fibers of π. Thus the arguments used in the reduced case (conjugacy of the holonomies implies conjugacy of the foliations) can be globalized here.

A great deal of other qualitative information about such $\mathcal{F}$ can now be obtained by using our knowledge of the subgroups of $\operatorname{Diff}(\mathbb{C}, 0)$. For instance, Nakai's theorem [**34**], which describes the pseudo-orbits of a nonsolvable subgroup $G \subset \operatorname{Diff}(\mathbb{C}, 0)$, can be used to show the density (generic in the Krull topology) of the distinct leaves of the separatrices.

8.8. First integrals. Still in the setting of generalized curves $\mathcal{F}$ that are reduced after one blow-up in $\mathbb{C}^2$, we can derive the following facts from the results in [**32**] and [**17**]:

- If the projective holonomy group of $\mathcal{F}$ is finite, then $\mathcal{F}$ has a holomorphic first integral $f \in \mathcal{O}(\mathbb{C}^2, 0)$.

– If it is linearizable, then $\mathcal{F}$ has a first integral of type $\sum \lambda_i \log f_i$, $f_i \in \mathcal{O}(\mathbb{C}^2, 0)$.

IDEA OF THE PROOF. We choose a transversal $(\mathbb{C}, 0)$ (a Hopf fiber), based at a nonsingular point, to the exceptional divisor. We may, and do, assume that the image of Hol is a linear group (finite in the first case) $\langle \mu_1 z, \ldots, \mu_n z \rangle$; such a group leaves the logarithmic form $dz/z = \eta_0$ invariant. This allows us to push η_0 along the leaves to define a meromorphic closed form $\widetilde{\eta}$ away from the separatrices of $E^{-1}(\mathcal{F})$ such that $\mathcal{F}$ is given by $\widetilde{\eta}$. It remains to extend $\widetilde{\eta}$ meromorphically along the separatrices, which is innocuous. One can check that the form η such that $E^* \eta = \widetilde{\eta}$ is of type $\sum \lambda df_i/f_i$, and is suitable. □

Clearly, one can also work in dimension 2 with foliations whose desingularizations require more than one blow-up, but this is rather hard: [**17**], [**32**], [**36**], [**41**], [**27**], [**33**], ...

8.9. Dimension ≥ 3; extension. One often goes by way of dimension 2 in constructing first integrals, as in 8.5; for example [**18**]:

– If $\mathcal{F}$ is a holomorphic foliation at the origin in $\mathbb{C}^n$ and $i : (\mathbb{C}^2, 0) \to (\mathbb{C}^n, 0)$ is a generic embedding, then $\mathcal{F}$ has an integrating factor if and only if $i^{-1}\mathcal{F}$ does.

One direction is obvious. The other is not hard; if ω defines $\mathcal{F}$ and f_0 is such that $d(d\omega_0/f_0) = 0$, $\omega_0 = i^*\omega$, then f_0 can be extended to an integrating factor $\widetilde{f}$ of ω defined on $\Delta \times B$, where Δ is an annulus in $\mathbb{C}^2$ and B is a polydisk in $\mathbb{C}^{n-2}$ (we have identified i here with the embedding $\mathbb{C}^2 \to \mathbb{C}^2 \times \mathbb{C}^{n-2}$) such that $\widetilde{f} \circ i = f_0$. This is a consequence of the classical Frobenius theorem applied to the points of $\Delta \times \{0\}$. By Cauchy's formula, $\widetilde{f}$ can be extended to a neighborhood of 0.

8.10. Dimension ≥ 3; a brief return to 2.2. Let $\mathcal{F}$ be defined on $(\mathbb{C}^3, 0)$ by $\omega = \omega_\nu + \cdots$, where

$$\frac{\omega_\nu}{P_{\nu+1}} = \sum_{i=1}^{s} \lambda_i \frac{dP_i}{P_i}, \quad \sum \lambda_i \nu(P_i) = 1,$$

and the P_i have normal crossings; by Deligne-Fulton, the holonomy of the divisor $D = E^{-1}(0)$ is abelian (finite if $s = 1$). It will be linearizable if, for instance, some λ_i is not real. This will be preserved if we cut by a plane section $i : (\mathbb{C}^2, 0) \to (\mathbb{C}^3, 0)$; by conjugating 8.8 and 8.9 we obtain the result stated in 2.2: $\mathcal{F}$ is defined by a logarithmic form $f_1 \cdots f_s \sum \lambda_i df_i/f_i$.

9. Foliations of $\mathbb{C}^3$ That Are Reduced by Point Blow-ups; Foliations with Very Connected Singular Locus

We say that a foliation $\mathcal{F}$ that is non-dicritical at the origin in $\mathbb{C}^3$ is reduced by point blow-up if, as in dimension 2, we can construct a reduction $\underline{E} : M(\mathcal{F}) \to \mathbb{C}^3$ by a finite composition of point blow-ups:

$$\mathbb{C}^3 = X_0 \overset{E_1}{\longleftarrow} X_1 \longleftarrow \cdots \overset{E_N}{\longleftarrow} X_N = M(\mathcal{F}).$$

Let m be a smooth point of $\operatorname{Sing} \mathcal{F}_N$; we know that, at m, $\mathcal{F}_N$ has a product structure "$\mathcal{G}_m \times \mathbb{C}$", where $\mathcal{G}_m$ is a reduced 2-dimensional foliation. We will say that $\mathcal{F}$ (or $\mathcal{F}_N$) is *of generalized Poincaré type* if each $\mathcal{G}_m$ is given by $\lambda_1 x dy + \lambda_2 y dx + \cdots$, with $\lambda_1/\lambda_2 \notin \mathbb{R}$.

The following statement (cf. [**10**]) generalizes the one presented in 8.10:

If $\mathcal{F}$ can be desingularized by point blow-ups into $\mathcal{F}_N$, which is of generalized Poincaré type, then $\mathcal{F}$ is given by a logarithmic form $f_1 \cdots f_p \sum \lambda_i df_i/f_i$.

The proof is by induction on N; the statement is true for $N = 0$ (description of simple singularities). Note that the foliation $\mathcal{F}_1 = E_1^{-1}\mathcal{F}$ satisfies the hypothesis at each of its singular points, which are reduced after fewer than N blow-ups; in particular, at each m_1 we can construct $f_{m_1} \in \mathcal{O}(E_1^{-1}(\mathbb{C}^3), m_1)$ such that $E_1^*\omega/f_{m_1}$ is closed, where ω defines $\mathcal{F}$. By the Poincaré hypothesis, f_{m_1} is unique up to scalar multiplication, and the df_{m_1}/f_{m_1} are well defined and glue together into a meromorphic closed form $\widetilde{\eta}_0$ defined in a neighborhood of $E_1^{-1}(0) \cap \operatorname{Sing}\mathcal{F}_1$ (the tangent cone to $\mathcal{F}$); a little topological argument lets us extend $\widetilde{\eta}_0$ to $\widetilde{\eta}$, which is meromorphic in a neighborhood of $E_1^{-1}(0)$. We then verify that if η satisfies $E_1^*\eta = \widetilde{\eta}$, then it is of type $\eta = df/f$, where $f \in \mathcal{O}(\mathbb{C}^3, 0)$ and ω/f is closed. The rest of the proof is easy. Note that we have made no direct use of holonomy arguments.

The foliations that can be reduced by point blow-ups are part of a more general class that consists of foliations with very connected singular locus. These are the foliations for which there exists a reduction

$$\mathbb{C}^3 = X_0 \xleftarrow{\pi_1} X_1 \longleftarrow \dots \xleftarrow{\pi_N} X_N = M(\mathcal{F})$$

such that, at each singular point m_i of $\mathcal{F}_i = (\pi_i \circ \cdots \circ \pi_1)^{-1}(\mathcal{F})$, the singular locus of $(\pi_N \circ \cdots \circ \pi_{i+1})^{-1}(\mathcal{F}_i, m_i)$ is connected. This notion can depend on the choice of blow-up strategy. For instance, we can consider the foliation associated in $\mathbb{C}^3$ with $f(x, y, z) = xy(x - y)$. We can reduce this by blowing up the z-axis, and the singular locus consists of three disjoint lines. But we can also reduce it by following the procedure sketched in 3.7, which produces a connected singular locus. An interesting example is that of the foliations $\mathcal{F}$ with separatrix a homogeneous cone $P(x, y, z) = 0$. Now we reduce $P = 0$ by following the procedure of 3.7, and require that $\mathcal{F}$ also be reduced; this almost always occurs for those $\mathcal{F}$ that are the natural extension of generalized curves (no saddle node "transversely"). If we impose a generalized Poincaré-type condition, then $\mathcal{F}$ will again be defined by a logarithmic form $f_1 \cdots f_p \sum \lambda_i df_i/f_i$; if we relax this condition by requiring only that, along the terminal singular locus, there be no holomorphic first integrals at the smooth points, then we will construct a formal integrating factor for $\mathcal{F}$. These hypotheses are not innocuous, and at present there is no generic description (in the Krull topology) of the singular holomorphic foliations in dimension 3.

10. Other Applications

The results described above in the complex setting can of course be applied elsewhere. The reader may consult the work of Cano-Lion-Moussu [**11**], for instance, which describes the boundaries of Pfaffian hypersurfaces. Here is a problem that one should be able to attack by resolution of singularities. It is known that a formal series in n real variables is the infinite jet of a C^∞ function at the origin in $\mathbb{R}^n$; this is a theorem of Borel. This result extends to ideals (Tougeron): an ideal $\widehat{I}$ of the ring of formal series is the infinite jet of a closed ideal of the ring of C^∞ functions at the origin in $\mathbb{R}^n$. The question is the following: *Is an integrable* 1*-form* $\widehat{\omega}$ *the infinite jet of an integrable* C^∞ 1*-form?* Since the reduction of singularities can be applied to formal foliations, at least in dimension 3, and the desired result is clear for (real) simple singularities, we can in fact think about constructing, by gluing, a C^∞ foliation that is defined on the reduction manifold of $\widehat{\omega}$ and whose infinite jet at each point is prescribed by the pullback of $\widehat{\omega}$. We mention a related question: Is a Lie algebra of formal vector fields the infinite jet of a C^∞ field?

Blow-up techniques and the study of holonomy groups allow us, in the general setting, to approach problems of computing relative cohomology (à la Malgrange), of describing periodic solutions of (real) differential equations, etc. In 4.12 I mentioned an application to second-order differential equations. This perspective does not seem to me to have been sufficiently exploited, and I would like to expand on it a bit.

In general, singular vector fields—except in dimension 2 (Camacho-Sad)—have no invariant hypersurfaces, or even invariant analytic curves passing through the singularity. Such an example is constructed in dimension 3 by Gomez-Mont and Luengo in [**23**]. Nor is there any global reduction strategy for singularities of vector fields in dimension 3 (cf. [**12**] for a local strategy and [**4**] for a special case). Let X be a holomorphic vector field with an isolated singularity at the origin in $\mathbb{C}^3$. I claim that if X has no invariant analytic curve passing through 0, then it is not tangent to a codimension-one holomorphic foliation $\mathcal{F}$. If it were, we would have $\operatorname{Sing}\mathcal{F} = \{0\}$ (because if γ were a one-dimensional branch of $\operatorname{Sing}\mathcal{F}$, it would have to be invariant under X). Let $\omega = \sum^3 a_i dx_i$ define $\mathcal{F}$, and let $X = \sum X_i \partial/\partial x_i$. Using the relation $0 = \omega(X) = \sum X_i a_i$, the fact that X has an isolated singularity, and de Rham–Saito, we can write $\omega = i_X i_Y(\text{volume})$, where Y is a holomorphic vector field (here we identify X with a 1-form and ω with a 2-form, a manipulation peculiar to dimension 3). But an argument with minors shows that $\operatorname{Sing}\omega$ must have codimension exactly 2. However, certain special vector fields X that appear in specific problems are naturally tangent to a codimension-one foliation. As an example (still in dimension 3), consider a holomorphic vector field X with Jordan decomposition $X = X_S + X_N$, where X_S is semisimple, X_N is nilpotent, and $[X_S, X_N] = 0$. When X_S and X_N are nonzero and noncollinear, we recover a foliation defined by $\omega = i_{X_S} i_{X_N}(\text{volume})$, which admittedly is only formal in general. Of course, pairs of commuting vector fields generalize this example, and the existence of invariant hypersurfaces (the non-dicritical case) will be useful in classifying them.

Now consider a non-dicritical foliation $\mathcal{F}$ given by the meromorphic 1-form $\omega = a dx + b dy + dz$. If F is the meromorphic function defined by $F(x, y, z) = -(a(x, y, z) + z b(x, y, z))$, consider the differential equation

$$y'' = F(x, y, y').$$

(This is singular when F is truly meromorphic.) There exists a surface S with equation $H(x, y, z) = 0$ such that, if $(x_0, y_0, y_0' = z_0)$ is an initial condition on S, then the solution $x \to y(x)$ realizing this initial condition stays on S: $H(x, y(x), y'(x)) = 0$. Indeed, the vector field

$$Z = \frac{\partial}{\partial x} + z\frac{\partial}{\partial y} + F(x, y, z)\frac{\partial}{\partial z},$$

which is naturally associated with our second-order equation $y'' = F$ by setting $y' = z$, is tangent to the foliation $\mathcal{F}$, which has a separatrix S.

Of course, we've cheated a bit here because, given an equation $y'' = F$, there is no reasonable procedure for deciding whether or not it can be "embedded" in a non-dicritical foliation $\mathcal{F}$, though the discussion above seems to suggest otherwise. But when one consults the classic books (Ince [**24**], for example), one is surprised to encounter this phenomenon abnormally often. Along these lines, consider, for instance, the second-order equation $x \cdot y'' = A(x, y, y')$, where A is holomorphic and not divisible by x in a neighborhood of the origin in $\mathbb{C}^3$, and vanishes there. As before, we associate with it the meromorphic vector field

$$\frac{\partial}{\partial x} + z\frac{\partial}{\partial y} + \frac{A}{x}\frac{\partial}{\partial z},$$

which is collinear with

$$X = x\frac{\partial}{\partial x} + xz\frac{\partial}{\partial y} + A\frac{\partial}{\partial z}.$$

X has nontrivial Jordan decomposition $X = X_S + X_N$ (i.e., X_S and X_N are nonzero and noncollinear), which produces a (formal) foliation $\omega = i_{X_S} i_{X_N}(\text{volume})$, the situation encountered above. It seems to me that it would be worthwhile to carry out a systematic study, local as well as global, of the vector fields contained in a codimension-one holomorphic foliation. On the subject of second-order equations, the reader may consult the monograph of Georges Reeb [**39**], one of the fathers of the theory of foliations, whose "wishes" were realised by Okamoto in [**35**], and Ince's book [**24**], which, of all the references, is still ... the most modern.

Bibliography

[1] P. F. Baum and R. Bott, *On the zeroes of meromorphic vector fields*, Essays on Topology and Related Topics (Mémoires dédiés à Georges de Rham), Springer, New York, 1970, pp. 29–47.

[2] A. D. Brjuno, *Analytic form of differential equations*, Trans. Moscow. Math. Soc. **25** (1971), 131–282.

[3] M. Brunella, *Feuilletages holomorphes sur les surfaces complexes compactes*, Ann. Sci. École Norm. Sup. (4) **30** (1997), 569–594.

[4] C. Camacho, F. Cano, and P. Sad, *Absolutely isolated singularities of holomorphic vector fields*, Invent. Math. **98** (1989), 351–369.

[5] C. Camacho and A. Lins Neto, *The topology of integrable differential forms near a singularity*, Inst. Hautes Études Sci. Publ. Math. **55** (1982), 5–35.

[6] C. Camacho, A. Lins Neto, and P. Sad, *Topological invariants and equidesingularization for holomorphic vector fields*, J. Differential Geom. **20** (1984), 143–174.

[7] ———, *Minimal sets of foliations on complex projective spaces*, Inst. Hautes Études Sci. Publ. Math. **68** (1988), 187–203.

[8] C. Camacho and P. Sad, *Invariant varieties through singularities of holomorphic vector fields*, Ann. of Math. (2) **115** (1982), 579–595.

[9] F. Cano and D. Cerveau, *Desingularization of nondicritical holomorphic foliations and existence of separatrices*, Acta Math. **169** (1992), 1–103.

[10] F. Cano, D. Cerveau, and B. Scárdua, To appear in J. Pure Appl. Algebra.

[11] F. Cano, J.-M. Lion, and R. Moussu, *Frontière d'une hypersurface pfaffienne*, Ann. Sci. École Norm. Sup. (4) **28** (1995), 591–646.

[12] F. Cano and J.-F. Mattei, *Hypersurfaces intégrales des feuilletages holomorphes*, Ann. Inst. Fourier (Grenoble) **42** (1992), 49–72.

[13] M. M. Carnicer, *The Poincaré problem in the nondicritical case*, Ann. of Math. (2) **140** (1994), 289–294.

[14] H. Cartan, *Sur le premier problème de Cousin*, C. R. Acad. Sci. Paris Ser. I Math. **207** (1938), 558–560.

[15] D. Cerveau and A. Lins Neto, *Holomorphic foliations in $\mathbb{CP}(2)$ having an invariant algebraic curve*, Ann. Inst. Fourier (Grenoble) **41** (1991), 883–903.

[16] ———, *Irreducible components of the space of holomorphic foliations of degree two in $\mathbb{CP}(n)$, $n \geq 3$*, Ann. of Math. (2) **143** (1996), 577–612.

[17] D. Cerveau and J.-F. Mattei, *Formes intégrables holomorphes singulières*, Astérisque, vol. 97, Société Mathématique de France, Paris, 1982.

[18] D. Cerveau and R. Moussu, *Extension de facteurs intégrants et applications*, C. R. Acad. Sci. Paris Sér. I Math. **294** (1982), 17–19.

[19] ———, *Groupes d'automorphismes de $(\mathbb{C}, 0)$ et équations différentielles $ydy + \cdots = 0$*, Bull. Soc. Math. France **116** (1988), 459–488.

[20] H. Dulac, *Intégrales d'une équation differentielle dans le voisinage d'un point singulier*, Ann. Univ. Grenoble **XVII** (1905), 1–51.

[21] J. Ecalle, *Les fonctions résurgentes.* Publications Mathématiques d'Orsay, 1981.

[22] E. Ghys, *Feuilletages holomorphes de codimension un sur les espaces homogènes complexes*, Ann. Fac. Sci. Toulouse Math. (6) **5** (1996), 493–519.

[23] X. Gómez-Mont and I. Luengo, *Germs of holomorphic vector fields in $\mathbb{C}^3$ without a separatrix*, Invent. Math. **109** (1992), 211–219.

[24] E. L. Ince, *Ordinary differential equations*, Dover Publications, New York, 1944.

[25] J. P. Jouanolou, *Équations de Pfaff algébriques*, Lecture Notes in Mathematics, vol. 708, Springer, Berlin, 1979.
[26] M. Klughertz, *Feuilletages holomorphes à singularité isolée ayant une infinité de courbes intégrales*, Thesis, Univ. Paul Sabatier, Toulouse, 1988.
[27] F. Loray, *Feuilletages holomorphes à holonomie résoluble*, Thesis, Université de Rennes I, 1994.
[28] B. Malgrange, *Frobenius avec singularités. I. Codimension un*, Inst. Hautes Études Sci. Publ. Math. (1976), 163–173.
[29] J. Martinet, *Normalisation des champs de vecteurs holomorphes (d'après A.-D. Brjuno)*, Bourbaki Seminar, Vol. 1980/81, Lecture Notes in Math., vol. 901, Springer, Berlin, 1981, pp. 55–70.
[30] J. Martinet and J.-P. Ramis, *Problèmes de modules pour des équations différentielles non linéaires du premier ordre*, Inst. Hautes Études Sci. Publ. Math. **55** (1982), 63–164.
[31] ———, *Classification analytique des équations différentielles non linéaires résonnantes du premier ordre*, Ann. Sci. École Norm. Sup. (4) **16** (1983), 571–621.
[32] J.-F. Mattei and R. Moussu, *Holonomie et intégrales premières*, Ann. Sci. École Norm. Sup. (4) **13** (1980), 469–523.
[33] J.-F. Mattei and E. Salem, *Complete systems of topological and analytical invariants for a generic foliation of* $(\mathbb{C}^2, 0)$, Univ. Toulouse, 1996.
[34] I. Nakai, *Separatrices for nonsolvable dynamics on* $(\mathbb{C}, 0)$, Ann. Inst. Fourier (Grenoble) **44** (1994), 569–599.
[35] K. Okamoto, *Sur les feuilletages associés aux équations du second ordre à points critiques fixes de P. Painlevé*, Japan. J. Math. (N.S.) **5** (1979), 1–79.
[36] E. Paul, *Formes abéliennes, formes résolubles*, Thèse d'habilitation, Univ. Toulouse, 1996.
[37] R. Pérez Marco and J.-C. Yoccoz, *Germes de feuilletages holomorphes à holonomie prescrite*, Complex analytic methods in dynamical systems, Astérisque, vol. 222, 1994 (with a bibliography on diffeomorphisms), 345–371.
[38] H. Poincaré, *Œuvres complètes.*
[39] G. H. Reeb, *Feuilletages: résultats anciens et nouveaux (Painlevé, Hector et Martinet)*, Les Presses de l'Université de Montréal, Montreal, Que., 1974.
[40] K. Saito, *On a generalization of de Rham lemma*, Ann. Inst. Fourier (Grenoble) **26** (1976), 165–170.
[41] F. Touzet, *Équations différentielles admettant des solutions liouvilliennes*, Thesis, Univ. Rennes I, 1995.
[42] G. Valiron, *Équations fonctionnelles et applications*, Masson et Cie, Paris, 1945.

RIEMANN SURFACE LAMINATIONS

Étienne Ghys

Abstract: The theory of foliations has its main roots in the qualitative study of ordinary differential equations in the complex domain. In recent years the concept of Riemann surface laminations seems to be central to the theory of holomorphic dynamical systems. These laminations are generalized foliations, in the sense that the ambient space is not necessarily a manifold. The leaves are (typically noncompact) Riemann surfaces. In this article we describe this type of object, emphasizing the analogy with compact Riemann surfaces. In particular, we study the conformal type of the leaves and the existence of meromorphic functions.

1. Introduction

Riemann surface laminations are foliated generalizations of the classical Riemann surfaces one encounters in many geometric and dynamical situations. The goal of this article is to describe some general results concerning these laminations, centering the discussion around some fundamental theorems on Riemann surfaces:

Topological classification. A compact Riemann surface is determined up to homeomorphism by its genus.

Uniformization theorem. Every simply connected Riemann surface is biholomorphically equivalent to the complex projective line $\mathbb{CP}^1$, the complex affine line $\mathbb{C}$, or the unit disk $\mathbb{D}$.

Riemann's theorem. Every Riemann surface has nonconstant meromorphic functions.

To what extent do these basic theorems generalize to Riemann surface laminations? We will see that there are some encouraging positive results but the general situation is not so simple ...

This article contains few proofs and, except in Section 7, no new results. We draw our inspiration from many papers, which we will cite as we go along, and we have not hesitated to copy from our own work, adapting parts of [**23**]. This text does not differ much from that distributed to participants in the session "État de la recherche" of January 1997; we have added 6.4, following an idea of Rick Kenyon, and corrected some statements that were sometimes a bit "optimistic" in the first version.

2. Examples

We begin with the definition of Riemann surface laminations. Consider a *compact* metric space M, covered by open sets U_i (which we will call *distinguished open sets*) and equipped with homeomorphisms h_i from U_i onto $\mathbb{D} \times T_i$, where $\mathbb{D}$ is the unit disk in $\mathbb{C}$ and T_i is some topological space. We say that these open sets define an *atlas* of the structure of a Riemann surface lamination on M if the *transition functions* $h_{ij} = h_j \circ h_i^{-1}$, on their domain of definition, have the form

$$h_{ij}(z,t) = (f_{ij}(z,t), \gamma_{ij}(t)),$$

where $f_{ij}(z,t)$ depends holomorphically on the variable z and continuously on the variable t. Two atlases are equivalent if their union is an atlas. A *Riemann surface lamination* is a compact space M equipped with an equivalence class $\mathcal{L}$ of atlases.

We call a set of the form $h_i^{-1}(\mathbb{D} \times \{t\})$ a *plaque*. The *leaves* of $\mathcal{L}$ are the smallest connected sets such that any plaque that intersects one of them lies completely inside it.

We sometimes write "lamination" for "Riemann surface lamination".

A subset F of M is called *saturated* if it is a union of leaves. If F is closed, then the restriction of the lamination to F defines a lamination structure on F.

A closed set F contained in M is called a *minimal set* if it is saturated, nonempty, and minimal among all closed sets with these properties. This amounts to saying that every leaf contained in F is dense in F. By Zorn's lemma, the closure of each leaf contains a minimal set. A lamination is called *minimal* if all its leaves are dense; that is, if the whole ambient space M is a minimal set.

We will now give a series of examples. We begin by observing that a connected compact Riemann surface is a lamination that has an atlas for which all the transverse spaces T_i reduce to points and that has only one leaf. The problem discussed in this article is whether this "trivial example" is sufficiently general.

2.1. Two-dimensional foliations. In this case, M is a compact differentiable manifold, and the leaves are given by the integral surfaces of an orientable, completely integrable, two-dimensional plane field. We will explain how the choice of a Riemannian metric on M allows us to consider this foliation as a Riemann surface lamination.

To do this, we recall a local theorem first proved by Gauss in the real-analytic case, then progressively improved up to relatively weak regularity hypotheses. Let g be a Riemannian metric on a connected oriented surface S. In a neighborhood of each point p of S we can introduce a system of *isothermal coordinates*; that is, a *conformal* diffeomorphism ϕ from a neighborhood of p onto an open set in the Euclidean plane. Of course, two such diffeomorphisms ϕ differ by a conformal diffeomorphism of an open set in the Euclidean plane; that is, by a holomorphic diffeomorphism of an open subset of $\mathbb{C}$ (if ϕ is required to preserve orientation). In other words, every Riemannian metric on an oriented surface determines a *Riemann surface* structure in a natural way. This theorem depends continuously on the metric; that is, if we have a family of Riemannian metrics on a surface that depend continuously on a parameter, then the isothermal coordinates depend continuously on this parameter [**3**]. Hence, *from a Riemannian metric on the tangent bundle to the leaves of an oriented two-dimensional foliation, one naturally defines the structure of a Riemann surface lamination* on this foliation: it suffices to apply the theorem cited above to the distinguished open sets for the foliation.

There are many methods of constructing foliations. We refer the reader to [**24, 30**] for examples and will be satisfied here with a few constructions, in order to illustrate the complexity of the situation.

Among the simplest examples are linear foliations on tori. Starting with the foliation of $\mathbb{R}^3$ whose leaves are planes parallel to a given plane Π, we pass to the quotient by integer translations (the quotient by $\mathbb{Z}^3$), which clearly preserve this foliation. On the quotient torus, we obtain a foliation whose leaves are all homeomorphic to planes if Π is "totally irrational".

A very general method of constructing examples is *suspension*. Let S be a manifold (which will be a compact Riemann surface in our case), and let T be a compact manifold. Consider a homomorphism h from the fundamental group Γ of S to the group of homeomorphisms of T. The group Γ operates diagonally on the product $\widetilde{S} \times T$ of the universal cover of S and T, and preserves the trivial foliation whose leaves are the sets $\widetilde{S} \times \{\star\}$. Passing to the quotient, we obtain a manifold M that fibers over S with fibers homeomorphic to T, equipped with a foliation $\mathcal{F}$ transverse to this fibration. The leaves of this foliation are covers of the base space S and intersect the fibers T in the orbits of the group $h(\Gamma)$.

For example, choose S to be a compact Riemann surface of genus 2, and let π be a surjection from its fundamental group onto a free group on two generators, $L(\alpha, \beta)$. By choosing two homeomorphisms a and b of T, one defines a homomorphism from $L(\alpha, \beta)$ to the group of homeomorphisms of T that sends α to a and β to b. Composing with π gives a homomorphism h from Γ to the group of homeomorphisms of T, and hence a foliation on a bundle over S. We make this more specific by choosing T to be the projective line $\mathbb{CP}^1$ and α and β to be two linear fractional transformations that generate a Kleinian group $G \subset \mathrm{PSL}(2, \mathbb{C})$ whose limit set $\Lambda \subset \mathbb{CP}^1$ is a Cantor set. In this case the suspension is a complex surface M that fibers over S with fibers $\mathbb{CP}^1$, and whose structure group reduces to G. It follows from Kodaira's embedding theorem that M is an algebraic surface; that is, it can be holomorphically embedded in $\mathbb{CP}^N$ (see [**27**]). Moreover, to the G-invariant subset Λ of $\mathbb{CP}^1$ there corresponds a compact set $X \subset M$ that is saturated by the foliation and intersects each fiber in a Cantor set. This compact set X, equipped with the restriction $\mathcal{L}$ of $\mathcal{F}$, is a typical example of a minimal Riemann surface lamination on a space that is not a manifold. (Minimality follows from the fact that all the orbits of the limit set of a Kleinian group are dense in the limit set.) There exists an embedding of X in $\mathbb{CP}^N$ that is continuous and holomorphic when restricted to the leaves. By choosing a generic projection, one can even show that X can be embedded in $\mathbb{CP}^3$. To summarize: *there exist minimal laminations that do not reduce to a Riemann surface and are holomorphically embedded in* $\mathbb{CP}^3$.

Of course, if we project this last example generically to a projective plane $\mathbb{CP}^2$, the leaves will not necessarily be embedded; they will only be immersed, and we do not obtain a lamination in $\mathbb{CP}^2$. *We do not know whether there exist minimal laminations that are holomorphically embedded in* $\mathbb{CP}^2$ *and do not reduce to a compact Riemann surface.* This question is a strong version of the question of the existence of an "exceptional minimal set" for polynomial differential equations in $\mathbb{C}^2$, which we recall here because it is one of the motivations for the systematic study of laminations. Let P and Q be relatively prime polynomials in two complex variables, and consider the following ordinary differential equation in $\mathbb{C}^2$:

$$\frac{dx}{dt} = P(x,y); \quad \frac{dy}{dt} = Q(x,y).$$

Away from the common zeros of P and Q, this equation defines a holomorphic foliation whose leaves are the complex solutions. When we compactify $\mathbb{C}^2$ to obtain $\mathbb{CP}^2$, it is not hard to see that this foliation extends to a holomorphic foliation of $\mathbb{CP}^2$ away from finitely many singularities. The question (called the "exceptional minimal set problem") is whether *the closure of every leaf contains a singular point.* This problem has been studied extensively and seems to be difficult (see [**6, 8, 9**]); a positive answer would give a complex analogue of the classical Poincaré-Bendixson theorem, which describes the limit sets of vector fields on the real 2-sphere. If a leaf of a polynomial foliation of $\mathbb{CP}^2$ did not accumulate at any singularity, its closure would be an embedded lamination, and one could consider a minimal sublamination $\mathcal{L}$ contained in this closure. It is easy to check that $\mathcal{L}$ cannot reduce to a compact Riemann surface: the normal bundle of a leaf of a foliation has a flat connection given by the holonomy, and it follows that if a foliation of the preceding type in $\mathbb{CP}^2$ had a compact leaf, then that leaf would be a Riemann surface embedded in $\mathbb{CP}^2$ and with zero self-intersection. But of course this contradicts Bezout's theorem. Thus, *the question of the existence of a nontrivial minimal lamination embedded in $\mathbb{CP}^2$ is stronger than that of an "exceptional minimal set".*

Here is another example of a two-dimensional foliation, due to M. Hirsch. In a solid torus $\overline{\mathbb{D}} \times \mathbb{S}^1$, we remove the interior of a solid torus, a neighborhood of a two-strand braid, for instance (see Figure 1). We thus obtain a three-dimensional manifold whose boundary consists of two tori. This manifold is naturally foliated by "pairs of pants", that is, spheres minus three disks. This foliation is transverse to the boundary and, on each boundary component, induces a trivial foliation, that is, a foliation of the torus whose leaves are circles. Gluing together the two connected components of the boundary by a suitable diffeomorphism, we obtain a closed 3-manifold equipped with a two-dimensional foliation. The leaves of this foliation are each homeomorphic to a sphere minus a Cantor set, except for those leaves that correspond to the "periodic points" of the gluing, which have nonzero genus.

2.2. One-dimensional dynamical systems. Here we describe a method, due to D. Sullivan, that allows us to associate laminations with certain dynamical systems in such a way that the dynamics are conjugate if and only if the associated laminations are isomorphic [**44**].

This idea of encoding dynamics by a geometric object is not new, and we start with an elementary example. Let $f : (\mathbb{C}, 0) \to (\mathbb{C}, 0)$ be a germ of a contracting holomorphic diffeomorphism. We assign to it the elliptic curve E that is the quotient of a sufficiently small deleted neighborhood of the origin by the action of f. This complex curve E is marked, in the sense that it has a distinguished homotopy class (the one that corresponds to f). It is very easy to check that two germs are holomorphically conjugate if and only if the marked elliptic curves are isomorphic. One can proceed in the same way in higher dimensions; thus one can show that studying the dynamics of germs of contracting biholomorphisms is equivalent to studying the geometry of Hopf manifolds [**29**]. Along these lines, the reader may consult [**32, 33**], which study certain local holomorphic dynamical systems by way of naturally associated complex manifolds.

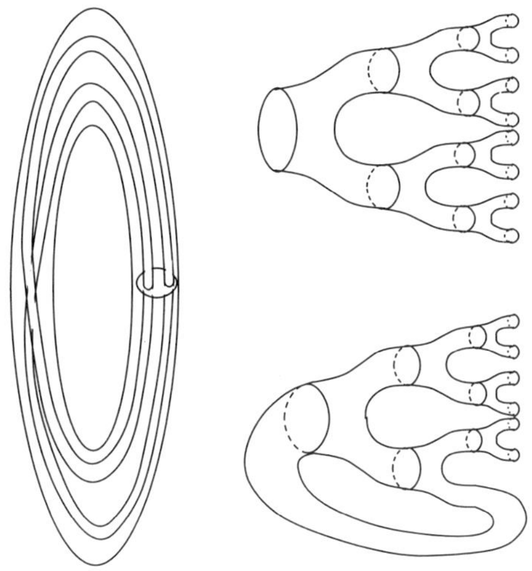

FIGURE 1

To explain D. Sullivan's construction, we begin by observing that the oriented affine line $\mathbb{R}$ can be regarded in a natural way as the boundary of the Poincaré upper half-plane $\mathbb{H} = \{z \in \mathbb{C} : \Im z > 0\}$; that is, every affine map $x \in \mathbb{R} \mapsto ax + b \in \mathbb{R}$, with $a > 0$, extends to a biholomorphism $z \in \mathbb{H} \mapsto az + b \in \mathbb{H}$. Another way of expressing the same thing is to say that the set of nonzero, positively oriented vectors on an oriented affine line can be naturally identified with the Poincaré upper half-plane.

Now suppose we have a diffeomorphism f of a compact manifold V that preserves an oriented foliation $\mathcal{F}$ of real dimension 1, with the following properties:

(1) f expands the leaves of $\mathcal{F}$.
This means that if V is equipped with a suitable Riemannian metric, the differential of f expands the tangent vectors to $\mathcal{F}$.

(2) f acts affinely on the leaves of $\mathcal{F}$.
This means that we assume the leaves are identified with affine lines and these affine structures are preserved by f. We will discuss this condition later and will see that it actually follows from condition (1).

We can consider the space V' consisting of those tangent vectors to $\mathcal{F}$ that are nonzero and positively oriented. This is a noncompact manifold, equipped with a foliation $\mathcal{F}'$ by copies of the half-plane $\mathbb{H}$ and on which the differential f acts holomorphically (and isometrically) in the leaves. Condition (1) implies that the action of f on V' is free and proper, so the quotient of V' by f is a compact manifold $\overline{V}$ equipped with a foliation (and hence a lamination) $\overline{\mathcal{F}}$ whose leaves are quotients of $\mathbb{H}$. More precisely, we distinguish two cases. If a leaf of $\mathcal{F}$ is not preserved by any power of f, then the corresponding leaf of $\overline{\mathcal{F}}$ is isomorphic to $\mathbb{H}$. On the other hand, if a leaf of $\mathcal{F}$ is preserved by f^n (and by no f^i with $0 < i < n$), then f^n acts on this leaf as a homothety of ratio $\lambda > 1$, and the corresponding leaf of $\overline{\mathcal{F}}$ is the quotient of $\mathbb{H}$ by this homothety: it is a cylinder that is isomorphic, as a Riemann surface, to the annulus with modulus $\log(\lambda)$ defined by $\{w \in \mathbb{C} : 1 < |w| < \lambda\}$.

We will not show in detail how condition (1) implies condition (2). A Riemannian metric on V lets us parametrize the leaves of $\mathcal{F}$ by arc length; in particular, this defines an affine structure on the leaves of $\mathcal{F}$. Of course, this structure does not satisfy condition (2) because f does not necessarily multiply the lengths of tangent vectors to $\mathcal{F}$ by a constant. The idea is to iterate the affine structure by powers of f and show that this sequence of structures converges to an affine structure satisfying (2). Concretely, defining an affine structure on a one-dimensional manifold is essentially equivalent to giving a definition of barycenter—to defining the midpoint of two points, for instance. If a and b are two points on the same leaf of $\mathcal{F}$, consider the points $f^n(c_n)$, where c_n is the midpoint of the segment joining $f^{-n}(a)$ to $f^{-n}(b)$, calculated with the auxiliary Riemannian metric. One shows that c_n converges, then defines the midpoint of the segment $[a,b]$ to be the limit of this sequence; this defines the desired invariant affine structure. The expansion condition clearly plays a significant role in the proof of convergence. For the precise proof and the required differentiability hypotheses, we refer the reader to [**34**].

We still have to describe the situations where such diffeomorphisms that expand a one-dimensional foliation arise. The most obvious case, and the most studied, is that of an Anosov diffeomorphism with a one-dimensional invariant foliation. Many examples can be obtained in this way.

The case studied by D. Sullivan, however, is different. Consider an expanding map $g : \mathbb{S}^1 \to \mathbb{S}^1$ from the circle to itself; that is, a map with derivative greater than 1 everywhere. Of course, such a map cannot be a diffeomorphism; it is a covering of the circle. To turn the non-invertible dynamics of g into invertible dynamics, we consider its *natural extension*, which is defined as follows. Let V be the set of sequences $(x_n)_{n\in\mathbb{Z}}$ of points on the circle that are orbits of g; that is, such that $g(x_n) = x_{n+1}$. This space, equipped with the topology induced by the product topology, is a compact space that has a natural bijection f defined by $f((x_n)) = ((x_{n+1}))$ and a projection $\pi : (x_n) \in V \mapsto x_0 \in \mathbb{S}^1$. The fibers of π are Cantor sets, and clearly $g \circ \pi = \pi \circ f$. There exists a "one-dimensional foliation" on V that is invariant under f and expanded by f. Indeed, being given a point in V amounts to being given a point x_0 on the circle and choosing a sequence of successive preimages under g. If the point x_0 moves continuously on the circle, we can follow these choices of preimages by continuity and thus describe a curve in V. These curves are the leaves of the foliation $\mathcal{F}$. We can also define these leaves as the unstable manifolds of f: two points a, b in V are in the same leaf of $\mathcal{F}$ if and only if the distance between $f^{-n}(a)$ and $f^{-n}(b)$ approaches 0 as n approaches $+\infty$.

We are in exactly the same situation as before except that V is not a manifold. But we can carry out the same construction: we produce an invariant affine structure on the leaves of $\mathcal{F}$, then construct a Riemann surface lamination $\overline{\mathcal{F}}$ as before. *In summary, we have associated a Riemann surface lamination with every expanding map g of the circle.*

To each periodic point x of g with period n, there corresponds a leaf of this lamination, which is an annulus with modulus $\log((g^n)'(x))$. Since the modulus of an annulus is a holomorphic invariant, we see that if two expanding maps of the circle define isomorphic laminations, then the derivatives of these two maps are the same at their periodic points. This observation underlies D. Sullivan's theorem, according to which *two expanding maps of the circle of class C^r $(r \geq 2)$ are conjugate by a diffeomorphism of class C^r if and only if their associated laminations are holomorphically equivalent.*

We can also proceed in a similar way with polynomial maps of one complex variable. More generally, let U be a simply connected open set in $\mathbb{C}$, and let $F : U \to F(U)$ be a proper holomorphic map such that $\overline{U} \subset F(U)$. For such *polynomial-like maps* (see [**16**]), the *filled Julia set* K is defined to be the intersection of the $F^{-n}(U)$ for $n \geq 0$. It is a compact set, which we will assume to be connected. We then consider the set V' of sequences of points $(z_n)_{n\geq 0} \in U \setminus K$ such that $F(z_n) = z_{n-1}$ for all $n > 0$. We agree to call two sequences z_n and z'_n equivalent if there exists an integer k such that $z_n = z'_{n+k}$ for all sufficiently large n. The quotient V of V' by this equivalence relation is a compact set equipped with a Riemann surface lamination. The leaves of this lamination are obtained by moving z_0 in $F(U) \setminus K$ and following the choices of the preimages z_n by continuity. *This allows us to associate a Riemann surface lamination with every polynomial-like map.*

In fact, these last two constructions are related. We can send the complement of K to the complement of the unit disk in $\mathbb{C}$ by conformal representation. The map F conjugated by this conformal representation is then defined in an annulus of the form $\{w \in \mathbb{C} : 1 < |w| < 1 + \varepsilon\}$ $(\varepsilon > 0)$ and extends by Schwarz's reflection principle to the unit circle. It can be shown that this extension is an expanding map of the circle and that the lamination associated with the latter is isomorphic to the lamination associated with F. For details and a great deal of supplementary information, see [**34**].

2.3. Polygonal tilings, Tits buildings. Here we describe other examples of laminations that arise naturally as abstract objects, not embedded in manifolds—in other words, for which the transverse spaces T_i are not manifolds.

Consider finitely many polygons $P_1, P_2, \dots, P_k$ in the Euclidean plane $\mathbb{R}^2$. Suppose these polygons tile the plane by translation; that is, $\mathbb{R}^2$ can be covered by tiles that are translates of the P_i, in such a way that two of these translates that are not disjoint intersect along a union of sides. Let $\mathcal{P}$ be the set of tilings of this type. We emphasize that two tilings that differ by a translation of $\mathbb{R}^2$ are a priori different: two tilings are considered to be identical only if they have the *same* tiles. We will turn $\mathcal{P}$ into a compact metric space equipped with a natural lamination. To illustrate the situation, we begin with the trivial case, where there is only one polygon: the unit square in $\mathbb{R}^2$. Up to translation, there is only one possible tiling, which is preserved by integer translations. Thus the space $\mathcal{P}$ is identified in this case with $\mathbb{R}^2/\mathbb{Z}^2$; that is, with the two-dimensional torus corresponding to the position of the origin with respect to the grid of the tiling.

More generally, if Π_1 and Π_2 denote two tilings in $\mathcal{P}$, we denote by $R(\Pi_1, \Pi_2)$ the largest nonnegative real number R (possibly infinite) such that the two tilings Π_1 and Π_2 coincide when restricted to the disk centered at the origin and with radius R. If $v \in \mathbb{R}^2$ and $\Pi \in \mathcal{P}$, we denote by $\Pi + v$ the tiling obtained by starting with Π and translating by the vector v. Let $\varepsilon, \varepsilon'$ be (small) positive real numbers, N a (large) positive real number, and Π a tiling in $\mathcal{P}$. We denote by $U_{\varepsilon,\varepsilon',N}(\Pi) \subset \mathcal{P}$ the set of Π' in $\mathcal{P}$ such that there exist vectors v, v' with $\|v\| < \varepsilon$, $\|v'\| < \varepsilon'$, and $R(\Pi + v, \Pi' + v') > N$. These sets form a basis for a topology that turns $\mathcal{P}$ into a (metrizable) compact space. Two tilings are close if, by translations by small vectors, they can be made to coincide on a large disk centered at the origin. Compactness follows from a classical diagonal argument and the fact that in a disk of given radius there are only finitely many possibilities for a tiling of which one of the vertices is a prescribed point. Details are left to the reader. Of course, $\mathbb{R}^2$ acts by translations on the compact set $\mathcal{P}$. The orbits define a lamination structure on $\mathcal{P}$. The leaves are Euclidean planes, cylinders, or tori; they are in one-to-one correspondence with the types of tilings, where we now identify two tilings if they differ by a translation. The compact orbits correspond to periodic tilings and, in general, the stabilizer of a point in $\mathcal{P}$ is the symmetry group of this tiling; that is, the group of translations that preserve it. Thus *we have associated with the family of polygons $P_1, \ldots, P_k$ a lamination whose dynamical structure precisely describes the space of tilings considered.*

There are now many examples of games with polygons P_i that tile the plane by translations but do not tile it periodically. The most famous examples are those of Penrose; we refer the reader to [**28**] for a description of these tilings. Here is an example (taken from [**37**]) of three "polyominoes" that tile the plane but do not tile it periodically. They thus provide examples of laminations without compact leaves. (To be precise, we would have to add to these three polygons their images under the isometries $(x, y) \mapsto (\pm x, \pm y)$ or $(\pm y, \pm x)$.)

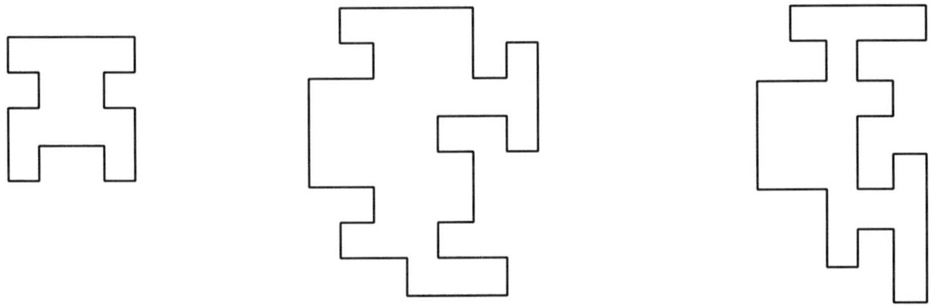

FIGURE 2

We can proceed in a similar way if we are interested in tilings whose tiles are not translates of P_i but where rigid motions of the Euclidean plane $\mathbb{R}^2$ are also permitted. The definition of distance between tilings will then have to be modified: two tilings will be close if they can be made to coincide on a large disk centered at the origin by moving them by isometries close to the identity. The corresponding space $\mathcal{P}$ is again compact, and equipped with an action by the group of motions: the orbits of the translation group then define a lamination on $\mathcal{P}$.

We can also modify the construction by starting with polygons P_i in the Poincaré disk $\mathbb{D}$ equipped with its hyperbolic metric. We then use the group of orientation-preserving isometries of $\mathbb{D}$, which is isomorphic to $\mathrm{PSL}(2, \mathbb{R})$ (since the stabilizer of a point is isomorphic to $\mathrm{SO}(2)$). We obtain a compact space $\mathcal{P}$, equipped with an action of $\mathrm{PSL}(2, \mathbb{R})$ whose orbits are three-dimensional. The action of the subgroup $\mathrm{SO}(2)$ is free except in the very special case where certain tilings are invariant under certain rotations. In the quotient space by this $\mathrm{SO}(2)$ action, we obtain a lamination with two-dimensional leaves; the leaf passing through a tiling Π is the quotient of $\mathbb{D}$ by the group of isometries of $\mathbb{D}$ that preserve the tiling. The situation where the $\mathrm{SO}(2)$ action is not free is similar to that of the quotient of a Riemann surface by a finite group, which leads, as is well known, to "orbifolds"; their theory is very similar to that of classical Riemann surfaces.

In the same vein, one can construct interesting examples of laminations by starting with certain *Tits buildings*. We refer the reader to [**7**] for an excellent exposition of the general theory, and will be satisfied here with an example. A Tits building of type $\widetilde{A}_2$ is a special two-dimensional simplicial complex; its faces are triangles called *chambers*, which are equipped with a flat metric that turns them into equilateral triangles with sides of length 1. We consider (the geometric realization of) this complex as a metric space for which the distance between two points is the length of the shortest path joining them. Moreover, certain subcomplexes are distinguished and called *apartments*; each is isomorphic to a Euclidean plane tiled with equilateral triangles. The conditions imposed on these apartments and chambers are as follows. First, any two chambers lie in a common apartment. Moreover, if two apartments have nonempty intersection, there must exist an isomorphism of the union that permutes the two apartments and is the identity on the intersection. One of the main justifications for studying these objects is that they play the role of symmetric spaces for p-adic Lie groups: for every prime number p, for instance, there exists a building I_p of type $\widetilde{A}_2$ on which the group $\mathrm{SL}(3, \mathbb{Q}_p)$ acts naturally. Every edge is incident to $p+1$ chambers, and the local situation in a neighborhood of a vertex is described as follows: there is an edge issuing from a vertex for each point and each line of the projective plane on the finite field $\mathbb{F}_p$, and two such edges are in the same chamber if they correspond to a point and a line that are incident. Every subgroup Γ of $\mathrm{SL}(3, \mathbb{Q}_p)$ acts on I_p, and there are many interesting arithmetic examples of groups Γ for which the action is free and the quotient I_p/Γ is a finite polyhedron P. The study of these finite quotients is similar to that of quotients of the Poincaré disk by arithmetic Fuchsian groups. Here is a very concrete example of such a quotient P for $p = 2$, taken from [**5**]. In Figure 3, edges with the same number are to be identified in threes; the resulting polyhedron P has one vertex, 7 faces, and 7 edges.

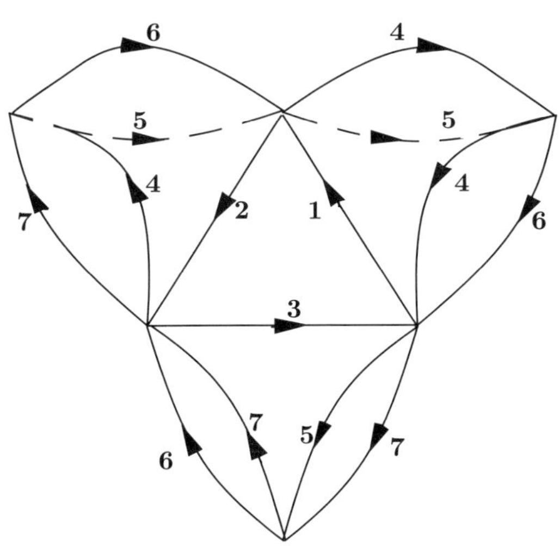

FIGURE 3

We show how to associate a lamination with such a situation. Let $\widetilde{E}$ be the space of isometric embeddings of the Euclidean plane $\mathbb{R}^2$ in I_p (the image is then an apartment), equipped with the topology of uniform convergence on compact sets. The group Γ acts naturally on $\widetilde{E}$ by composition in the range, and the quotient is a compact space E. The additive group $\mathbb{R}^2$ acts by translations in the domain; its orbits define a lamination on E. The dynamical study of these laminations is very interesting: it can be shown, for instance, that the union of the compact leaves is dense in E.

In the same way, one can construct many laminations whose leaves are not flat surfaces but surfaces with curvature -1. It suffices to consider buildings of another type, whose apartments are now Poincaré disks tiled by hyperbolic polygons. Many examples can be found in the work of M. Gromov, N. Benakli, and F. Haglund.

Here is another example of a lamination obtained by the same sort of process. Let S be a compact Riemann surface. One can consider the family of finite coverings (étales) of S. This forms a natural projective system: the projective limit of this system is a lamination $\mathcal{L}_S$ on a compact space M_S. Concretely, a point m of M_S is a map that assigns to each finite covering $\pi : \Sigma \to S$ a point $m(\pi)$ of Σ in such a way that if $p : \Sigma' \to \Sigma$ is a finite covering, then $p(m(\pi \circ p)) = m(\pi)$. Understanding the dynamics of such a lamination is not easy ...

We will not elaborate on these examples. Our only goal is to emphasize that the study of laminations need not be limited to "classical" dynamical systems.

3. Transverse Measures, Harmonic Measures

3.1. Fundamental class. We fix a lamination $\mathcal{L}$ on a compact space M and an atlas $h_i : U_i \to \mathbb{D} \times T_i$ whose transition functions are of the form $h_{ij}(z,t) = (f_{ij}(z,t), \gamma_{ij}(t))$.

With the goal of trying to generalize the theory of Riemann surfaces, we are led to introduce differential forms on M.

In a distinguished open set of type $\mathbb{D} \times T$, we say that a *differential k-form* is a family of real differential k-forms (of class C^∞, say) on the plaques $\mathbb{D} \times \{t\}$ that depends continuously on the transverse parameter t (in the C^∞ topology).

A differential k-form on the lamination $\mathcal{L}$ is given by differential k-forms on the distinguished open sets of an atlas that are compatible on the intersections in an obvious sense. We denote by $A^k(\mathcal{L})$ the space of k-forms on $\mathcal{L}$; this is a topological vector space. The differentiation operator along the leaves defines an operator $d : A^k(\mathcal{L}) \to A^{k+1}(\mathcal{L})$. Differential forms of class C^r, for $0 \leq r \leq \infty$, are defined similarly.

The complex structure on the leaves can be exploited in exactly the same way to define spaces $A^{p,q}(\mathcal{L})$ of (complex-valued) forms of type (p,q) and operators ∂, $\overline{\partial}$ from $A^{p,q}(\mathcal{L})$ to $A^{p+1,q}(\mathcal{L})$ and $A^{p,q+1}(\mathcal{L})$, respectively.

More generally, it is not hard to extend most of the classical definitions concerning Riemann surfaces: in each situation, one considers objects that depend continuously on the point and are smooth or holomorphic on the leaves, depending on the cases considered. The derivatives along the leaves are always assumed to be continuous on M.

The situation most closely related to that of Riemann surfaces is that in which there is a *fundamental class*: here one is concerned with being able to integrate a 2-form. A 2-form is called (strictly) *positive* if it is (strictly) positive when restricted to the leaves.

DEFINITION 3.1. A foliation cycle for the lamination $\mathcal{L}$ is a continuous linear operator $I : A^2(\mathcal{L}) \to \mathbb{R}$ that is strictly positive on the strictly positive forms and zero on the exact forms.

The most obvious example is that given by the current of integration on a compact leaf. This concept was introduced in a more general form by D. Sullivan in [**43**] and is a continuation of work of many authors, among them J. Plante [**39**]. The important result is that these foliation cycles correspond to *invariant transverse measures.*

DEFINITION 3.2. An invariant transverse measure μ for the lamination $\mathcal{L}$ is defined by giving a positive measure μ_i on each transverse space T_i such that, if $B \subset T_i$ is a Borel subset of the domain of definition of γ_{ij}, then $\mu_i(B) = \mu_j(\gamma_{ij}(B))$

It is not hard to see that being given an invariant transverse measure for an atlas produces another such measure for any equivalent atlas.

Let μ be an invariant transverse measure and f_i a partition of unity subordinate to the cover by the U_i (it is easy to check the existence of such partitions of unity, smooth on the leaves). Let ω be a 2-form on $\mathcal{L}$. If the support of ω is contained in a distinguished open set U_i, we may consider ω as a form on $\mathbb{D} \times T_i$. Integrating over the plaques $\mathbb{D} \times \{t\}$, we obtain a function on T_i that can in turn be integrated with respect to the measure μ_i. We then write $I(\omega)$ for the resulting number. When ω is not supported in an open set U_i, we decompose ω into $\sum_i f_i\omega$ and define $I(\omega)$ to be the sum $\sum_i I(f_i\omega)$. This defines a linear operator $I : A^2(\mathcal{L}) \to \mathbb{R}$, and it is easy to check that this is independent of the choices made (atlas, partition of unity) and that it is indeed a foliation cycle (i.e., it is zero on the exact forms). Thus every invariant transverse measure gives rise to a foliation cycle. It turns out that the converse is also true, so the two points of view are equivalent [**43**].

One of the major difficulties of this approach is that very many laminations have no foliation cycles. The Hahn-Banach theorem implies:

PROPOSITION 3.3 ([**43**]). *The following conditions are equivalent:*

- *The lamination $\mathcal{L}$ has no foliation cycle.*
- *There exists an exact 2-form that is strictly positive.*

In practice, it is easier to verify that there exists an invariant transverse measure. We encourage the reader to determine, among the examples we have described, which are the laminations that have invariant transverse measures. For a lamination defined by the suspension of a homomorphism from the fundamental group of a surface to the group of homeomorphisms of a compact set T, the existence of an invariant transverse measure is equivalent to the existence of a finite measure on T that is invariant under the group action. Thus we see that in general the existence of a foliation cycle is a very strong hypothesis.

However, there are geometric hypotheses on the leaves —related to their growth, for instance—that imply the existence of a foliation cycle (see [**26, 39**]).

A more general notion than that of foliation cycles is that of harmonic currents, introduced by L. Garnett in a more general context [**19**]. Note first that $\sqrt{-1}\partial\bar{\partial}$ is a real operator from $A^0(\mathcal{L})$ to $A^2(\mathcal{L})$ whose image is contained in the space of exact forms.

DEFINITION 3.4. A (foliation) harmonic current is a linear operator $I : A^2(\mathcal{L}) \to \mathbb{R}$ that is continuous, strictly positive on strictly positive forms, and zero on forms of the type $\sqrt{-1}\partial\bar{\partial}u$, for every function u that is continuous on M and smooth on the leaves.

A foliation cycle is clearly a harmonic current. The advantage of this notion comes from the following result:

PROPOSITION 3.5 ([**19**]). *Every lamination has at least one harmonic current.*

PROOF. By the Hahn-Banach theorem, what must be shown is that a strictly positive 2-form ω cannot be a limit of forms of type $\sqrt{-1}\partial\bar{\partial}u_n$. Thus it suffices to show that $\sqrt{-1}\partial\bar{\partial}u$ cannot be strictly positive. Since M is compact, u attains its maximum at some point x. The restriction of u to the leaf through x also attains its maximum at x, and $\sqrt{-1}\partial\bar{\partial}u$ is therefore less than or equal to zero at this point (in local coordinates $x+\sqrt{-1}y$, we have $\sqrt{-1}\partial\bar{\partial}u = 1/2(\partial^2u/\partial x^2+\partial^2u/\partial y^2)dxdy$). □

A less canonical way of defining these harmonic currents consists of choosing a Hermitian metric on the tangent bundle to $\mathcal{L}$. This defines an area element along the leaves and thus allows us to identify $A^2(\mathcal{L})$ with the space of functions on M. Hence the current I defines an element of the dual of the space of functions on M, and the positivity condition shows that I extends to the space of continuous functions; that is, we have a positive measure μ on M. From this point of view, the harmonicity condition on the measure μ can be expressed as follows. The integral $\int \Delta u\, d\mu$ is zero for every smooth function u on M, where Δ denotes the Laplacian along the leaves for the chosen Hermitian metric. For this reason we will often speak of harmonic measure, even if this assumes a non-canonical choice of a Hermitian metric along the leaves.

The local nature of harmonic currents is easy to describe [**19**]. Consider a harmonic current I and its restriction to a distinguished open set $U_i \simeq \mathbb{D}\times T_i$. There exist a probability measure ν_i on T_i and a function $\phi_i : \mathbb{D}\times T_i \to \mathbb{R}^+$ that, for ν_i-almost every point t in T_i, is defined and harmonic on the whole plaque $\mathbb{D}\times\{t\}$, such that if ω is a 2-form with support in U_i, then

$$I(\omega) = \int_{T_i}\left(\int_{\mathbb{D}\times\{t\}} \phi_i\omega\right) d\nu_i(t).$$

This local expression for I is not unique. However, if we have two local expressions using two systems ν_i, ϕ_i and ν_i', ϕ_i', then there exists a function $\delta_i : T_i \to \mathbb{R}_\star^+$, defined ν_i-almost everywhere, such that

$$\nu_i' = \delta_i^{-1}\nu_i \qquad \phi_i'(z,t) = \delta_i(t)\phi_i(z,t).$$

It follows that the harmonic functions $\phi_i(z,t)$ defined on the plaques are compatible on the intersections of the plaques, up to a multiplicative constant, and this allows us to define positive harmonic functions $\widetilde{\phi}$ on the universal cover of almost every leaf (in fact, on an abelian cover).

Harmonic currents that are foliation cycles are characterized by the fact that the functions $\phi_i(z,t)$ are independent of the variable z; in other words, $\widetilde{\phi}$ is a constant. Since every positive harmonic function on $\mathbb{C}$ is constant, we obtain the following corollary:

COROLLARY 3.6. *Let μ be a harmonic measure for a lamination. If the universal cover of μ-almost every leaf is conformally equivalent to $\mathbb{C}$, then μ is in fact a foliation cycle.*

3.2. The ergodic theorem, the topological type of the leaves. As an application of the notion of harmonic measure, we briefly describe an invariant of a lamination that is analogous to the genus of a Riemann surface. It is clear that classifying laminations up to homeomorphism is not reasonable. For instance, a theorem of J. Cantwell and L. Conlon states that for every connected noncompact surface L and every compact 3-manifold, there is a two-dimensional foliation on M one of whose leaves is homeomorphic to L [**11**]. Note that there exists an uncountable family of homeomorphism types of noncompact surfaces ... On the other hand, the more modest attempt to study the topology of almost all the leaves, with respect to a harmonic measure, yields the following theorem.

THEOREM 3.7 ([**22**]). *Let $(M, \mathcal{L})$ be a lamination and μ a harmonic measure. Then, for μ-almost every point x in M, the leaf of $\mathcal{L}$ that passes through x is either a compact Riemann surface or homeomorphic to one of the following six surfaces:*

(1) *the plane,*
(2) *the plane with infinitely many handles attached (the "Loch Ness monster"),*
(3) *the cylinder,*
(4) *the cylinder to which are attached infinitely many handles that accumulate at the two ends ("Jacob's ladder"),*
(5) *the sphere minus a Cantor set,*
(6) *the sphere minus a Cantor set, to which are attached infinitely many handles that accumulate at all the ends.*

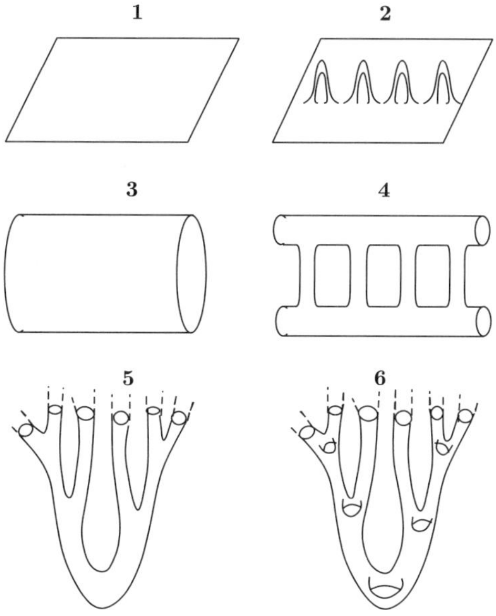

FIGURE 4

We give the idea of a special case rather than going into the details of the proof. This will let us show how being given a harmonic measure makes possible an ergodic study of laminations.

Let L be a Riemannian manifold, $x \in L$ a base point, and $\Omega_x(L)$ the set of continuous maps γ from $\mathbb{R}^+$ to L such that $\gamma(0) = x$. If we use the heat kernel on the manifold L, the theory of Brownian motion lets us construct a natural probability measure w_x, called the Wiener measure, on $\Omega_x(L)$.

This construction can be applied, in particular, to the leaf L_x of the lamination $\mathcal{L}$ that passes through the point x of M. The union Ω of the $\Omega_x(L_x)$ is the space of continuous maps from $\mathbb{R}^+$ to M that have image contained in a leaf. Given a finite measure μ on M, we can integrate all these measures w_x to produce a measure $\overline{\mu}$ on Ω. To be precise, if B is a Borel set in Ω, we let

$$\overline{\mu}(B) = \int_M w_x(B \cap \Omega_x(L_x))d\mu.$$

The advantage of this space Ω is that it is endowed with dynamics: if $\tau > 0$ and $\gamma \in \Omega$, we set $S_\tau(\gamma)(s) = \gamma(s+\tau)$. These transformations S_τ of Ω form a semigroup, i.e. $S_{\tau_1+\tau_2} = S_{\tau_1} \circ S_{\tau_2}$, and it is easy to show that S_τ preserves $\overline{\mu}$ if and only if μ is harmonic. To summarize: *a harmonic measure for a lamination leads to a one-parameter semigroup of transformations that preserve a finite measure to which one can apply the classical methods of ergodic theory*—Birkhoff's ergodic theorem, for instance:

THEOREM 3.8. *Let μ be a harmonic measure for a lamination $(M, \mathcal{L})$ equipped with a Hermitian metric. Let B be a Borel subset of M. Then, for μ-almost every point x in M and w_x-almost every path γ in $\Omega_x(L_x)$, the limit*

$$\ell(x, \gamma) = \lim_{t\to\infty} \frac{1}{t} m\{\tau \in [0, t] : \gamma(t) \in B\}$$

exists. Here m denotes Lebesgue measure. Moreover, there exists a measurable function $\ell : M \to \mathbb{R}$, constant on the leaves of $(M, \mathcal{L})$, such that $\ell(x) = \ell(x, \gamma)$ for μ-almost every x and w_x-almost every γ. Finally, $\int \ell(x)d\mu = \mu(B)$.

To illustrate how the ergodic theorem is used, we will show that there is no lamination $(M, \mathcal{L})$ all of whose leaves are homeomorphic to a sphere minus three points. Of course, this is a very special case of Theorem 3.7. In fact, to simplify the situation further, we also assume that all the leaves have constant curvature -1. We will discuss this hypothesis in Subsection 5.2 and will see that it is actually innocuous.

On a surface L homeomorphic to a sphere minus three points, there are three homotopy classes of simple closed curves that are not homotopic to a point. If a Riemann surface with negative curvature is a leaf of a lamination, it is complete, and each of these homotopy classes contains a unique closed geodesic. Indeed, in a lamination, a sufficiently short closed curve contained in a leaf is contained in a plaque; hence it is homotopic to a point in this leaf. The three geodesics of L bound a compact domain $c(L) \subset L$, the "convex core" of L.

Now suppose there exists a lamination all of whose leaves are of this type, and consider the union B of the convex cores of all the leaves. It is not hard to see that B is a Borel set, so we can apply the ergodic theorem to B, for some harmonic measure μ. The measure $\mu(B)$ is certainly nonzero because, by the local description of harmonic measures, a Borel set of measure zero intersects almost every leaf in a Borel set of zero area (with respect to the area element of the leaves). The ergodic theorem implies that for μ-almost every point x and w_x-almost every path $\gamma : \mathbb{R}^+ \to L_x$, the average time spent by γ in the core $c(L_x)$ exists and is nonzero. It is intuitively clear, and easy to prove, that this is impossible: for w_x-almost every path $\gamma : \mathbb{R}^+ \to L_x$, the path γ approaches infinity in L_x, and in particular the average time spent in $c(L_x)$ is zero. This is the desired contradiction.

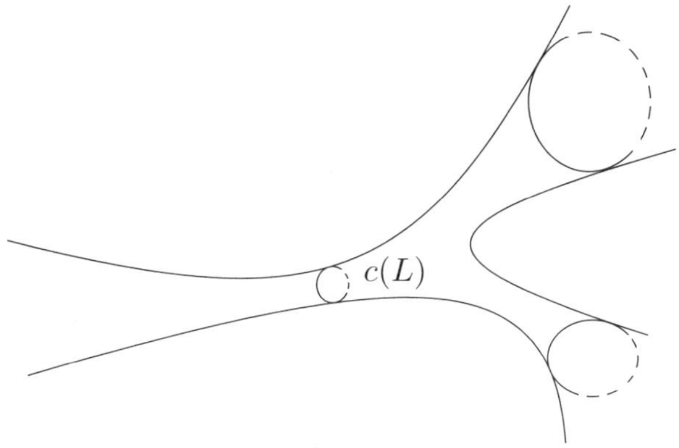

FIGURE 5

There are more elementary proofs of the fact that there is no lamination all of whose leaves are homeomorphic to spheres minus three points. We sketched this one, however, because essentially the same idea is what makes the proof of Theorem 3.7 work: one shows that an analogue of the convex core can be defined for every surface that is not one of the six described above.

J. Cantwell and L. Conlon have very recently obtained a topological analogue of Theorem 3.7 (see [**12**]). They show, in particular, that for every *minimal* lamination $(M, \mathcal{L})$ there exists a dense G_δ set $X \subset M$ that is a union of leaves that are homeomorphic to one of the six leaves described in Theorem 3.7.

4. The Gauss-Bonnet and Riemann-Roch Theorems

4.1. Some counterexamples. It is tempting to try to generalize the cohomological formalism and theorems like those of de Rham, Gauss-Bonnet, and Riemann-Roch to laminations. We begin by giving some examples showing that things aren't so simple!

First we consider the de Rham cohomology $H^\star(\mathcal{L})$, obtained by taking the quotient of the closed forms by the exact forms. The difficulty here is that this space may be infinite-dimensional even in the most elementary cases.

The simplest example where this phenomenon occurs is that of a product lamination $S \times T$ of a Riemann surface S by a compact space T. A closed (resp. exact) form on $\mathcal{L}$ is just a continuous map from T to the space of closed (resp. exact) forms on S. Hence the de Rham cohomology of $\mathcal{L}$ is the (generally infinite-dimensional) space of continuous maps from T to the de Rham cohomology of S.

This example may seem artificial because it does not correspond to a minimal lamination. The example of linear foliations on tori is more convincing. We will just sketch it because it is now well known (see also [**17**] for other examples of computations). Consider the torus $\mathbb{T}^2 = \mathbb{R}^2/\mathbb{Z}^2$, which has fundamental group generated by two commuting elements a and b, and let h be the homomorphism from this group to the group of diffeomorphisms of the circle $\mathbb{R}/\mathbb{Z}$ that sends a and

b to rotations through angles α and β, respectively. The foliation $\mathcal{L}$ obtained by suspension is linear on the torus $\mathbb{T}^3$. The circle $\mathbb{R}/\mathbb{Z}$ can be regarded as embedded in $\mathbb{T}^3$, transverse to the foliation. If x is a point of $\mathbb{R}/\mathbb{Z}$ and we lift a loop homotopic to a (resp. b) in the leaves, with the point x as origin, then the endpoint of the path a_x (resp. b_x) thus obtained is $x+\alpha$ (resp. $x+\beta$). We assume that α and β are numbers that are linearly independent over $\mathbb{Q}$, so that all the leaves of $\mathcal{L}$ are dense and homeomorphic to planes.

We evaluate the first cohomology group $H^1(\mathcal{L})$. Let ω be a closed 1-form. The integral of ω along a_x (resp. b_x) defines a continuous function A (resp. B): $\mathbb{R}/\mathbb{Z} \to \mathbb{R}$. Since ω is closed,

$$A(x) + B(x+\alpha) = B(x) + A(x+\beta). \tag{1}$$

If ω is the differential of a function $f : \mathbb{T}^3 \to \mathbb{R}$, then the restriction F of f to the fiber $\mathbb{R}/\mathbb{Z}$ over the base point in $\mathbb{T}^2$ satisfies

$$A(x) = F(x+\alpha) - F(x), \qquad B(x) = F(x+\beta) - F(x). \tag{2}$$

Conversely, it is easy to prove that $H^1(\mathcal{L})$ is isomorphic to the space of pairs of functions A, B satisfying (1) modulo those of the form (2). It is then natural to expand A, B, and F in Fourier series:

$$A(x) = \sum_n a_n \exp(2\sqrt{-1}\pi n x), \qquad B(x) = \sum_n b_n \exp(2\sqrt{-1}\pi n x)$$

$$F(x) = \sum_n f_n \exp(2\sqrt{-1}\pi n x).$$

Condition (1) can be written as

$$a_n(1 - \exp(2\sqrt{-1}\pi\beta n)) = b_n(1 - \exp(2\sqrt{-1}\pi\alpha n)). \tag{3}$$

Relation (2) becomes

$$a_n = (\exp(2\sqrt{-1}\pi n\alpha) - 1) f_n, \quad b_n = (\exp(2\sqrt{-1}\pi n\beta) - 1) f_n. \tag{4}$$

Thus, for a pair A, B to correspond to a coboundary, a_0 and b_0 must be zero. Conversely, if a_0 and b_0 are zero, we can use formula (4) to compute the f_n in two ways, which are compatible by (3).

$$f_n = a_n(\exp(2\sqrt{-1}\pi n\alpha) - 1)^{-1} = b_n(\exp(2\sqrt{-1}\pi n\beta) - 1)^{-1}.$$

However, the Fourier series with coefficients f_n may diverge even if the Fourier series with coefficients a_n and b_n converge. This is the well-known phenomenon of *small divisors*. If α (or β) does not satisfy a diophantine condition, $(\exp(2\sqrt{-1}\pi n\alpha) - 1)$ may be small for values of n that are not too large ... Thus we see that the space of pairs A, B for which the Fourier series with coefficients f_n converges may have infinite codimension.

By a suitable choice of α and β, we can obtain examples for which $H^1(\mathcal{F})$ is infinite-dimensional. Note moreover that $H^1(\mathcal{F})$ is not Hausdorff as a quotient topological vector space. A priori, we could just as well have defined the cohomology $H^1(\mathcal{L})$ by using less regular differential forms, which (for instance) would depend only measurably on the point in the transverse direction. This would indeed change the group $H^1(\mathcal{F})$ but would not eliminate the fact that the dimension could be infinite.

In connection with this, *the Teichmüller space of a lamination may be infinite-dimensional.* To give some meaning to this last assertion, we make the following definitions. A quasiconformal homeomorphism $f : M \to M$ is a homeomorphism that preserves each leaf of $\mathcal{L}$ and induces a (uniformly) quasiconformal homeomorphism on it [**2**]. The Teichmüller space of the lamination $(M, \mathcal{L})$ is the space $\mathcal{T}(M, \mathcal{F})$ of quasiconformal homeomorphisms f, where we identify f_1 and f_2 if $f_1 \circ f_2^{-1}$ is holomorphic on the leaves and homotopic to the identity by a homotopy that consists of quasiconformal homeomorphisms preserving each leaf.

Consider a product lamination $S \times \mathbb{S}^1$, where S is a Riemann surface that is not the sphere. The Teichmüller space of this lamination is clearly the space of continuous maps from $\mathbb{S}^1$ to the Teichmüller space of S. It is thus infinite-dimensional, but the lamination is not minimal.

A far more interesting example of such a situation is given by Sullivan's construction, which was described in 2.2. We have seen that every expanding map g of the circle has a corresponding lamination $\mathcal{L}_g$. It is not hard to check that all these laminations associated with maps g of the same topological degree are quasiconformally equivalent, and we have pointed out that they are holomorphically equivalent only if the corresponding maps g are differentiably conjugate. But the space of conjugacy classes of expanding maps of the circle with a given degree is infinite-dimensional. Indeed, an expanding map of the circle has infinitely many periodic points; if we fix finitely many of them, then we can arbitrarily prescribe the derivatives (> 1) at these points.

4.2. The index theorem. We will state the index theorem of A. Connes for foliations, restricting on the one hand to Riemann surface laminations and on the other to the case where there exists a foliation cycle. For the general theory, the reader is referred to [**13, 14**].

Fix a lamination $(M, \mathcal{L})$ equipped with a foliation cycle μ, which we will also think of as an invariant transverse measure. Choose a Hermitian metric along the leaves. Fix an integer $\ell = 0, 1$, or 2.

For each point x of M, the leaf $L(x)$ passing through x is in general a noncompact Riemannian manifold, so the space of harmonic forms of degree ℓ on $L(x)$ is in general infinite-dimensional. Let $\mathcal{H}^\ell(L(x))$ denote the Hilbert space of square-integrable harmonic ℓ-forms on $L(x)$. One thus obtains a measurable Hilbert space bundle over M. Of course, the fibers over two points in the same leaf are canonically identified, so it is better to think of this fiber as lying over the "space of leaves", even if the latter is not Hausdorff in general.

Actually, for technical reasons we have to consider the holonomy cover $\overline{L}(x)$ of the leaf $L(x)$ for each point x. We let $\mathcal{H}^\ell(L(x))$ denote the space of square-integrable ℓ-forms on $\overline{L}(x)$ (rather than $L(x)$)—see [**24**], for example. It turns out that in most cases, almost every leaf has trivial holonomy; hence $L(x) = \overline{L}(x)$ for almost every x, so the two definitions are the same almost everywhere.

Here is another way to construct such a "fiber bundle over the space of leaves". Let $B \subset M$ be a Borel set *transverse* to $\mathcal{L}$; that is, B intersects each leaf in a finite or countably infinite set. This allows us to define the transverse measure $\mu(B)$ unambiguously. For each point x in M, we can consider the Hilbert space $\ell^2(B \cap L(x))$ of square-integrable functions on this countable set. This is also a Hilbert space bundle, "constant" on the leaves. The number $\mu(B)$ is called the *Murray–von Neumann dimension* of this Hilbert space bundle. This definition is justified by the following lemma (see [**13, 14**]).

LEMMA 4.1. *Let B_1 and B_2 be Borel sets transverse to $\mathcal{L}$. Suppose there exists a measurable map that is constant on the leaves and assigns to μ-almost every point x of M an isometry Φ_x between $\ell^2(B_1 \cap L(x))$ and $\ell^2(B_2 \cap L(x))$. Then $\mu(B_1) = \mu(B_2)$.*

We return to our Hilbert space bundle $\mathcal{H}^\ell(L(x))$. A. Connes proved that it is isomorphic to a bundle of the form $\ell^2(B \cap L(x))$ for some transverse Borel set, so we can define its Murray–von Neumann dimension. The (nonnegative real) number thus obtained is called the ℓth Betti number of the lamination relative to the foliation cycle μ and is denoted by $\beta_\ell(\mathcal{L}, \mu)$. It is easy to check that this number is independent of the choice of the metric along the leaves.

Let k be the curvature of the metric; depending on one's point of view, k can be regarded as a 2-form on $\mathcal{L}$ or as a function on M. We can now state the Gauss-Bonnet theorem for foliations, which evaluates $\int_M k\, d\mu$, the integral of k on the foliation cycle μ.

THEOREM 4.2 ([**13, 14**]). $\int_M k\, d\mu = \beta_0(\mathcal{L}, \mu) - \beta_1(\mathcal{L}, \mu) + \beta_2(\mathcal{L}, \mu)$

This number, obtained by starting either with a metric along the leaves or with the Betti numbers for foliations, is of course called the *Euler-Poincaré characteristic for foliations* and denoted by $\chi(\mathcal{L}, \mu)$. See [**38, 44**] for a concrete interpretation of this number as the average of the Euler-Poincaré characteristics of large domains contained in the leaves.

The numbers $\beta_0(\mathcal{L}, \mu)$ and $\beta_2(\mathcal{L}, \mu)$ are of course equal and easy to evaluate: there can exist a nonzero square-integrable harmonic function on $\overline{L}(x)$ only if this manifold is compact. Although $\beta_1(\mathcal{L}, \mu)$ is harder to evaluate, it is nonnegative... The next result follows from this.

COROLLARY 4.3 ([**13, 14**]). *If the union of the compact leaves of $(M, \mathcal{L})$ has μ-measure zero, then the integral $\int_M k\, d\mu$ is nonpositive.*

The reader will recognize that Theorem 4.2 is valid in a much more general situation, which includes all elliptic operators that are defined along the leaves. We mention here only the Riemann-Roch theorem for foliations.

Consider a holomorphic line bundle E over the lamination $(M, \mathcal{L})$. This is simply a line bundle that is trivial over the distinguished open sets $U_i \simeq \mathbb{D} \times T_i$ and that has transition functions of the form

$$(z, t, \zeta) \in \mathbb{D} \times T_i \times \mathbb{C} \longmapsto (f_{ij}(z,t), \gamma_{ij}(t), g_{ij}(z,t)\zeta) \in \mathbb{D} \times T_j \times \mathbb{C},$$

where $g_{ij}(z,t) \in \mathbb{C}^\star$ depends continuously on (z,t) and holomorphically on z.

Just as in the classical case, the choice of a Hermitian metric $\| \, \|$ on the fibers of E allows us to define a "curvature" 2-form, which can be written in local coordinates as

$$\gamma_i = \frac{1}{2\sqrt{-1}\pi} \partial\bar{\partial}(\log \|(z,t,1)\|^2).$$

Since $g_{ij}(z,t)$ is holomorphic in z, these 2-forms are compatible on the intersections of the U_i. Hence they define a 2-form γ on $\mathcal{L}$.

When we change the Hermitian metric, the ratio of the two metrics is a positive function on M of type $\exp u$, and the form γ becomes $\gamma + 1/2\sqrt{-1}\pi\partial\bar{\partial}u$. In particular, γ is well defined modulo exact forms. The class in $H^2(\mathcal{L})$ thus defined is of course called the *Chern class* of E and denoted by $c(E)$.

It is important, however, to note that we are better off considering γ as well defined modulo forms of type $\sqrt{-1}\partial\overline{\partial}u$; this is much more precise than the simple fact of being defined modulo exact forms. This observation, due to A. Candel, shows in particular that *the pairing $\langle c(E), \mu\rangle$ of the Chern class with a harmonic current can be unambiguously defined even if the current is not a foliation cycle.*

For $\ell = 0, 1$ and x a point in M, we denote by $\mathcal{H}^\ell(L(x), E)$ the Hilbert space of square-integrable holomorphic ℓ-forms on $L(x)$ (or, more precisely, on its holonomy cover $\overline{L}(x)$) with values in E. In the same way as before, this Hilbert-space bundle has a Murray–von Neumann dimension, denoted by $h^\ell(E, \mu)$.

The version of the Riemann-Roch theorem proved by A. Connes becomes:

THEOREM 4.4 ([**14**]). $h^0(E,\mu) - h^1(E,\mu) = \langle c(E), \mu\rangle + \frac{1}{2}\chi(\mathcal{L}, \mu)$.

A detailed description of this theorem, as well as interesting examples, can be found in [**31**].

The following lemma will be used only in Section 7, but its proof can serve to illustrate the ideas that we just introduced.

LEMMA 4.5. *Let E be a holomorphic line bundle over a lamination $(M, \mathcal{L})$, and let $c(E)$ be its Chern class. Suppose that there exists a foliation cycle μ and that E has a holomorphic section s that is not identically zero on any leaf of $\mathcal{L}$. Then $\langle c(E), \mu\rangle \geq 0$. If the set of leaves that intersect the zero locus of s does not have μ-measure zero, then $\langle c(E), \mu\rangle > 0$.*

PROOF. Let $\|\ \|$ be a Hermitian metric on E, and let γ be its curvature form. Observe that the function $\log(\|s\|) : M \to \mathbb{R} \cup \{-\infty\}$ is locally integrable on each leaf.

We must evaluate $\langle c(E), \mu\rangle = \langle \gamma, \mu\rangle$. To do this, we choose a cover by open sets U_i that simultaneously trivialize the bundle and the lamination. Let s_i be a nonvanishing holomorphic section over U_i, and let f_i be the holomorphic function on U_i such that $s = f_i \cdot s_i$. Choose a partition of unity ϕ_i subordinate to this cover.

Then $\langle \gamma, \mu\rangle = \sum_i \langle \phi_i\gamma, \mu\rangle$, and each term of this sum can be computed in an open set $U_i \simeq \mathbb{D}\times T_i$ for which we have a local description of the foliation cycle. This involves integrating the 2-form $\phi_i\gamma$ on each plaque $P_t = \mathbb{D} \times \{t\}$, then integrating this function of t against a certain measure ν_i on T_i.

On each plaque P_t of U_i, we have $\gamma = \dfrac{1}{2\sqrt{-1}\pi}\partial\overline{\partial}(\log\|s_i\|^2)$. Since $s = f_i \cdot s_i$ on each plaque, it follows that

$$\gamma = \frac{1}{2\sqrt{-1}\pi}\partial\overline{\partial}(\log\|s_i\|^2) + [D_{P_t}]$$

in the sense of distributions, where $[D_{P_t}]$ denotes the sum of the Dirac measures at those points of P_t where the section s vanishes, counting multiplicities (which are positive integers). In our one-dimensional case, this formula, called the Poincaré-Lelong formula, is just a version of the Cauchy-Green-Stokes ... formula (see [**15**], for example). In particular, applying this formula to the test function ϕ_i restricted to the plaque P_t gives

$$\int_{P_t}\phi_i\gamma = \int_{P_t}\log\|s\|\cdot\frac{1}{\sqrt{-1}\pi}\partial\overline{\partial}\phi_i + \sum m_\alpha\phi_i(z_\alpha) \geq \int_{P_t}\log\|s\|\cdot\frac{1}{\sqrt{-1}\pi}\partial\overline{\partial}\phi_i,$$

where the z_α are the zeros of s on the plaque P_t, and m_α their multiplicities. Integrating against the transverse measure ν_i on T_i, we have

$$\langle \phi_i \gamma, \mu \rangle \geq \left\langle \log \|s\| \cdot \frac{1}{\sqrt{-1}\pi} \partial\bar{\partial}\phi_i, \mu \right\rangle .$$

The right-hand side makes sense because $\log \|s\|$ is integrable. Since $\sum_i \phi_i = 1$, summing over i gives

$$\langle c(E), \mu \rangle = \langle \gamma, \mu \rangle \geq 0.$$

This is the first part of the lemma. Note moreover that the inequality is strict unless, for every U_i, ν_i-almost every plaque does not intersect the zero locus of s. In other words, if the set of leaves that intersect the zero locus of s does not have μ-measure zero, then $\langle c(E), \mu \rangle > 0$. □

5. Uniformization of Laminations

5.1. The conformal type of the leaves. One of the main results of the theory of Riemann surfaces is the uniformization theorem, according to which the universal cover of any (connected) Riemann surface is isomorphic to $\mathbb{CP}^1$ (*elliptic* type), $\mathbb{C}$ (*parabolic* type), or $\mathbb{D}$ (*hyperbolic* type). In other words, any Riemann surface can be obtained from one of these three examples by passing to the quotient by a discrete group Γ that acts holomorphically, freely, and properly. We will say that a Riemann surface is elliptic, parabolic, or hyperbolic according to the type of its universal cover. This terminology does not agree with that used in potential theory. One should also note that an elliptic curve, i.e. a complex torus, is of parabolic type because its universal cover is $\mathbb{C}$!

In the hyperbolic case, the automorphisms of $\mathbb{D}$ are also the orientation-preserving isometries of the Poincaré disk. Riemann surfaces of this type are thus obtained by taking the quotient of $\mathbb{D}$ by a discrete, torsion-free subgroup of the isometry group of this disk, which is isomorphic to $\mathrm{PSL}(2,\mathbb{R})$. Riemann surfaces of hyperbolic type are thus naturally endowed with a complete Hermitian metric with constant curvature -1. This metric is *unique* because the Poincaré metric is the unique complete Hermitian metric with curvature -1 on the disk.

In the parabolic case, the automorphisms of $\mathbb{C}$ are the complex affine transformations, and only the translations act without fixed points. The Riemann surfaces of parabolic type are therefore, up to isomorphism, the plane $\mathbb{C}$, the quotient $\mathbb{C}/\mathbb{Z} \simeq \mathbb{C}^\star$, and the elliptic curves (quotients of $\mathbb{C}$ by lattices). Note that the complete flat Hermitian metric on $\mathbb{C}$ is invariant, so these parabolic Riemann surfaces are also flat surfaces. This flat Hermitian metric is unique *up to a constant multiplicative factor* because the canonical Hermitian metric on $\mathbb{C}$ is the unique complete flat metric on $\mathbb{C}$, up to a constant multiplicative factor.

In the elliptic case, every isomorphism of $\mathbb{CP}^1$ is a linear fractional transformation and hence has a fixed point. It follows that $\mathbb{CP}^1$ is the only Riemann surface of elliptic type. Of course, $\mathbb{CP}^1$ has a Hermitian metric with constant curvature $+1$.

If we consider a Riemann surface lamination $(M, \mathcal{L})$, we can apply this uniformization theorem to each leaf. The question we will discuss in this section is whether the uniformizations of the leaves depend continuously on the leaves. A more naive and basic problem is to study the partition of M by this type of leaves. The elliptic case is "easy": it is an instance of Reeb's stability theorem (see [**24**], for example).

THEOREM 5.1. *The union of the elliptic leaves of the lamination $\mathcal{L}$ is open. In this open set U, the leaves of $\mathcal{L}$ are the fibers of a locally trivial fibration.*

PROOF. Reeb's stability theorem states that in an arbitrary foliation, the union of the simply connected compact leaves is an open set and each of these leaves has a neighborhood that is trivially foliated. This theorem applies to the general case of the laminations we are considering (that is, on spaces that are not necessarily manifolds), although, strictly speaking, we know of no reference for this general statement... (However, see [**10**].) □

It follows that removing the leaves of elliptic type from a lamination yields a new compact space, hence a new lamination without elliptic leaves. From now on, *we assume that the lamination under consideration has no elliptic leaves.*

The case of parabolic leaves is harder, but we can state the following proposition, which will be proved later.

PROPOSITION 5.2. *The union of the parabolic leaves is a G_δ in the sense of Baire, that is, a countable intersection of open sets.*

There exist laminations that have both parabolic and hyperbolic leaves. Starting with a foliation on a 3-manifold, for instance, one can modify it by "Reeb turbularization" (see [**24**], for example). This result of this is to introduce a solid torus foliated by planes of parabolic type and bounded by a torus, also of parabolic type. In general, the topology of the leaves outside the solid torus is not that of the plane, the cylinder, or the torus, so these leaves are of hyperbolic type. These examples are clearly unsatisfactory because the types are not really mixed. L. Mosher and U. Oertel have constructed an example of a lamination such that the union of the parabolic leaves is not closed [**36**]. In this example, there exists a unique minimal set consisting of hyperbolic leaves. In 6.4 we will describe an example where the leaves are *all* dense and there is a mixture of conformal types.

We have already observed that a Riemannian metric along the leaves of a two-dimensional foliation defines the structure of a Riemann surface lamination. Note that the conformal types of the Riemann surfaces thus obtained are independent of the Riemannian metric because the disk, the plane, and the sphere are not quasiconformally isomorphic to each other.

5.2. Uniformization of hyperbolic laminations. Here we describe the result of A. Candel, which completely settles the hyperbolic case. Recall that a Hermitian metric on a lamination is defined by giving smooth Hermitian metrics on the leaves that (together with their derivatives) depend continuously on the point in M.

THEOREM 5.3 ([**10**]). *Let $(M, \mathcal{L})$ be a Riemann surface lamination. There exists a Hermitian metric on $(M, \mathcal{L})$ with constant curvature -1 if and only if, for every foliation cycle μ, the Euler-Poincaré characteristic $\chi(\mathcal{L}, \mu)$ is negative.*

The necessary condition (that is, the negativity of $\chi(\mathcal{L}, \mu)$) clearly follows from Theorem 4.2 (the Gauss-Bonnet theorem for foliations).

Note that this theorem includes the (quite general) case where there is no foliation cycle ...

Before sketching the proof of Theorem 5.3, we begin with some general properties. Fix a Hermitian metric g on $(M, \mathcal{L})$ and let $k : M \to \mathbb{R}$ denote the curvature function on the leaves. Let $u : M \to \mathbb{R}$ be a smooth function along the leaves, and let g' denote the Hermitian metric $\exp(2u)g$. The formula giving the curvature k' of g' is well known:

$$k' = \exp(-2u)(k - \Delta u), \tag{5}$$

where Δ denotes the Laplacian along the leaves.

We know that on every leaf $L(x)$ of hyperbolic type, there exists a unique smooth function $u : L(x) \to \mathbb{R}$ such that the metric $\exp(2u)g$ on $L(x)$ is complete and has curvature -1. We set $u = -\infty$ on any leaf $L(x)$ of parabolic type. This defines a global function $M \to \mathbb{R} \cup \{-\infty\}$ that satisfies

$$k = \Delta u - \exp(2u)$$

wherever it is finite.

PROPOSITION 5.4 ([**20, 45**]). *For every lamination (with no leaves of elliptic type), the function u is upper semicontinuous and its gradient along the leaves is bounded.*

SKETCH OF THE PROOF. Let $x \in M$ be such that the leaf $L(x)$ is of hyperbolic type, and let $f : \mathbb{D} \to L(x)$ be a holomorphic covering such that $f(0) = x$. By definition of the function u, the norm of the derivative of f at 0 is $\exp(-u(x))$. Let $0 < r < 1$, and consider the restriction of f to the disk D_r with center 0 and radius r. Since D_r is simply connected and relatively compact in $\mathbb{D}$, there exist a pointed space $(Q, *)$ and a map $F : D_r \times Q \to M$ such that

- $F(z, *) = f(z)$;
- F is a local homeomorphism in a neighborhood of x;
- the restriction of F to $D_r \times \{q\}$ is a holomorphic map into a leaf.

Now, using Schwarz's lemma in the leaves, we see that the norm of the derivative at z of $F(z, q)$ is less than or equal to $\exp(-u(F(z, q))$ when the norm is computed in the Poincaré metric in the domain and in the metric g in the range. Thus we have found locally defined, continuous functions that are upper bounds for the function u. Letting r go to 1, it is not hard to see that u is the greatest lower bound of these continuous functions; this proves semicontinuity. The proof that the gradient of u is bounded is elementary and reduces to Koebe's distortion lemma: if $j : \mathbb{D} \to \mathbb{C}$ is an injective holomorphic map with $j(0) = 0$ and $j'(0) = 1$, then $|j''(0)| \leq 2$. For more details, see [**20**]. □

Observe in particular that this proposition shows that the set of points where $u = -\infty$ is a G_δ in the sense of Baire: this is Proposition 5.2. It was this proposition that also allowed us, in [**20**], to extend Corollary 4.3:

COROLLARY 5.5 ([**20**]). *Let μ be an arbitrary harmonic measure. Then the integral $\int k \, d\mu$ is nonpositive.*

SKETCH OF THE PROOF. We have

$$\int k \, d\mu = \int \Delta u \, d\mu - \int \exp(2u) \, d\mu.$$

By the definition of a harmonic measure, the first integral on the right-hand side would vanish if u were continuous. It turns out that semicontinuity and the bound on the gradient are enough to ensure vanishing (see [**20**]). The sign of the curvature integral follows. □

See [**18, 42**] for other proofs of Corollary 5.5 and interesting ramifications.

We return to A. Candel's theorem. The proof splits into two parts. The first, in the same vein as Corollary 3.6, shows that if $\chi(\mathcal{L},\mu) < 0$ for every foliation cycle μ, then all the leaves are of hyperbolic type.

PROPOSITION 5.6. *If a lamination $(M,\mathcal{L})$ has a leaf L of parabolic type, then there exists a foliation cycle μ (supported in the closure of L) such that $\chi(\mathcal{L},\mu) = 0$.*

SKETCH OF THE PROOF. Suppose, for example, that there exists a holomorphic diffeomorphism $f : \mathbb{C} \to L$; the case of the cylinder is similar (and that of the torus is trivial). Let $|df|$ denote the norm of the differential of f, computed in the Euclidean metric on $\mathbb{C}$ and in the metric g on L. Let D_r be the disk of radius $r > 0$ in $\mathbb{C}$, and $L_r = f(D_r)$. The area of L_r in the metric g is the integral over D_r of $|df|^2$, and the length of the boundary ∂L_r is the integral of $|df|$ over the circle of radius r. Using a classical lemma due to Ahlfors (which reduces to an application of the Cauchy-Schwarz inequality), we conclude that

$$\liminf_{r\to\infty} \frac{\text{length}(\partial L_r)}{\text{area}(L_r)} = 0.$$

This means that the L_r make up an *averaging sequence*; that is, if c_r is the current of integration on the disk L_r, there exists a subsequence c_{r_n}, where r_n goes to infinity, that converges to a foliation cycle μ (because the boundary of L_r "disappears" as r goes to infinity; see [**26**]). It remains to show that $\chi(\mathcal{L},\mu) = 0$. This is done by using the classical Gauss-Bonnet theorem to integrate the curvature over L_r, and showing that the boundary effect disappears as r goes to infinity. □

The second part of the proof of Theorem 5.3 is the following:

PROPOSITION 5.7. *If all the leaves of a lamination are of hyperbolic type, then there exists a Hermitian metric of curvature -1.*

SKETCH OF THE PROOF. What must be shown is that, if the function u does not assume the value $-\infty$, it is continuous. By Proposition 5.4, it suffices to show that u is lower semicontinuous. Let $\mathcal{K}$ be the space of holomorphic maps from the disk $\mathbb{D}$ to (a leaf of) $(M,\mathcal{L})$, equipped with the topology of uniform convergence on compact sets. One first shows that $\mathcal{K}$ is compact. To do this, one considers a sequence of holomorphic maps $f_n : \mathbb{D} \to M$ with no convergent subsequence. Dilating their domains of definition by suitable homotheties and taking subsequences, one constructs a nonconstant holomorphic map from $\mathbb{C}$ to a leaf of $\mathcal{L}$ (Brody's lemma). This contradicts the fact that all the leaves are of hyperbolic type. Finally, to prove the lower semicontinuity of u, it suffices to write it as

$$\exp(u(x)) = \inf_{f\in\mathcal{K}} \{|df(0)|^{-1} \in \mathbb{R}^+_\star : f(0) = x\}$$

because the function $\mathcal{K} \ni f \mapsto |df(0)| \in \mathbb{R}^+_\star$ is continuous on the compact set $\mathcal{K}$. □

6. Uniformization of Parabolic Laminations

Now that we have managed to uniformize laminations all of whose leaves are hyperbolic, we will study laminations *all* of whose leaves are parabolic. The question is now whether there exists a flat Hermitian metric along the leaves.

6.1. Approximate uniformization and a counterexample.

THEOREM 6.1 ([**23**]). *Let* $(M, \mathcal{L})$ *be a lamination all of whose leaves are parabolic. Then there exists a sequence of smooth Hermitian metrics* g_n *on* $(M, \mathcal{L})$ *such that the curvature* form *of* g_n *approaches* 0 *uniformly as* n *approaches infinity.*

We emphasize that there would be no advantage in letting the curvature *function* approach 0. When a metric g is multiplied by a constant that approaches infinity, the curvature obviously approaches 0 ... But the area element da is multiplied by the square of this constant, so the curvature form $k\, da$ is preserved! This is why Theorem 6.1 considers the curvature form.

PROOF. Fix a lamination $(M, \mathcal{L})$ all of whose leaves are parabolic, as well as a Hermitian metric g. Let $k : M \to \mathbb{R}$ be the curvature function. Another way of writing formula (5), which gives the change of curvature under a conformal change of metric, is as follows. If $k\, da$ denotes the curvature form of g, then the curvature form $k'\, da'$ of $g' = \exp(2u)g$ is

$$k'\, da' = k\, da - 2\sqrt{-1}\partial\bar{\partial}u.$$

To prove the theorem, we must find a sequence of smooth functions u_n such that $2\sqrt{-1}\partial\bar{\partial}u_n$ approaches $k\, da$ uniformly.

Let $\mathcal{E}$ be the Banach space of 2-forms on $\mathcal{L}$ that are continuous on M, and $\mathcal{H}$ the subspace of forms of type $\sqrt{-1}\partial\bar{\partial}u$, with u smooth. By definition, the harmonic measures are the elements of the topological dual of $\mathcal{E}$ that vanish on $\mathcal{H}$ and assume positive values on positive elements. A continuous linear form on $\mathcal{E}$ that vanishes on $\mathcal{H}$ is thus the difference of two harmonic measures. By the Hahn-Banach theorem, we may conclude that an element is in the closure of $\mathcal{H}$ if and only if it vanishes on all the harmonic measures.

Thus, to prove the theorem it suffices to show that $\int k\, d\mu$ is zero for every harmonic measure. By Corollary 3.6, any harmonic measure comes from an invariant transverse measure, and the conclusion follows from Theorem 4.2. □

It is easy to construct laminations all of whose leaves are parabolic and for which there is no smooth Riemannian metric that is flat on the leaves. The simplest example is the usual Reeb foliation of the sphere $\mathbb{S}^3$. If there were a metric on the sphere such that all the leaves were flat, the noncompact leaves would be isometric to a Euclidean plane. Hence the growth of the areas of disks in the noncompact leaves would be quadratic, contradicting the fact that this growth is obviously linear. But this example is unsatisfying for two reasons. First, the leaves are not dense, and it is more reasonable to consider minimal laminations. Moreover, we observe in the example of the Reeb foliation that there exists a *measurable* Riemannian metric that is smooth on the leaves and flat. This metric is actually continuous in the complement of the compact leaf, and approaches infinity as one approaches this compact leaf. In what follows, we will restrict our attention to minimal laminations and look for measurable metrics.

THEOREM 6.2 ([**23**]). *There exists a lamination* $(M, \mathcal{L})$ *such that*

- *all the leaves of* $\mathcal{L}$ *are parabolic;*
- *all the leaves of* $\mathcal{L}$ *are dense;*
- *there is no Hermitian metric that is* measurable *on* M *and differentiable on the leaves, and also complete and flat on the leaves.*

We settle here for constructing this counterexample, and refer the reader to [**23**] for the proof that it actually has all the required properties.

First, consider the three 1-complex-parameter subgroups of $\mathrm{SL}(2,\mathbb{C})$ defined as follows:

$$d^t = \begin{pmatrix} \exp(t) & 0 \\ 0 & \exp(-t) \end{pmatrix}, \quad h_+^s = \begin{pmatrix} 1 & s \\ 0 & 1 \end{pmatrix}, \quad h_-^s = \begin{pmatrix} 1 & 0 \\ s & 1 \end{pmatrix}.$$

These 1-parameter groups satisfy

$$d^t h_\pm^s d^{-t} = h_\pm^{\exp(\pm 2t)s}.$$

Let $\widetilde{\mathcal{F}}$ be the foliation of $\mathrm{SL}(2,\mathbb{C})$ whose leaves are the left cosets of $\{h_+^s\}$. Its leaves are parametrized by $\mathbb{C}$ and hence equipped with a natural flat metric. The relation above shows that left translation by d^t globally preserves the leaves of $\widetilde{\mathcal{F}}$ and acts on them as a similarity.

Let Γ be a discrete subgroup of $\mathrm{SL}(2,\mathbb{C})$ such that the quotient $M = \mathrm{SL}(2,\mathbb{C})/\Gamma$ is compact. Of course, there are many examples of such groups associated with real-hyperbolic 3-manifolds. Since right translations commute with left translations, the foliation $\widetilde{\mathcal{F}}$ passes to the quotient as a foliation $\mathcal{F}$ on M. The leaves of $\mathcal{F}$ are the orbits of the natural holomorphic action of a 1-complex-parameter group, which we continue to denote by h_+^s, so all the leaves are equipped with a flat metric. Thus the foliation $\mathcal{F}$ is *not* the desired counterexample!

We modify this example as follows. We know that there exist real-hyperbolic 3-manifolds with nonzero first Betti number. Hence there are examples of groups Γ that have a nontrivial homomorphism $c : \Gamma \to \mathbb{Z}$. Let ε be a positive real number. Consider the right action of Γ on $\mathrm{SL}(2,\mathbb{C})$ defined by

$$(x,\gamma) \in \mathrm{SL}(2,\mathbb{C}) \times \Gamma \longmapsto d^{\varepsilon\cdot c(\gamma)} x\,\gamma \in \mathrm{SL}(2,\mathbb{C}).$$

It turns out that if ε is sufficiently small, this action is free, proper, and cocompact [**21**]. Let $M_{\varepsilon\cdot c}$ denote the quotient compact manifold. Note that this new action still globally preserves the foliation $\widetilde{\mathcal{F}}$, and hence defines a holomorphic foliation $\mathcal{L}$ on the manifold $M_{\varepsilon\cdot c}$. But the flat metric on the leaves of $\widetilde{\mathcal{F}}$ is no longer preserved, and the leaves of $\mathcal{L}$ are not equipped with a flat metric. On the other hand, the action is conformal on the leaves of $\widetilde{\mathcal{F}}$, so the leaves of $\mathcal{L}$ are naturally equipped with a conformal structure, which depends analytically on the point in $M_{\varepsilon\cdot c}$. In other words, there exists a real-analytic Riemannian metric g that is defined on the tangent bundle to the leaves on $M_{\varepsilon\cdot c}$ and that, when lifted to $\mathrm{SL}(2,\mathbb{C})$ and restricted to a leaf of $\widetilde{\mathcal{F}}$, is conformally equivalent to the natural flat metric on this leaf.

One shows that this really is a non-uniformizable foliation, in the sense that *there is no measurable function $u : M_{\varepsilon\cdot c} \to \mathbb{R}$ that is differentiable along the leaves and such that the metric $g' = \exp(2u)g$ is complete and flat on the leaves.* The proof is not very hard, though it uses delicate results on the ergodic theory of the horocyclic flow on compact manifolds with negative curvature—more precisely, on their cyclic covers [**23**].

6.2. Continuity of the affine structure. We have seen that it is impossible in general to construct a flat metric along the leaves of a lamination with parabolic leaves. Note, however, that there does exist a *flat affine structure* along the leaves that is continuous. Here is an explanation.

Let S be a parabolic Riemann surface. The conformal map ρ between the universal cover of S and $\mathbb{C}$ is unique, up to an affine map, so S is naturally equipped with a complex affine structure. For instance, if x, y, z are distinct points in S that are near each other, one can choose lifts $\widetilde{x}, \widetilde{y}, \widetilde{z}$ in the universal cover that are near each other, and the ratio $(\rho(\widetilde{x}) - \rho(\widetilde{y}))/(\rho(\widetilde{x}) - \rho(\widetilde{z})) \in \mathbb{C}$ is independent of the choice of lifts and the choice of ρ. We will denote this ratio by $(x-y)/(x-z)$.

THEOREM 6.3 ([**23**]). *Let $(M, \mathcal{L})$ be a lamination all of whose leaves are of parabolic type. Then the affine structure on the leaves is continuous, in the following sense. Let (x_i, y_i, z_i) $(i \geq 0)$ be a sequence of triples of distinct points contained in the same distinguished open set and such that, for each i, the points x_i, y_i, z_i are in the same plaque. Suppose that x_i, y_i, z_i converge respectively to the distinct points $x_\infty, y_\infty, z_\infty$ in this same distinguished open set. Then $(x_i - y_i)/(x_i - z_i)$ converges to $(x_\infty - y_\infty)/(x_\infty - z_\infty)$ as i approaches infinity.*

We refer the reader to [**23**] for the proof and restrict ourselves here to indicating the tools used. Fix a lamination $(M, \mathcal{L})$ all of whose leaves are parabolic, equipped with a smooth metric g.

Let $*$ be a base point in M. We denote by L_* the leaf through $*$. Since the universal cover of L_* is conformally equivalent to the complex line, we have a conformal covering $\psi : \mathbb{C} \to L_*$ such that $\psi(0) = *$. Let D_n be the disk in $\mathbb{C}$ with center 0 and radius n (a positive integer). Since D_n is simply connected, the immersed disk $\psi(D_n)$ can be lifted into neighboring leaves. This means that there exist local homeomorphisms

$$\Psi_n : D_n \times Q_n \to M,$$

where the Q_n form a decreasing family of open neighborhoods of a base point q_∞ in a space Q, such that

- the restriction of Ψ_n to $D_n \times \{q_\infty\}$ coincides with the restriction of ψ to D_n;
- the image of Ψ_n is an open neighborhood of $*$;
- Ψ_n immerses each $D_n \times \{q\}$ into a leaf of $\mathcal{L}$.

Let ε_n be a decreasing sequence of real numbers that converges to 0. Shrinking Q_n if necessary, we may assume that

- the restriction of Ψ_n to each $D_n \times \{q\}$ is a $(1+\varepsilon_n)$-quasiconformal immersion into a leaf of $\mathcal{L}$ (see [**2**]).

The proof of the theorem now uses two facts.

– The first follows easily from general theorems about quasiconformal maps [**2**]. Let D^a, D^b, D^c be (small) disjoint closed disks, contained in a closed disk $D \subset \mathbb{C}$ with center 0. For every real number $\eta > 0$, there exists an $\varepsilon > 0$ such that if $f : \mathbb{D} \to \mathbb{D}$ is a $(1+\varepsilon)$-quasiconformal homeomorphism that fixes the origin and a, b, c are points in D^a, D^b, D^c, respectively, then

$$\left| \frac{a-b}{a-c} - \frac{f(a)-f(b)}{f(a)-b(c)} \right| < \eta.$$

– The second is Koebe's distortion theorem (see [**41**]). Recall that the theorem implies, in particular, that if h is a univalent holomorphic function $\mathbb{D} \to \mathbb{C}$ and

$\zeta_1, \zeta_2, \zeta_3$ are three points in $\mathbb{D}$ with modulus less than $r < 1$, then

$$\left| \frac{h(\zeta_1) - h(\zeta_2)}{h(\zeta_1) - h(\zeta_3)} - \frac{\zeta_1 - \zeta_2}{\zeta_1 - \zeta_3} \right| < A(r), \tag{6}$$

where $A(r)$ is a universal function that approaches 0 as r approaches 0.

By the first fact, the identifications between nearby plaques are almost affine with respect to the affine structures given by their embedding into D_n, for large n. By the second fact, the embeddings of D_n into the universal covers of the leaves given by the Ψ_n are almost affine when restricted to a fixed D_i, for n approaching infinity. This leads to a fairly easy proof that the affine structure on the leaves is in fact continuous.

6.3. A problem concerning linear foliations of tori. The simplest example of a parabolic foliation is of course a codimension-one linear foliation $\mathcal{F}$ on the three-dimensional torus $\mathbb{T}^3 = \mathbb{R}^3/\mathbb{Z}^3$.

Let g be a smooth Riemannian metric along the leaves of $\mathcal{F}$. As we have seen, this allows us to define a structure of a Riemann surface lamination on $\mathcal{F}$.

We do not know whether it is always possible to find a continuous (or even measurable) function $u : \mathbb{T}^3 \to \mathbb{R}$ such that $\exp(u)g$ is complete and flat along the leaves.[1]

The only (little) result we have in this direction is the following. A real number α is said to satisfy a *diophantine condition* if there exist constants $C > 0$ and $\varepsilon > 0$ such that

$$\left| \alpha - \frac{p}{q} \right| \geq \frac{C}{q^{2+\varepsilon}}$$

for all integers (p, q) with $q > 0$.

PROPOSITION 6.4 ([**23**]). *Let $\mathcal{F}$ be the linear foliation of $\mathbb{R}^3/\mathbb{Z}^3$ with equation $dt = \alpha_1 dx + \alpha_2 dy$, and let g be a metric of class C^∞. If the subgroup of $\mathbb{R}$ generated by α_1 and α_2 contains a nonzero rational number or a number satisfying a diophantine condition, then there exists a function $u : \mathbb{R}^3/\mathbb{Z}^3 \to \mathbb{R}$ of class C^∞ such that $\exp(u)g$ is flat on the leaves.*

The fact that this theorem holds in both the rational and the diophantine irrational cases suggests that it may hold without any arithmetic hypothesis.

We reproduce here the proof given in [**23**]. Let $\mathcal{F}$ be the linear foliation of the torus $\mathbb{T}^3 = \mathbb{R}^3/\mathbb{Z}^3$ with equation $dt = \alpha_1 dx + \alpha_2 dy$ (where (x, y, t) denote the coordinates in $\mathbb{R}^3$), and let g be a metric of class C^∞ on $\mathbb{T}^3$. We may regard $\mathcal{F}$ as the quotient of $\mathbb{R}^2 \times \mathbb{R}/\mathbb{Z}$, foliated by planes $\mathbb{R}^2 \times \{*\}$, by the group generated by the following two commuting diffeomorphisms:

$$T_1 : (x, y, t) \in \mathbb{R}^2 \times \mathbb{R}/\mathbb{Z} \longmapsto (x + 1, y, t + \alpha_1) \in \mathbb{R}^2 \times \mathbb{R}/\mathbb{Z},$$

$$T_2 : (x, y, t) \in \mathbb{R}^2 \times \mathbb{R}/\mathbb{Z} \longmapsto (x, y + 1, t + \alpha_2) \in \mathbb{R}^2 \times \mathbb{R}/\mathbb{Z}.$$

For each $t \in \mathbb{R}/\mathbb{Z}$, the plane $\mathbb{R}^2 \times \{t\}$ is equipped with the metric $\widetilde{g}_t$ obtained by lifting g. There exists a unique conformal diffeomorphism ψ_t from $(\mathbb{R}^2 \times \{t\}, \widetilde{g}_t)$ to the complex plane $\mathbb{C}$ that sends the points $(0, 0, t)$ and $(0, 1, t)$ to 0 and 1, respectively. Since we are assuming that the metric g is of class C^∞, the metric

[1] A. A. Glutsuk has just given an affirmative answer to this question (October 1999).

$\widetilde{g}_t$ depends on t in a C^∞ way. The version with parameters of the uniformization theorem given in [**1, 2**] shows that the bijection

$$F : (x, y, t) \in \mathbb{R}^2 \times \mathbb{R}/\mathbb{Z} \longmapsto (\psi_t(x, y), t) \in \mathbb{C} \times \mathbb{R}/\mathbb{Z}$$

is a diffeomorphism of class C^∞.

Since T_1 and T_2 act by isometries with respect to $\widetilde{g}$, their conjugates by F must act conformally on the $\mathbb{C} \times \{*\}$. In other words, $T_1' = F \circ T_1 \circ F^{-1}$ can be written as

$$T_1'(\zeta, t) = (a_1'(t)\zeta + b_1'(t), t + \alpha_1),$$

where the functions $a_1', b_1' : \mathbb{R}/\mathbb{Z} \to \mathbb{C}^*$ are of class C^∞.

First we study the case where α_1 is zero.

Since T_1' acts freely on $\mathbb{C}\times\mathbb{R}/\mathbb{Z}$, we observe that $a_1'(t)$ must be identically equal to 1, so T_1' acts on each $\mathbb{C} \times \{t\}$ as a nontrivial translation by $b_1'(t)$. Conjugating T_1' by

$$G : (\zeta, t) \in \mathbb{C} \times \mathbb{R}/\mathbb{Z} \longmapsto (b_1'^{-1}(t)\zeta, t) \in \mathbb{C} \times \mathbb{R}/\mathbb{Z},$$

we obtain

$$T_1'' = G \circ T_1' \circ G^{-1} : (\zeta, t) \in \mathbb{C} \times \mathbb{R}/\mathbb{Z} \longmapsto (\zeta + 1, t) \in \mathbb{C} \times \mathbb{R}/\mathbb{Z}.$$

We now look for $T_2'' = (G \circ F) \circ T_2 \circ (G \circ F)^{-1}$ in the form

$$T_2''(\zeta, t) = (a_2''(t)\zeta + b_2''(t), t + \alpha_2).$$

The condition that T_1'' and T_2'' commute can be written as

$$a_2''(t) = 1;$$

that is, T_2'' also acts by translations. This means that the metric obtained by transporting the Euclidean metric on $\mathbb{C} \times \{*\}$ by $(G \circ F)^{-1}$ is invariant under both T_1 and T_2, and hence passes to the quotient on the torus $\mathbb{T}^3$. The metric on $\mathbb{T}^3$ thus obtained is smooth, flat on the leaves, and conformally equivalent to g. Theorem 6.4 has been proved in the special case where $\alpha_1 = 0$.

More generally, suppose that the subgroup of $\mathbb{R}$ generated by α_1, α_2, and 1 has rank less than or equal to 2. Then the leaves of the foliation $\mathcal{F}$ are all cylinders or all tori. Transforming the foliation by an appropriate linear diffeomorphism of $\mathbb{T}^3$, we reduce to the case just studied, where $\alpha_1 = 0$. Theorem 6.4 is thus proved in this case as well.

We now tackle the more interesting case where α_1 satisfies a diophantine condition.

The following proposition is well known; it is the most elementary result of the theory of small denominators (see [**4**], for example). It is proved by simply evaluating the Fourier coefficients and using the diophantine estimate to find an upper bound.

PROPOSITION 6.5. *If α_1 satisfies a diophantine condition of the type above for every function $v : \mathbb{R}/\mathbb{Z} \to \mathbb{R}$ of class C^∞, then there exist a constant $\overline{v}$ and a function $w : \mathbb{R}/\mathbb{Z} \to \mathbb{R}$ of class C^∞ such that*

$$v(t) = w(t + \alpha_1) - w(t) + \overline{v}.$$

We return to the study of

$$T_1'(\zeta,t) = (a_1'(t)\zeta + b_1'(t), t+\alpha_1).$$

Applying Proposition 6.5 to the function $v(t) = \log|a_1(t)|$, we find that there exist a function $w : \mathbb{R}/\mathbb{Z} \to \mathbb{R}$ of class C^∞ and a constant $k > 0$ satisfying the homological equation

$$|\exp(w(t+\alpha_1))\exp(-w(t))| = k|a_1(t)|.$$

Now we can consider the diffeomorphism

$$G(\zeta,t) \in \mathbb{C}\times\mathbb{R}/\mathbb{Z} \longmapsto (\exp(-v(t))\zeta, t) \in \mathbb{C}\times\mathbb{R}/\mathbb{Z}.$$

Setting $T_1'' = (G\circ F)\circ T_1 \circ (G\circ F)^{-1}$, we obtain

$$T_1'' : (\zeta,t) \in \mathbb{C}\times\mathbb{R}/\mathbb{Z} \longmapsto (a_1''(t)\zeta + b_1''(t), t+\alpha_1) \in \mathbb{C}\times\mathbb{R}/\mathbb{Z},$$

where

$$|a_1''(t)| = k.$$

We claim that $k=1$. This could be deduced from Theorem 6.1, but it is elementary in this case. Indeed, let b_1'' denote an upper bound for the modulus of $b_1''(t)$. Then we can show by induction that the nth power $(n>0)$ of T_1'' satisfies

$$T_1''^{\,n}(0,0) = (\zeta_n, t+n\alpha_1),$$

where

$$|\zeta_n| \le b_1''(1+k+\cdots+k^{n-1}).$$

Suppose, for instance, that $k<1$. Then the preceding formula shows that the points $T_1''^{\,n}(0,0)$ stay in a compact subset of $\mathbb{C}\times\mathbb{R}/\mathbb{Z}$. This is impossible because the abelian group generated by T_1 and T_2 acts properly on $\mathbb{C}\times\mathbb{R}/\mathbb{Z}$. Similarly, we show that k cannot be greater than 1 by considering the backward iterates of T_1''. Thus we have shown that $k=1$.

In other words, T_1'' acts by isometries on the $\mathbb{C}\times\{*\}$. Set $T_2'' = (G\circ F)\circ T_2\circ (G\circ F)^{-1}$. We claim that T_2'' must also act by isometries. Writing T_2'' in the form

$$T_2'' : (\zeta,t) \in \mathbb{C}\times\mathbb{R}/\mathbb{Z} \longmapsto (a_2''(t)\zeta + b_2''(t), t+\alpha_2) \in \mathbb{C}\times\mathbb{R}/\mathbb{Z}$$

and writing out the commutation relation between T_1'' and T_2'', we obtain

$$a_1''(t)a_2''(t+\alpha_1) = a_2''(t)a_1''(t+\alpha_2)$$

$$a_1''(t+\alpha_2)b_2''(t) + b_1''(t+\alpha_2) = a_2''(t+\alpha_1)b_1''(t) + b_2''(t+\alpha_1).$$

Since α_1 is irrational and the modulus of $a_1''(t)$ is 1, the modulus of $a_2''(t)$ must be constant. Exactly as we showed above that $k=1$, we show here that the modulus of $a_2''(t)$ is in fact 1.

In summary, we have shown that if we conjugate both T_1 and T_2 by $G\circ F$, we obtain diffeomorphisms that act isometrically on the $\mathbb{C}\times\{*\}$. In other words, the Euclidean metric on the complex lines $\mathbb{C}\times\{*\}$ is invariant under T_1'' and T_2''. Hence this Euclidean metric, transported by $G\circ F$, passes to the quotient in the torus $\mathbb{T}^3$. We have obtained a smooth metric that is flat on the leaves and conformally equivalent to g, as claimed.

When the group generated by α_1, α_2, and 1 contains a number satisfying a diophantine condition, the situation reduces by a change of basis in the lattice $\mathbb{Z}^3$ to the case where α_1 satisfies a diophantine condition. Since we just analyzed this case, the proof of Theorem 6.4 is complete.

We point out a problem that would probably be worth careful study. Let $(M, \mathcal{L})$ be a compact lamination all of whose leaves are dense and parabolic (we have seen that this is independent of the choice of metric). *Does there exist a smooth (or even measurable)* Riemannian *metric that is flat on the leaves?* This question is different from the one studied in this subsection because the desired metric need not be Hermitian.

6.4. An example of a lamination where the conformal types are mixed. During the "État de la recherche" conference, we mentioned the problem of the existence of a minimal lamination with some leaves of parabolic type and others of hyperbolic type. Richard Kenyon showed us a nice idea for constructing an example, which we will describe here. We thank him for his kind permission to include the example in this article.

THEOREM 6.6. *There exists a Riemann surface lamination $\mathcal{L}$ on a compact metric space $\widetilde{\mathbb{X}}$ that satisfies the following conditions.*

- *All the leaves of $\mathcal{L}$ are dense.*
- *One leaf of $\mathcal{L}$ is conformally equivalent to a sphere minus four points (and hence is hyperbolic). All the other leaves are conformally equivalent to either the plane $\mathbb{C}$ or the parabolic cylinder $\mathbb{C}/\mathbb{Z} \simeq \mathbb{C}^\star$.*
- *All the leaves of $\mathcal{L}$ have trivial holonomy and polynomial growth.*

The last point is not the most important: for the notions of growth and holonomy, see [**24**].

We begin by describing a very general metric space that is analogous to those we described in 2.3. Consider the graph G whose vertices are the integer points $\mathbb{Z} \times \mathbb{Z}$ in the plane $\mathbb{R}^2$ and whose edges are the horizontal segments $[i, i+1] \times \{j\}$ or the vertical segments $\{i\} \times [j, j+1]$, where $(i, j) \in \mathbb{Z} \times \mathbb{Z}$. Let $\mathbb{A}$ be the set of (finite or infinite) subgraphs of G that contain the origin $(0, 0)$ and are trees (i.e. connected and with no cycles). The *distance* between two vertices of a tree is the minimal length of a path that joins them in this graph. If T is a tree in $\mathbb{A}$, $p = (i, j)$ one of its vertices, and n an integer, we will denote by $T(p, n)$ the ball with center p and radius n; that is, the subtree whose vertices are those vertices whose distance from p is less than or equal to n. We will turn $\mathbb{A}$ into a compact metric space. Let T_1 and T_2 be two elements of $\mathbb{A}$. By definition, the distance between T_1 and T_2 is $\exp(-n)$, where n is the largest integer (possibly $+\infty$) such that $T_1((0, 0), n) = T_2((0, 0), n)$. One can show by a diagonal argument that $\mathbb{A}$ is compact when equipped with this metric. The important point is of course that a limit of trees is a tree.

There is a natural equivalence relation on $\mathbb{A}$, which consists of changing the base point. More precisely, two trees T_1 and T_2 in $\mathbb{A}$ will be called equivalent if there exists an integer translation of the plane that sends T_1 to T_2. The corresponding equivalence classes are countable.

We will now modify the space $\mathbb{A}$ slightly, in order to construct a lamination. Our goal is to construct a compact metric space $\widetilde{\mathbb{A}}$ equipped with a lamination $\mathcal{L}$ and containing a copy of $\mathbb{A}$, in such a way that the leaves of $\mathcal{L}$ intersect $\mathbb{A}$ in the orbits of the equivalence relation we just defined on $\mathbb{A}$.

Consider a tree T in $\mathbb{A}$. There are 16 different possibilities for the ball with center the origin and radius 1 in T; these depend on the vertices of T at distance 1 from the origin that can form an arbitrary subset of the set with four elements

$\{(\pm 1, \pm 1)\}$. Thus $\mathbb{A}$ is the disjoint union of 16 subsets $\mathbb{A}_P$ indexed by the subsets P of $\{(\pm 1, \pm 1)\}$. These $\mathbb{A}_P$ are open and closed subsets of $\mathbb{A}$. For each subset P, consider a compact surface with boundary Σ_P that is homeomorphic to a sphere minus a number of open disks equal to the cardinality of P. The boundary components of Σ_P are indexed by the elements of P: we denote by $\partial_p \Sigma_P$ the boundary component of Σ_P indexed by $p \in P$. We also choose Riemannian metrics on these Σ_P such that all the connected boundary components have open neighborhoods isometric to a Euclidean annulus $\mathbb{S}^1 \times [0, 1)$, and fix such isometries. Finally, we choose an arbitrary base point $\star_P$ in each Σ_P. In Figure 6 we have indicated 4 of the 16 possibilities.

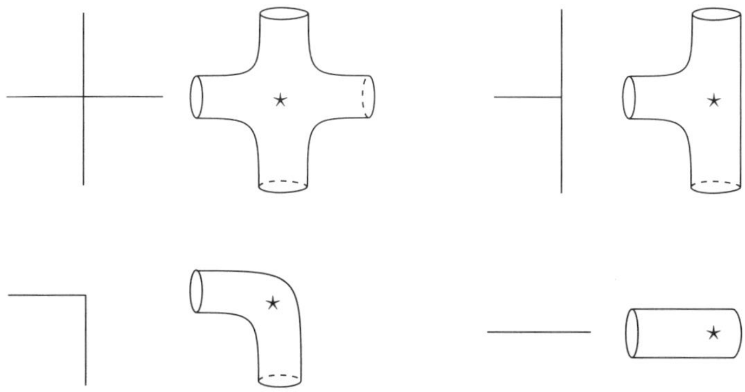

FIGURE 6

The desired space $\widetilde{\mathbb{A}}$ will be obtained by gluing the boundaries of the union of the $\widetilde{\mathbb{A}}_P = \mathbb{A}_P \times \Sigma_P$ in a suitable way. To be precise, let T be an element of $\mathbb{A}$, let P be the subset such that $T \in \mathbb{A}_P$, and choose an element $p \in \{(\pm 1, \pm 1)\}$ of P. The translate $T + p$ of T by the vector p is another element of $\mathbb{A}$, and the subset Q such that $T + p \in \mathbb{A}_Q$ contains the vector $-p$. Both the boundaries $\partial_p(\Sigma_P)$ and $\partial_{-p}(\Sigma_Q)$ are identified with the circle $\mathbb{R}/\mathbb{Z} \simeq \mathbb{S}^1 \subset \mathbb{S}^1 \times \{0\} \subset \mathbb{S}^1 \times [0, 1)$. For every $\theta \in \mathbb{R}/\mathbb{Z}$, the point θ on one of the boundaries is identified with the point $-\theta$ on the other boundary. These gluings are done for every tree $T \in \mathbb{A}$. One can check that the space thus obtained is compact; the important point is that, for each pair of subsets (P, Q) such that there exists an element $p \in \{(\pm 1, \pm 1)\}$ with $p \in P$ and $-p \in Q$, the set of trees T such that $T \in \mathbb{A}_P$ and $T + p \in \mathbb{A}_Q$ is an open and closed subset of $\mathbb{A}$. The embedding of each $\mathbb{A}_P \simeq \mathbb{A}_P \times \{\star_P\}$ in $\mathbb{A}_P \times \Sigma_P$ gives an embedding of $\mathbb{A}$ in $\widetilde{\mathbb{A}}$.

The various gluings described above are clearly compatible with the trivial laminations of $\mathbb{A}_P \times \Sigma_P$ by the surfaces with boundary Σ_P. We thus obtain a lamination $\mathcal{L}$ on $\widetilde{\mathbb{A}}$. The leaves of this lamination are equipped with a Riemannian metric, so $\mathcal{L}$ can be regarded as a Riemann surface lamination. Each leaf of $\mathcal{L}$ intersects $\mathbb{A}$ in an equivalence class of the relation on $\mathbb{A}$ described above. Each leaf L of $\mathcal{L}$ corresponds to a tree T in $\mathbb{A}$ that is defined up to integer translations. The tree T is in fact embedded in L as a "skeleton"; that is, every point of L is at uniformly bounded distance from T. Thus, studying the geometry of a leaf reduces essentially to studying the geometry of the corresponding tree. The case where P is empty is rather uninteresting; the only tree in $\mathbb{A}_\emptyset$ is the one that reduces to the origin, and since $\Sigma_\emptyset$ is a sphere, this leads to a spherical leaf that is isolated in the lamination $(\widetilde{\mathbb{A}}, \mathcal{L})$.

This lamination $(\widetilde{\mathbb{A}}, \mathcal{L})$ is unsatisfying from our point of view because $\mathbb{A}$ contains far too many trees, so the lamination is not minimal and contains many parabolic leaves and many hyperbolic leaves (as well as isolated elliptic leaves, which correspond to the finite trees in $\mathbb{A}$). The lamination $(\widetilde{\mathbb{X}}, \mathcal{L})$ that we are looking for will actually be the restriction of $\mathcal{L}$ to the closure of a very special leaf L_∞, which we will construct. To do this, we construct a special tree T_∞ in $\mathbb{A}$.

T_∞ is constructed recursively, as an increasing union of finite trees T_n in $\mathbb{A}$. The tree T_1 is indicated in Figure 7. Suppose we have constructed the tree T_n whose intersection with the vertical axis is the interval $\{0\} \times [-2^n, 2^n]$ and whose intersection with the horizontal axis is the interval $[-2^n+1, 2^n-1] \times \{0\}$. We then construct T_{n+1} as follows. First we translate T_n vertically by a vector $(0, 2^n)$, then consider the union of the images of this translate under the four rotations centered at the origin and through angles 0, $\pm\pi/2$, π. One can check that the graph thus obtained is a tree. Finally, we lop off the two extreme edges on the horizontal axis so that the tree T_{n+1} finally obtained actually intersects this axis in the interval $[-2^{n+1}+1, 2^{n+1}-1] \times \{0\}$, as required. An induction argument shows that each T_{n+1} extends T_n in such a way that the union T_∞ of the T_n is a tree in $\mathbb{A}$.

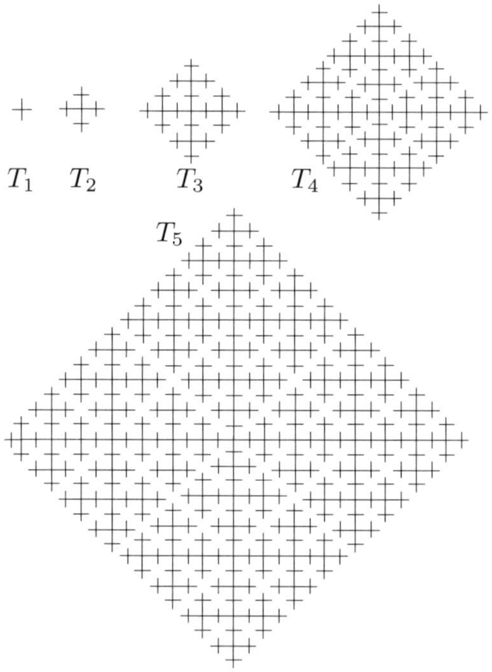

FIGURE 7

It can also be verified by induction that the vertices of T_∞ are the points with coordinates (i,j) such that $(i,j) = (0,0)$ or the 2-adic valuations of i and j are different.

Recall that the number of ends of a locally finite graph is the least upper bound (possibly infinite) of the number of infinite connected components of the complement of a finite subset. Note that the tree T_∞ has exactly 4 ends; it consists of the two coordinate axes onto which are grafted finite trees, which do not change the ends of T_∞. Hence the topology of the leaf of $\mathcal{L}$ that corresponds to T_∞ is that of a sphere minus four points, even if, from a geometric point of view, one has to imagine this sphere as being very "lumpy"...

Let $\mathbb{X}$ be the closure of the equivalence class of T_∞ in $\mathbb{A}$. For the moment, we assume the following two lemmas.

LEMMA 6.7. *For every tree T in $\mathbb{X}$, the closure of the equivalence class of T in $\mathbb{X}$ is dense in $\mathbb{X}$.*

LEMMA 6.8. *Let T be a tree in $\mathbb{X}$ that is not in the equivalence class of T_∞. Then T has one or two ends.*

To the closed subset $\mathbb{X}$ of $\mathbb{A}$ there corresponds a closed subset $\widetilde{\mathbb{X}}$ of $\widetilde{\mathbb{A}}$ that is saturated by the lamination $\mathcal{L}$. We will show that this lamination $(\widetilde{\mathbb{X}}, \mathcal{L})$ satisfies all the constraints of Theorem 6.6.

By Lemma 6.7, all the leaves of $\mathcal{L}$ are dense in $\widetilde{\mathbb{X}}$. Since the trees in $\mathbb{A}$ are embedded in the lattice of integer points of $\mathbb{R}^2$, it is clear that the number of vertices of a tree in $\mathbb{A}$ at distance n from the origin grows at most quadratically in n. Hence all the leaves of $\mathcal{L}$ have at most quadratic growth.

Consider a tree T in $\mathbb{A}$ with exactly two ends. There is exactly one infinite path that joins these two ends and has no backtrackings; thus T can be regarded as a line to which one attaches a collection of finite trees. The corresponding leaf L of $\mathcal{L}$ is then obtained by starting with an infinite cylinder and taking the connected sum of (lumpy!) spheres; that is, this leaf L is homeomorphic to a cylinder. We claim that this cylinder is conformally equivalent to $\mathbb{C}/\mathbb{Z} \simeq \mathbb{C}^\star$. It suffices to observe that L contains an infinite collection of disjoint open cylinders that are not homotopic to zero and are conformally equivalent to $\mathbb{S}^1 \times (0,1)$, so the modulus of L, considered as a Riemann surface, is necessarily infinite. This implies that L is indeed conformally isomorphic to $\mathbb{C}^\star$.

Similarly, a tree T in $\mathbb{A}$ that has only one end can be obtained by starting with a ray and grafting finite trees onto it. Thus the corresponding leaf L is topologically a plane. To show that it is in fact conformally isomorphic to $\mathbb{C}$, we need only observe that L contains a union of nested annuli that are all isomorphic to $\mathbb{S}^1 \times (0,1)$ and hence all have the same modulus.

It follows from these remarks and Lemma 6.8 that all the leaves of $(\widetilde{\mathbb{X}}, \mathcal{F})$ except for the leaf L_∞ corresponding to T_∞ are planes or cylinders, conformally equivalent to $\mathbb{C}$ or $\mathbb{C}^\star$. A similar argument shows that the leaf L_∞ is conformally equivalent to a sphere minus 4 points.

To prove Theorem 6.6, it remains to prove the two lemmas... We begin with two simple preliminary remarks whose proofs are left to the reader. They can easily be verified by using the definition of T_∞. One can also use the description we gave of the vertices of T_∞ in terms of 2-adic valuations, and observe that if i is a nonzero integer multiple of 2^n and a is a nonzero integer with absolute value less than 2^n, then the 2-adic valuations of a and $i + a$ are the same.

REMARK 1. For every ball $B = T_\infty((i,j), 2^{n+1})$, there exists an integer translation τ such that the image $\tau(T_\infty((0,0), 2^n))$ is contained in B and coincides with the ball $T_\infty(\tau(0,0), 2^n)$.

REMARK 2. For every ball $B = T_\infty((i,j),2^n)$, there exists a vertex (i',j') of T_∞ with $|i'| \leq 2^n$ and $|j'| \leq 2^n$ such that the translation by $(i'-i, j'-j)$ sends B to the ball $T_\infty((i',j'),2^n)$.

We first prove Lemma 6.7. Consider a tree T in the closure $\mathbb{X}$ of the equivalence class of T_∞. It must be shown that the equivalence class of T is also dense in $\mathbb{X}$. Clearly it suffices to show that T_∞ is in the closure of the class of T. In other words, it must be shown that for every integer n, there exists an integer translation (i,j) such that $T_\infty((0,0),2^n)+(i,j)$ is the ball $T((i,j),2^n)$.

Since T is in the closure of the class of T_∞, there exists an integer translation (i_0,j_0) such that $T((0,0),2^{n+1})+(i_0,j_0)$ is the ball $T_\infty((i_0,j_0),2^{n+1})$. By Remark 1, there exists (i_1,j_1) such that $T_\infty((0,0),2^n)+(i_1,j_1)$ is contained in $T((0,0),2^{n+1})+(i_1,j_1)$ and is the ball $T_\infty((i_1,j_1),2^n)$. It follows that the translation (i_1-i, j_1-j) sends the ball $T_\infty((0,0),2^n)$ onto the ball $T((i_1-i,j_1-j),2^n)$. This is what had to be shown.

We now prove Lemma 6.8. Let T be a tree in $\mathbb{A}$ that has at least three ends. Then T contains a Y, that is, a vertex p from which issue three paths γ_1, γ_2, γ_3 that are infinite and disjoint, and have neither double points nor backtrackings.

Let T be a tree in $\mathbb{X}$ with more than two ends. We will show that T is a translate of T_∞. To do this, we choose a vertex p and three paths as above. After a translation, we may always assume that p is the origin of T. We will analyze the nature of the intersections of these paths with $T((0,0),2^n)$, which thus consists of three disjoint paths γ_1^n, γ_2^n, γ_3^n issuing from the origin and of length 2^n. By Remark 2, there is a translation (i,j) with $|i| \leq 2^n$ and $|j| \leq 2^n$ that takes $T((0,0),2^n)$ onto $T_\infty((i,j),2^n)$, which is contained in $T_\infty((0,0),2^{n+1})$. The finite tree $T_\infty((0,0),2^{n+1})$ has the following structure. It is a cross consisting of four branches on the coordinate axes, of length 2^{n+1}, onto which are grafted trees of diameter less than 2^n (see the figure). Since the three paths $\gamma_1^n+(i,j)$, $\gamma_2^n+(i,j)$, $\gamma_3^n+(i,j)$ are disjoint and have length 2^n, it is clear that the only possibility is that these paths start at the origin; that is, $(i,j)=(0,0)$. It follows that for every n, the balls $T((0,0),2^n)$ and $T_\infty((0,0),2^n)$ coincide. We have proved that the tree T is actually T_∞, and this concludes the proof of Theorem 6.6.

7. Meromorphic Functions on Laminations

On any compact Riemann surface, there exist nonconstant meromorphic functions. More precisely, given two distinct points on such a surface, there exists a meromorphic function that assumes different values at these points. This can be used to show that all Riemann surfaces are algebraic; that is, they can be holomorphically embedded in a complex projective space. It is this result of Riemann, one of the most important results in the classical theory, that we will try to generalize.

Let $(M,\mathcal{L})$ be a Riemann surface lamination and U an open set in M. A *holomorphic function* on U is a continuous function $f: U \to \mathbb{C}$ that is holomorphic on each leaf. This defines a sheaf of rings $\mathcal{O}$ on M.

A *meromorphic function* is usually defined as a function that is locally the quotient of two holomorphic functions. Several difficulties appear when one tries to generalize this definition to laminations.

The first is that the ring of germs of functions that are holomorphic in a neighborhood of a point in a lamination is not necessarily an integral domain. Indeed, it suffices to observe that this ring contains the subring of germs of continuous functions that are constant on the plaques, which is isomorphic to the ring of germs of functions that are continuous on a transversal, in a neighborhood of one of its points. The latter ring is not an integral domain if the point considered is not isolated.

The second difficulty is illustrated by the following example. Consider the trivial lamination $\mathbb{D} \times [0,1]$, and let z_1, z_2 be continuous functions from $[0,1]$ to $\mathbb{D}$ such that $z_1(t) = z_2(t)$ for $t \leq 1/2$ and $z_1(t) \neq z_2(t)$ for $t > 1/2$. The function

$$F : (z,t) \in \mathbb{D} \times [0,1] \longmapsto (z - z_1(t))/(z - z_2(t))$$

is the quotient of two holomorphic functions, and should therefore be called meromorphic. But it is hard to define the "divisor of poles" of F. Indeed, for $t \leq 1/2$ the function F equals 1 and hence has no poles, whereas the curve $(z_2(t), t)$ is contained in "the poles" of F for $t > 1/2$. The divisor of poles thus has a "stopping point".

Recall that we have already defined the notion of a line bundle over M. If we denote by $\mathcal{O}^\star$ the sheaf of germs of nonzero holomorphic functions, the line bundles over M can be identified with the elements of $H^1(M, \mathcal{O}^\star)$.

If U is an open subset of M, we denote by $\mathcal{H}(U)$ the set of functions that are holomorphic in U and not identically zero on any leaf of the restriction of $\mathcal{L}$ to U. This is a monoid under multiplication. The sheaf of monoids associated with this presheaf will be denoted by $\mathcal{H}$; it contains $\mathcal{O}^\star$ as a subsheaf. A global section of the quotient $\mathcal{H}/\mathcal{O}^\star$ will be called an *effective divisor* on M. In less precise terms, an effective divisor is defined locally by a holomorphic equation that is not identically zero on any plaque, and these local equations are well defined modulo multiplication by a nonvanishing holomorphic function. The *support* of an effective divisor is the closed subset of M where the local equations are zero.

As in the classical case, an effective divisor D naturally defines a line bundle equipped with a holomorphic section. Conversely, a holomorphic section of a line bundle defines an effective divisor: its "divisor of zeros".

A *meromorphic function* on $(M, \mathcal{L})$ is a function $f : M \to \mathbb{CP}^1$ such that there exist a line bundle E over M and two holomorphic sections u, v of this bundle such that $f = u/v$ and v is not the zero function on any leaf. As the notation suggests, the function f is undefined at the points where u and v are zero; everywhere else, the quotient of the two sections can actually be viewed as an element of $\mathbb{C} \cup \{\infty\}$. Of course, the line bundle E is not unique. The divisor of zeros of the section v is called *a* divisor of poles of f.

Even though a meromorphic function f is undefined at the common zeros of u and v, the restriction of f to each leaf is a meromorphic function in the usual sense and therefore extends to a continuous function with values in $\mathbb{CP}^1$. Note, however, that the function f thus defined on M is not necessarily continuous: this is the phenomenon we have already seen of zeros meeting poles.

To simplify our statements and avoid certain pathological phenomena, we will agree that a meromorphic function is *nonconstant* if it is not constant on any leaf.

LEMMA 7.1. *The support of a divisor of poles of a nonconstant meromorphic function intersects all the leaves of a lamination.*

PROOF. Let $f : M \to \mathbb{CP}^1$ be a nonconstant meromorphic function, and let X be the support of a divisor of poles. We begin by observing that by Rouché's theorem, if a plaque of $\mathcal{L}$ intersects X, then so do all nearby plaques. In other words, the union of the leaves that intersect X is open. If the complement were nonempty, it would be a lamination $\mathcal{L}'$ equipped with a continuous holomorphic function to $\mathbb{C}$. One could then consider a point of $\mathcal{L}'$ where f attained its maximum modulus, and this would contradict the maximum principle in the leaf containing this point. □

7.1. A counterexample. We begin by indicating a necessary condition for the existence of nonconstant meromorphic functions and giving examples of laminations that do not satisfy this condition.

By our definition, a foliation cycle is an element of the dual of $A^2(\mathcal{L})$ that is zero on the exact forms. Under reasonable conditions on the ambient space M, a foliation cycle can be considered as a homology class of dimension 2 on M that can be homologous to zero in certain cases, as we will see later. We will not try to find the most general setting in which this would make sense, and will restrict our attention to the case where M is a compact differentiable manifold and $\mathcal{L}$ is a smoooth foliation. By restriction, the differential 2-forms on M give elements of $A^2(\mathcal{L})$; a foliation cycle can thus be considered as a closed current on M in the usual sense, so it has a homology class.

LEMMA 7.2. *Let $(M, \mathcal{L})$ be a foliation by Riemann surfaces on an oriented manifold. Suppose $\mathcal{L}$ has a foliation cycle μ that is homologous to zero. Then the support of an effective divisor cannot intersect all the leaves. In particular, if a foliation cycle is homologous to zero, then there are no nonconstant meromorphic functions.*

PROOF. Let E be a holomorphic line bundle over M, and let s be a global holomorphic section of E. We will show that under the hypotheses of the lemma, the zero locus of s cannot intersect all the leaves.

Since E is a line bundle over M, it has a Chern class $c(E)$ that is an element of the de Rham cohomology group $H^2(M;\mathbb{R})$. Restricting to the leaves, we have a map $H^2(M;\mathbb{R}) \to H^2(\mathcal{L})$, which of course sends $c(E)$ to the Chern class defined in 4.2. Let γ be a closed 2-form on M that represents $c(E)$.

The foliation cycle μ can be viewed as a current on M. To say that the cycle is homologous to zero means that this current is zero on closed forms. In particular, $\langle c(E), \mu\rangle = \langle \gamma, \mu\rangle = 0$, and by Lemma 4.5 μ-almost every leaf does not intersect the zero locus of s. □

It is easy to construct a foliation that does not satisfy this criterion. Consider a Hopf surface, a quotient of $\mathbb{C}^2 \setminus \{(0,0)\}$ by a homothety of ratio 2. We have a fibration $\pi : M \to \mathbb{CP}^1$ that assigns to each point the complex line in $\mathbb{C}^2$ passing through the point. The fibers of π are elliptic curves, which define a lamination on M. Since M is diffeomorphic to $\mathbb{S}^3 \times \mathbb{S}^1$, its second homology group is zero, so the fundamental class of the leaves is zero in the homology of M. Thus, *there are no nonconstant meromorphic functions on this lamination on M.* This generalizes the well-known fact that Hopf surfaces are not algebraic.

Of course, the preceding example could be criticized for not being minimal. We return to the notation of 6.1. The one-parameter group of left translations h_+^s ($s \in \mathbb{C}$) on the homogeneous space $\mathrm{SL}(2,\mathbb{C})/\Gamma$ generates a lamination $\mathcal{L}$ on $M = \mathrm{SL}(2,\mathbb{C})/\Gamma$, which is equipped with a flat Hermitian metric in which the associated holomorphic vector field is of length 1. This lamination has a foliation cycle (unique up to a multiplicative constant) defined as follows: if ω is a 2-form on

$\mathcal{L}$, it can be evaluated on the field of unit 2-vectors carried by $\mathcal{L}$ to give a function on M, for which one can compute the integral $I(\omega)$ on M with respect to the Haar measure on $\mathrm{SL}(2,\mathbb{C})/\Gamma$. It is easy to check that this defines a foliation cycle. On the other hand, the complex flow d^t satisfies $d^t h_+^s d^{-t} = h_+^{\exp(2t)s}$. This means that d^t preserves $\mathcal{L}$ globally and expands the flat Hermitian metric along the leaves by a constant not equal to 1. In particular, the foliation cycle described above is also multiplied by a constant. Since a flow acts trivially on homology, it follows that the homology class of this foliation cycle is zero. *Thus we have constructed a minimal lamination with no nonconstant meromorphic functions.*

Given a lamination $(M,\mathcal{L})$, the question of the existence of an effective divisor whose support intersects all the leaves clearly precedes the question of the existence of meromorphic functions; the former is a purely topological question.

7.2. Construction of meromorphic functions. Here we describe three results on the existence of meromorphic functions for laminations all of whose leaves are of the same type.

A *total transversal* for a lamination $(M,\mathcal{L})$ is a *closed set* $\mathcal{T} \subset M$ that intersects all the leaves of $\mathcal{L}$ and whose intersection with any distinguished open set $U_i \simeq \mathbb{D}\times T_i$ is the disjoint union of finitely many graphs of the form $\{(\sigma(t),t)\}$, where the σ are continuous and defined in open subsets of T_i. Such a transversal is clearly an example of an effective divisor, if we agree to give it multiplicity 1.

Let $(M,\mathcal{L})$ be a lamination all of whose leaves are parabolic. We say that $\mathcal{L}$ satisfies the *growth condition* if there exist constants $C,\ell > 0$ with the following property. Let $\phi : \mathbb{C} \to L$ be a holomorphic covering of a leaf L, and let $|d\phi|$ be the norm of the derivative of ϕ (computed with an auxiliary metric on $\mathcal{L}$). Then

$$C^{-1}(1+|z|)^{-\ell}|d\phi(0)| \leq |d\phi(z)| \leq C(1+|z|)^{\ell}|d\phi(0)|.$$

Of course, this condition is satisfied if the lamination is uniformizable because then $|d\phi|$ is bounded.

THEOREM 7.3. *Let $(M,\mathcal{L})$ be a lamination of finite topological dimension, all of whose leaves are of elliptic type. Then the meromorphic functions on $(M,\mathcal{L})$ separate the points of M.*

THEOREM 7.4. *Let $(M,\mathcal{L})$ be a lamination of finite topological dimension all of whose leaves are of parabolic type. If $(M,\mathcal{L})$ has a total transversal and $\mathcal{L}$ satisfies the growth condition, then the meromorphic functions on $(M,\mathcal{L})$ separate the points of M.*

THEOREM 7.5. *Let $(M,\mathcal{L})$ be a lamination of finite topological dimension all of whose leaves are of hyperbolic type. If $(M,\mathcal{L})$ has a total transversal, then the meromorphic functions on $(M,\mathcal{L})$ separate the points of M.*

Before sketching the proofs, we make a few remarks on the hypotheses.

(1) We already pointed out in the proof of Proposition 5.6 that a lamination with parabolic leaves has an averaging sequence, but we do not know whether such a lamination necessarily satisfies the growth condition.
(2) One can prove the same result (with essentially the same proof), assuming only the existence of an effective divisor whose support intersects all the leaves instead of assuming the existence of a total transversal. But we

chose the weaker statement because we are convinced that this kind of topological hypothesis is temporary and will soon be eliminated...

(3) For codimension-one oriented foliations on closed 3-manifolds, it is known that there exists a total closed transversal if and only if no foliation cycle is homologous to zero (one then says that the foliation is *taut*). More precisely, there exists no total transversal if and only if the foliation contains finitely many toroidal compact leaves whose (oriented) union bounds a submanifold with boundary (see [**43, 25**]). In particular, if all the leaves of a codimension-one foliation on a 3-manifold are of hyperbolic type, then there exists a total transversal and one can then apply Theorem 7.5.

The *proof of Theorem 7.3* is of course the easiest. Let $(M, \mathcal{L})$ be a lamination all of whose leaves are elliptic. We saw in 5.1 that the leaves of $\mathcal{L}$ are the fibers of a fibration $\pi : M \to Q$. One can proceed as follows. Let N be a positive integer, and consider the line bundle over $\mathbb{CP}^1$ that is the Nth power of the tangent bundle. In other words, the global sections of this bundle are the holomorphic tensors that can be written in each affine chart on $\mathbb{CP}^1$ as

$$g(z)\left(\frac{\partial}{\partial z}\right)^N.$$

The space of holomorphic sections is a space E_N of complex dimension $2N+1$ on which $\mathrm{PGL}(2,\mathbb{C})$ acts naturally; it corresponds to the functions g that are polynomials of degree less than or equal to $2N$. As we have seen, the "quotient" of two nonzero elements of E_N can be viewed as a rational map on $\mathbb{CP}^1$. We can construct the vector bundle of rank $2N+1$ over Q that is associated with E_N. Since M is assumed to have finite topological dimension, a vector bundle of sufficiently high rank over M has nonvanishing sections. Thus, by choosing N sufficiently large and considering the quotient of two nonvanishing sections, we can obtain meromorphic functions on $(M, \mathcal{L})$. This construction is sufficiently general to produce meromorphic functions that separate the points of M. This proves Theorem 7.3.

The *proof of Theorem* 7.4 consists of adapting the classical construction of elliptic functions. We first consider a very special case: suppose that $(M, \mathcal{L})$ is a lamination with parabolic leaves such that the leaves of $\mathcal{L}$ are the orbits of a complex flow ϕ^z ($z \in \mathbb{C}$) acting holomorphically on the leaves. Let $\mathcal{T} \subset M$ be a total transversal. If x is a point in M, then the set $Z(x) = \{z \in \mathbb{C} : \phi^z(x) \in \mathcal{T}\}$ is discrete in $\mathbb{C}$, and there exists $\varepsilon > 0$ (independent of x) such that the disks of radius ε centered at the points of $Z(x)$ are disjoint. It follows that for some integer $N > 2$, the series

$$f(x) = \sum_{z \in Z(x)} z^{-N} \tag{7}$$

converges outside $\mathcal{T}$ and defines a holomorphic function on $(M, \mathcal{L})$, with values in $\mathbb{CP}^1 \simeq \mathbb{C} \cup \{\infty\}$. Note that if $s : \mathcal{T} \to \mathbb{R}$ is a continuous function, then the image of the map $t \in \mathcal{T} \mapsto \phi^{s(t)}(t) \in M$ is another total transversal. We can use this to show that if x and y are two distinct points, then there exists a total transversal that contains x but not y. We can then construct meromorphic functions, with values in $\mathbb{CP}^1 \simeq \mathbb{C} \cup \{\infty\}$, that have a pole at x and are finite at y; this proves Theorem 7.4 in the special case of a holomorphic flow.

We now consider the general case of Theorem 7.4: a lamination with parabolic leaves, satisfying the growth condition and with a total transversal $\mathcal{T}$. Let x be a point in M. By hypothesis, there exists a holomorphic covering $\phi : \mathbb{C} \to L(x)$ of the leaf through x. Let $Z(x)$ be the set of $z \in \mathbb{C}$ such that $\phi(z) \in \mathcal{T}$. This is still a discrete set, and it is still true that there exists an $\varepsilon > 0$ such that the disks centered at the points of $Z(x)$ and with radius ε are disjoint, but the disks must be considered in the metric on $\mathbb{C}$ that is the pullback under ϕ of a Hermitian metric on $(M, \mathcal{L})$. For this reason, the growth condition ensures that the expression (7) is convergent for sufficiently large N. However, the sum of this series depends on the choice of the covering ϕ, which is unique only up to an affine map. The object that is well defined, independently of the choice of ϕ, is the expression

$$\rho(x) = \sum_{z \in Z(x)} z^{-N} dz^N.$$

In other words, ρ defines a meromorphic differential of order N along the leaves, which has a pole along $\mathcal{T}$. The continuity of ρ on M follows from the continuity of the affine structure of the leaves, which we proved in 6.3. If we proceed in this way with two transversals and consider the ratio of the two differentials ρ thus obtained, we obtain a meromorphic function; this proves Theorem 7.4. To be precise, we would have to say that the two differentials ρ constructed in this way can be considered as holomorphic sections of a line bundle: it suffices here to consider the product of the Nth power of the cotangent bundle to the leaves and the Nth powers of the two bundles defined by the two transversals.

Before proving Theorem 7.5, we recall how H. Poincaré constructed *Fuchsian functions* [**40**]. Let Γ be a Fuchsian group, that is, a discrete subgroup of $\mathrm{PSL}(2,\mathbb{R})$. Let R be a rational map of one complex variable z, and consider the quadratic differential form $\rho = R(z)dz^2$. Since Γ acts on the Riemann sphere (preserving the disk $\mathbb{D}$), it acts on the quadratic differentials. We now set

$$\sigma_R = \sum_{\gamma \in \Gamma} \gamma^\star \rho.$$

Since Γ is discrete and acts properly on the disk $\mathbb{D}$, this series converges in the whole disk $\mathbb{D}$ (minus the orbit under Γ of the poles of R, of course). This follows from the fact that if x is a point in the Poincaré disk, then there is an $\varepsilon > 0$ such that the hyperbolic disks of radius ε with centers on the orbit Γx are pairwise disjoint, so the series of their Euclidean areas converges. Of course, the quadratic differential σ_R is invariant under Γ. The quotient $\sigma_{R_1}/\sigma_{R_2}$ of two rational maps R_1 and R_2 is a meromorphic function on $\mathbb{D}$ that is invariant under the Fuchsian group Γ; that is, a meromorphic function f on the Riemann surface $\mathbb{D}/\Gamma$. However, one has to check that f is not constant. H. Poincaré did this by arranging that the poles of R_1 and R_2 do not lie on the same orbit of Γ, which ensures that f has a zero and a pole and is therefore nonconstant.

The proof of Theorem 7.5 consists of imitating Poincaré's construction in the setting of laminations. Let $(M, \mathcal{L})$ be a lamination all of whose leaves are of hyperbolic type. We know that there exists a Hermitian metric g such that all the leaves have curvature -1. Let $T\mathcal{L}$ be the tangent bundle to the lamination. Considering the exponential map along the leaves with respect to the metric g, we define a map $\exp : T\mathcal{L} \to M$ whose restriction to each tangent plane is a covering because each leaf is negatively curved. We can thus equip each tangent plane with the pullback metric, and from now on we consider $T\mathcal{L}$ as a bundle over M whose fibers are copies of the Poincaré disk $\mathbb{D}$. The map exp is holomorphic and isometric on the fibers.

Let $T\mathcal{T}$ be the subset of $T\mathcal{L}$ that lies over the total transversal $\mathcal{T}$. We may consider $T\mathcal{T}$ as a noncompact lamination whose leaves are all Poincaré disks and that is mapped holomorphically to $\mathcal{L}$ by $\exp : T\mathcal{T} \to M$. This map is surjective and has discrete fibers. To construct a quadratic differential on M, we simply take the "pushforward" of a quadratic differential on $T\mathcal{T}$.

Let ρ be a quadratic differential on $T\mathcal{T}$. This means that ρ continuously assigns to each point x in $\mathcal{T}$ a quadratic differential $\rho(x)$ on the corresponding tangent plane, which is isomorphic to $\mathbb{D}$. We require that ρ be rational, that is, of the form $R_x(z)dz^2$, where $R_x(z)$ depends rationally on z. Moreover, we choose ρ so that each $\rho(x)$ has a pole (not necessarily simple) at the origin of the corresponding disk. We can choose $R_x(z)$ to be of the form $z^{-N}P_x(z)$, where N is a sufficiently large integer and P_x is a nonzero polynomial of degree less than or equal to N. Indeed, since M is finite-dimensional by hypothesis, a vector bundle over M of sufficiently high finite dimension must have a continuous nonvanishing section.

We now set $\sigma = \exp_\star(\rho)$, but we have to explain what we mean by this "unnatural" pushforward ...

The universal cover $\widetilde{L}$ of a leaf L of $\mathcal{L}$ is isomorphic to the disk $\mathbb{D}$. Choose an isomorphism such that there is a holomorphic immersion $i : \mathbb{D} \to M$ with image L. The inverse image $i^{-1}(\mathcal{T})$ is a discrete subset of $\mathbb{D}$, and there is a greatest lower bound for the Poincaré hyperbolic distance between two of its points. For each $t \in i^{-1}(\mathcal{T})$, there is an exponential map $\exp_t : \mathbb{D} \to \widetilde{L} \simeq \mathbb{D}$ that sends the origin to the point t and is an isometry. For each t there is also a rational quadratic differential ρ_t on $\mathbb{D}$. Using exactly the same reasoning as H. Poincaré, we verify that the sum $\sum_t (\exp_t^{-1})^\star(\rho_t)$ converges away from its poles. This quadratic differential on $\widetilde{L}$ passes naturally to the quotient on L and defines a quadratic differential on each leaf of L; this is the meromorphic quadratic differential we denoted earlier by σ. The continuous dependence in the transverse direction has to be checked, but this follows easily from the uniform convergence of the series under consideration. We have indeed constructed a meromorphic differential on $(M, \mathcal{L})$ that has a pole along the transversal $\mathcal{T}$.

Note that if s is a section of the bundle $T\mathcal{T}$ over $\mathcal{T}$, then the image of the map $t \in \mathcal{T} \mapsto \exp(s(t)) \in M$ is another total transversal. This allows us to show that if x and y are distinct points, then there exists a total transversal containing x but not y.

To construct nonconstant meromorphic functions on $(M, \mathcal{L})$, it suffices to consider the quotient of two of these quadratic differentials, corresponding to the choice of two total transversals. This ratio is not constant because it has poles. Here again, to be precise, we must consider these two meromorphic differentials as holomorphic sections of the same line bundle. Given two distinct points in M, we have indeed constructed a meromorphic function on M that assumes different values at these points. This completes the proof of Theorem 7.5.

Bibliography

[1] W. Abikoff, *The real analytic theory of Teichmüller space*, Lecture Notes in Mathematics, vol. 820, Springer, Berlin, 1980.

[2] L. V. Ahlfors, *Lectures on quasiconformal mappings*, Wadsworth & Brooks/Cole Advanced Books & Software, Monterey, CA, 1987.

[3] L. Ahlfors and L. Bers, *Riemann's mapping theorem for variable metrics*, Ann. of Math. (2) **72** (1960), 385–404.

[4] V. Arnol′d, *Chapitres supplémentaires de la théorie des équations différentielles ordinaires*, "Mir", Moscow, 1980, Translated from the Russian by D. Embarek.

[5] S. Barré, *Polyèdres de rang 2*, Thesis, ENS Lyon, 1996.

[6] C. Bonatti, R. Langevin, and R. Moussu, *Feuilletages de $\mathbb{CP}(n)$: de l'holonomie hyperbolique pour les minimaux exceptionnels*, Inst. Hautes Études Sci. Publ. Math. **75** (1992), 123–134.

[7] K. S. Brown, *Buildings*, Springer-Verlag, New York, 1989.

[8] C. Camacho, *Problems on limit sets of foliations on complex projective spaces*, Proceedings of the International Congress of Mathematicians (Kyoto, 1990), Math. Soc. Japan, 1991, pp. 1235–1239.

[9] C. Camacho, A. Lins Neto, and P. Sad, *Minimal sets of foliations on complex projective spaces*, Inst. Hautes Études Sci. Publ. Math. **68** (1988), 187–203.

[10] A. Candel, *Uniformization of surface laminations*, Ann. Sci. École Norm. Sup. (4) **26** (1993), 489–516.

[11] J. Cantwell and L. Conlon, *Every surface is a leaf*, Topology **26** (1987), 265–285.

[12] ———, *Generic leaves*, Comment. Math. Helv. **73** (1998), 306–336.

[13] A. Connes, *A survey of foliations and operator algebras*, Operator algebras and applications, Part I (Kingston, Ont., 1980), Proc. Sympos. Pure Math., vol. 38, Amer. Math. Soc., Providence, R.I., 1982, pp. 521–628.

[14] ———, *Noncommutative geometry*, Academic Press Inc., San Diego, CA, 1994.

[15] J.-P. Demailly, *Courants positifs et théorie de l'intersection*, Gaz. Math. **53** (1992), 131–159.

[16] A. Douady and J. H. Hubbard, *On the dynamics of polynomial-like mappings*, Ann. Sci. École Norm. Sup. (4) **18** (1985), 287–343.

[17] A. El Kacimi-Alaoui, *Sur la cohomologie feuilletée*, Compositio Math. **49** (1983), 195–215.

[18] B. Enriquez, *Sur le théorème de Gauss-Bonnet pour les feuilletages avec mesure harmonique*, C. R. Acad. Sci. Paris Sér. I Math. **309** (1989), 733–736.

[19] L. Garnett, *Foliations, the ergodic theorem and Brownian motion*, J. Funct. Anal. **51** (1983), 285–311.

[20] E. Ghys, *Gauss-Bonnet theorem for 2-dimensional foliations*, J. Funct. Anal. **77** (1988), 51–59.

[21] ———, *Déformations des structures complexes sur les espaces homogènes de* $\mathrm{SL}(2,\mathbb{C})$, J. Reine Angew. Math. **468** (1995), 113–138.

[22] ———, *Topologie des feuilles génériques*, Ann. of Math. (2) **141** (1995), 387–422.

[23] ———, *Sur l'uniformisation des laminations paraboliques*, Integrable systems and foliations (Montpellier, 1995), Progr. Math., vol. 145, Birkhäuser Boston, Boston, MA, 1997, pp. 73–91.

[24] C. Godbillon, *Feuilletages*, Progress in Mathematics, vol. 98, Birkhäuser Verlag, Basel, 1991.

[25] S. E. Goodman, *Closed leaves in foliations of codimension one*, Comment. Math. Helv. **50** (1975), 383–388.

[26] S. E. Goodman and J. F. Plante, *Holonomy and averaging in foliated sets*, J. Differential Geom. **14** (1979), 401–407.

[27] P. Griffiths and J. Harris, *Principles of algebraic geometry*, John Wiley & Sons Inc., New York, 1994.
[28] B. Grünbaum and G. C. Shephard, *Tilings and patterns*, W. H. Freeman and Company, New York, 1989.
[29] A. Haefliger, *Deformations of transversely holomorphic flows on spheres and deformations of Hopf manifolds*, Compositio Math. **55** (1985), 241–251.
[30] G. Hector and U. Hirsch, *Introduction to the geometry of foliations. Parts A and B*, Aspects of Mathematics, vol. E1, E3, Vieweg & Sohn, Braunschweig, 1981, 1983.
[31] S. Hurder, *The $\overline{\partial}$ operator*, Global analysis on foliated spaces, MSRI Publ., vol. 9, Springer-Verlag, New York, 1988, Appendix A.
[32] J. -J. Loeb and M. Nicolau, *Holomorphic flows and complex structures on products of odd-dimensional spheres*, Math. Ann. **306** (1996), 781–817.
[33] S. López de Medrano and A. Verjovsky, *A new family of complex, compact, non-symplectic manifolds*, Bol. Soc. Brasil. Mat. (N.S.) **28** (1997), 253–269.
[34] W. de Melo and S. van Strien, *One-dimensional dynamics*, Ergebnisse der Mathematik und ihrer Grenzgebiete (3), vol. 25, Springer-Verlag, Berlin, 1993.
[35] C. C. Moore and C. Schochet, *Global analysis on foliated spaces*, MSRI Publ., vol. 9, Springer-Verlag, New York, 1988.
[36] L. Mosher and U. Oertel, *Spaces which are not negatively curved*, Comm. Anal. Geom. **6** (1998), 67–140.
[37] R. Penrose, *Shadows of the mind*, Oxford University Press, Oxford, 1994.
[38] A. Phillips and D. Sullivan, *Geometry of leaves*, Topology **20** (1981), 209–218.
[39] J. F. Plante, *Foliations with measure preserving holonomy*, Ann. of Math. (2) **102** (1975), 327–361.
[40] H. Poincaré, *Sur les fonctions fuchsiennes*, Acta Math. **1** (1882), 193–294.
[41] C. Pommerenke, *Univalent functions*, Vandenhoeck & Ruprecht, Göttingen, 1975.
[42] J. Roe, *Remark on a paper of É. Ghys: "Gauss-Bonnet theorem for 2-dimensional foliations"*, J. Funct. Anal. **89** (1990), 150–153.
[43] D. Sullivan, *Cycles for the dynamical study of foliated manifolds and complex manifolds*, Invent. Math. **36** (1976), 225–255.
[44] ———, *Bounds, quadratic differentials, and renormalization conjectures*, AMS centennial publications, vol. II (Providence, RI, 1988), Amer. Math. Soc., Providence, RI, 1992, pp. 417–466.
[45] A. Verjovsky, *A uniformization theorem for holomorphic foliations*, The Lefschetz centennial conference (Mexico City, 1984), Contemp. Math., vol. 58, Amer. Math. Soc., Providence, RI, 1987, pp. 233–253.

DYNAMICS OF RATIONAL MAPS ON $\mathbb{P}^k$

Nessim Sibony

Abstract: We present the basics of a Fatou-Julia theory for rational maps on $\mathbb{P}^k$. We consider primarily those aspects that use pluripotential theory.

Given a dominant rational map $f : \mathbb{P}^k \longrightarrow \mathbb{P}^k$, we define the Julia set associated with f. We then introduce a closed, positive current T of bidegree $(1,1)$ whose properties of invariance, support, and regularity give information about the dynamics of f.

The second chapter is devoted to the study of regular polynomial biholomorphisms of $\mathbb{C}^k$: periodic points; entropy; ergodic measure defined as an "intersection" of currents; stable manifolds; and Fatou-Bieberbach domains.

In the last chapter we consider the case of holomorphic endomorphisms of $\mathbb{P}^k$. The support of T coincides with the Julia set. The measure $\mu := T^k$ is mixing and maximizes entropy.

Introduction

The dynamical study of rational maps of one complex variable initiated by Fatou, Julia, and Léau has undergone a significant revival of activity in the past two decades. This is due in part to the computer visualization of Julia sets. It is now a highly developed theory.

By comparison, the theory of iteration of rational maps of the projective space $\mathbb{P}^k$ is still in its infancy. The problems that motivate it are, however, quite natural.

Let $F = (P, Q)$ be a polynomial map of $\mathbb{C}^2$. Suppose we want to localize the zeros of F; that is, to approximate the solutions of the system of equations

$$\begin{cases} P(z,w) = 0 \\ Q(z,w) = 0. \end{cases}$$

Newton's method consists of iterating the map

$$(z,w) \longmapsto (z,w) - \left(F'(z,w)\right)^{-1} \circ F(z,w).$$

Passing to homogeneous coordinates in $\mathbb{P}^2$ leads to the study of iterates of a rational map from $\mathbb{P}^2$ to $\mathbb{P}^2$. One can verify that, generically in (P,Q), one obtains a meromorphic map from $\mathbb{P}^2$ to $\mathbb{P}^2$, which has points of indeterminacy. Note that in one variable, rational maps from $\mathbb{P}^1$ to $\mathbb{P}^1$ have no points of indeterminacy.

Another motivation is the study of real Hénon maps. Consider the polynomial diffeomorphism $h_{a,c}$ on $\mathbb{R}^2$ defined by

$$(x,y) \longmapsto (x^2 + c + ay, x).$$

It has been known since Hénon's experimental studies that, for certain values of the parameters (a,c), the *Hénon map* $h_{a,c}$ exhibits new dynamical phenomena. The existence of a "strange attractor" for certain parameter values is now well established through the work of Benedicks-Carleson [**12**]. It is tempting to view the maps $h_{a,c}$ as biholomorphisms of $\mathbb{C}^2$ and study their dynamics by using the tools of complex analysis. It is even useful to consider the Hénon maps in homogeneous coordinates

$$h_{a,c} : [z : w : t] \longmapsto [z^2 + ct^2 + awt : zt : t^2].$$

It is easy to see that $h_{a,c}$ has a point of indeterminacy $I_+ = [0 : 1 : 0]$ and that $h_{a,c}^{-1}$, considered as a meromorphic map on $\mathbb{P}^2$, has a point of indeterminacy $I_- = [1 : 0 : 0]$. The point I_- is attracting for $h_{a,c}$, and I_+ is attracting for $h_{a,c}^{-1}$. It is this property, $I_+ \cap I_- = \emptyset$, that distinguishes the generalized Hénon maps among the polynomial automorphisms of $\mathbb{C}^2$. The elementary maps

$$[z : w : t] \longmapsto [w^2 + ct^2 + azt : wt : t^2]$$

satisfy $I_+ = I_- = [1 : 0 : 0]$, and their dynamics are very simple (Chapter 2).

A basic tool in the iteration theory of rational maps of $\mathbb{P}^1$ is Montel's theorem: a family of holomorphic maps of the unit disk with values in $\mathbb{P}^1 \setminus \{0, 1, \infty\}$ is locally equicontinuous. The corresponding results for holomorphic maps with values in $\mathbb{P}^k$ minus hypersurfaces do not have the same adaptability.

These notes deal mainly with results that can be obtained by using techniques of pluripotential theory.

Given a dominant meromorphic map f from $\mathbb{P}^k$ to $\mathbb{P}^k$ (that is, one whose image contains an open set), we can define the Fatou set of f as the largest open set in which the family of iterates (f^n) is locally equicontinuous. The Julia set J is the complement of the Fatou set. The point here is to introduce dynamically interesting closed positive currents on J and analyze them.

We sketch the theory in the setting of polynomials in one complex variable. In this case, the currents considered are just positive measures. Let

$$f_a(z) = z^d + a_1 z^{d-1} + \cdots + a_d$$

be a polynomial on $\mathbb{C}$ of degree d, parametrized by $a \in \mathbb{C}^d$. Set

$$K_a = \{z \in \mathbb{C} : \{f_a^n(z)\} \text{ bounded}\}.$$

The Julia set J_a is the boundary of K_a. Let ε_w denote the Dirac measure at w. Brolin [**16**] showed that the harmonic measure μ_a of K_a with respect to the point at infinity has remarkable dynamical properties. On the one hand, it is mixing for f_a; on the other, except for at most one point $w \in \mathbb{C}$, the sequence of discrete measures

$$\frac{(f^n)^* \varepsilon_w}{d^n} = \frac{1}{d^n} \sum_{f^n(w_i) = w} \varepsilon_{w_i} \tag{1}$$

converges vaguely to μ_a.

Tortrat [**81**] has shown that the sequence of measures

$$\nu_n = \frac{1}{d^n} \sum_{f^n(z) = z} \varepsilon_z \tag{2}$$

also converges to μ.

I observed in 1981 ([**74**]) that one could treat some aspects of the elementary theory of dynamics of polynomials without recourse to the theorem of Montel mentioned above. What replaces it is a compactness theorem for subharmonic functions that are locally bounded above (see Theorem A.1.2). Set

$$G(z,a) = \lim_{n\to\infty} \frac{1}{d^n} \log^+ |f_a^n(z)|. \tag{3}$$

One shows that this limit exists and that it defines a continuous plurisubharmonic function on $\mathbb{C} \times \mathbb{C}^d$. It satisfies the functional equation

$$G(f_a(z), a) = dG(z,a). \tag{4}$$

Moreover,

$$K_a = \{z : G(z,a) = 0\}.$$

In addition, the function G is pluriharmonic (it is locally the real part of a holomorphic function) outside the closed set

$$\mathcal{K} = \{(z,a) \in \mathbb{C} \times \mathbb{C}^d : z \in K_a\}.$$

For fixed a, $G(\cdot, a)$ is the Green's function in $\mathbb{C}$ of the compact set K_a, with pole at infinity. If Δ_z denotes the Laplacian with respect to z, we have

$$\mu_a = \Delta_z(G(z,a)).$$

This is the harmonic measure considered by Brolin.

The Julia set appears as the support of μ_a. It is the set of boundary points of $\{G_a > 0\}$; the function G_a has no harmonic extension to a neighborhood of any of these points. Elementary results from potential theory imply that the Julia set is perfect.

The functional equation (4) can be used to show that the function $G(\cdot, a)$ is in fact locally Hölder continuous. From this, still because J_a is the support of μ_a, we can derive estimates for the local Hausdorff dimension of the Julia set.

The critical points of $G(\cdot, a)$ carry information about the connectivity of J_a. Differentiating equation (4) in $\{G(z,a) > 0\}$, where the function G is harmonic, yields the classical result that if the critical points of f_a are in K_a, then the boundary J_a of K_a is connected. For iteration theory in one complex variable, we refer the reader to the monograph of Carleson-Gamelin [**18**].

The continuity of the function $G(z,a)$ with respect to the parameter a shows that the measure μ_a varies continuously, although the Julia sets J_a do not vary continuously with respect to the Hausdorff distance on compact sets.

To determine the distribution of the preimages of a point w, one is led to study the convergence of the sequence of subharmonic functions

$$u_n(z) = \frac{1}{d^n} \log |f_a^n(z) - w|.$$

One shows that the sequence u_n converges to G_a. Similarly, the approximation of the measure μ_a by Dirac measures concentrated at the periodic points comes from studying the convergence to G_a of the sequence of subharmonic functions

$$v_n(z) = \frac{1}{d^n} \log |f_a^n(z) - z|.$$

These notions generalize to several variables. It was Hubbard [**50**] who had the idea of considering the Green's function of Hénon maps. For a Hénon map h, he defined

$$G^{\pm}(z,w) = \lim_{n\to\infty} \frac{1}{2^n} \log^+ |h^{\pm n}(z,w)|$$

and showed that G^+ and G^- are plurisubharmonic functions on $\mathbb{C}^2$, pluriharmonic outside the sets

$$K^{\pm} = \{(z,w) : \{h^{\pm n}(z,w)\} \text{ is bounded}\}.$$

The compact set

$$K = K^+ \cap K^-$$

is the set of points whose iterates h^n and h^{-n}, $n \geq 0$, are bounded.

Then $K^+ = \{G^+ = 0\}$ and $K^- = \{G^- = 0\}$. The function $G = \sup(G^+, G^-)$ vanishes exactly on K. The function G^+ measures the rate of convergence to infinity [**50**].

It was natural to consider the dynamical study of the closed positive currents

$$T_+ := dd^c G^+ \quad \text{and} \quad T_- := dd^c G^-$$

and the measure

$$\mu := (dd^c)^2 G.$$

Here $(dd^c)^2$ denotes the Monge-Ampère operator. The current T_+ is supported on the boundary of K^+, that is, on the set of points in any neighborhood of which the function G^+ is not pluriharmonic.

The basic properties of these currents and μ were established in joint work with E. Bedford. Some of them appear in [**7**]. The theory of complex Hénon maps was developed in a number of papers by Bedford-Smillie [**6, 7, 8, 9**], Bedford-Smillie-Lyubich [**3, 4**], Hubbard-Obersthe-Vorth [**51**], Fornaess and the author [**27, 29**].

Bedford-Smillie-Lyubich showed, in particular, that the invariant measure μ maximizes entropy and is the limit of Dirac measures at the hyperbolic periodic points.

The Julia set of h has a certain rigidity. It is shown in [**29**] that the only closed positive currents with support in K^+ are proportional to T_+. This uniqueness property can be understood heuristically by saying that the stable manifolds that laminate ∂K^+ force the current to equal T_+.

The study of Hénon maps in $\mathbb{C}^2$ is certainly the most highly developed aspect of the theory of holomorphic maps. The fact that the maps are algebraic is responsible for the rigidity of the objects constructed. Note, for instance, that for a transcendental automorphism of $\mathbb{C}^k$, the set

$$K_f = \{z \in \mathbb{C}^k : \{f^n(z)\} \text{ is bounded}\}$$

can be everywhere dense and not equal to $\mathbb{C}^k$. For the dynamical study of transcendental maps, we refer the reader to the survey by Fornaess [**26**].

One can ask: Why introduce closed positive currents into the study of the dynamics of rational maps?

Consider the very simple example of the endomorphism f of $\mathbb{P}^2$ defined by

$$f[z:w:t] = [z^2 : w^2 : t^2].$$

It has algebraic degree 2 and topological degree 4. It has three attracting fixed points: the points $[0:0:1]$, $[1:0:0]$, and $[0:1:0]$. The union of their basins of attraction is the Fatou set. Its boundary is the Julia set. We work in the chart $\{t \neq 0\}$. We are in $\mathbb{C}^2$, and the basin of attraction at the point $[0:0:1]$ is the polydisk $D^2 = \{|z| < 1, |w| < 1\}$. Its boundary consists of

$$A_1 = \{z = e^{i\theta}, |w| < 1\}, \quad A_2 = \{w = e^{i\varphi}, |z| < 1\}, \quad A_3 = \{z = e^{i\theta}, w = e^{i\varphi}\}.$$

The open subset A_1 of the Julia set is foliated by the stable manifolds corresponding to the points of the circle $|z| = 1$. The circle $\{w = 0, |z| = 1\}$ is invariant. The periodic points are dense in it, and the mixing measure that describes their distribution is Lebesgue measure. Each point has a stable manifold that is a union of disks $z = e^{i\theta}$. Thus the current

$$\frac{1}{2\pi}\int_0^{2\pi}[z = e^{i\theta}]d\theta = dd^c \log^+ |z|,$$

where $[z = e^{i\theta}]$ denotes the current of integration on the line $z = e^{i\theta}$, has a dynamical interpretation. This current describes the distribution of the stable manifolds. It is the analogue, in the setting of endomorphisms, of the dynamical currents introduced by Ruelle-Sullivan [**68**].

We show that if ω is the usual Kähler form on $\mathbb{P}^2$, the sequence of forms $(f^n)^*\omega/d^n$ converges to a closed positive current T whose restriction to A_1 we have described. In fact, this current T contains other information. If we define the wedge product $\mu := T \wedge T$ appropriately, we find the Lebesgue measure on the torus A_3; it is invariant under f. The torus A_3 contains all the repelling periodic points of f. For this example, it is easy to check that if $a \notin \{zwt = 0\}$, then

$$\frac{1}{d^{2n}}(f^n)^*\varepsilon_a \to \mu \quad \text{and} \quad \frac{1}{d^{2n}}\sum_{f^n(z)=z}(f^n)^*\varepsilon_z \to \mu.$$

These are the questions we will study for the endomorphisms of $\mathbb{P}^k$. Pluripotential theory is used for the proofs in the general case.

These notes are divided into three chapters and an appendix.

In Chapter 1 we introduce the basic notions of meromorphic maps on $\mathbb{P}^k$. Let $f : \mathbb{P}^k \to \mathbb{P}^k$ be a dominant meromorphic map; we define the Fatou set of f and its complement, the Julia set. Let ω be the Kähler form on $\mathbb{P}^k$; we show that if f is algebraically stable (Definition 1.4.4) and of degree $d \geq 2$, then the sequence of forms

$$\frac{1}{d^n}(f^n)^*\omega$$

converges to a closed positive current T that satisfies

$$f^*T = dT.$$

When f is normal (Definition 1.5.2), the support of T coincides with the Julia set J (Theorem 1.6.5). We give estimates for the Hausdorff dimension of J and for the current T. Finally, we prove a result on convergence to the current T (Theorem 1.10.1).

Chapter 2 is devoted to a generalization of Hénon maps from $\mathbb{C}^2$ to $\mathbb{C}^k$. We introduce the notion of a regular automorphism of $\mathbb{C}^k$. These are polynomial automorphisms f such that the indeterminacy sets I_+ (associated with f) and I_- (associated with f^{-1}) are disjoint. We restrict ourselves to this hypothesis to simplify the exposition, though a number of properties are true under more general hypotheses. The hypothesis $I_+ \cap I_- = \emptyset$ guarantees a relation between the degree of f and that of f^{-1}.

Regular automorphisms are normal. The current T_+ is not necessarily supported on K^+, as it is in dimension 2.

If $\dim I_- = 0$, then T_+ is supported on ∂K^+, and all the closed positive currents supported on K^+ are proportional to T_+. In particular, the stable manifolds are dense in ∂K^+ (a theorem of Bedford-Smillie on compositions of generalized Hénon maps).

We then discuss some classical results on the basins of attraction of attracting periodic points (Fatou-Bieberbach domains), as well as global properties of stable manifolds.

Under the hypothesis $\dim I_- = \ell - 1$, we show that the measure $\mu := T_+^\ell \wedge T_-^{k-\ell}$ is invariant. It is mixing for f if $\dim I_- = 0$. In particular, it is mixing for Hénon maps on $\mathbb{C}^2$ [**9**] and for every regular automorphism of $\mathbb{C}^3$.

In Chapter 3 we treat the case of holomorphic endomorphisms of $\mathbb{P}^k$ of algebraic degree $d \geq 2$. We attempt to give a version of the most elementary properties of the iteration theory of rational maps of $\mathbb{P}^1$. The Fatou components are Kobayashi hyperbolic (a theorem of Ueda). The support of T has some of the properties of the Julia set in $\mathbb{P}^1$. In particular, it is the complement of the Fatou set, but it is not necessarily contained in the set of nonwandering points. The support of $\mu := T^k$, denoted by J_k, is contained in the set of recurrent points. In fact, the measure μ is invariant under f, maximizes entropy, and is mixing (3.6). If U is a union of Fatou components and satisfies $f^{-1}(U) \subset U$, then $J_k \subset \partial U$.

The periodic points are dense in J_k, and the Lyapunov exponents of the measure μ are positive (a theorem of Briend-Duval).

We have collected in an appendix those properties of plurisubharmonic functions and currents that are used repeatedly.

Finally, let us point out that these notes make extensive use of joint work with J. E. Fornaess.

Julien Duval, Charles Favre, Vincent Guedj, and Monique Hakim contributed to this article. I am very grateful to all of them.

CHAPTER 1

Iteration of Rational Maps of $\mathbb{P}^k$

1.1. Definitions. Dominant Maps

Let $\mathbb{P}^k$ be the complex projective space of dimension k. We write

$$z = [z_0 : z_1 : \ldots : z_k]$$

for a point z of $\mathbb{P}^k$ defined by its homogeneous coordinates. A rational map of degree d is written as

$$f = [F_0 : F_1 : \ldots : F_k],$$

where the F_j are homogeneous polynomials of degree d without common factors. We assume in what follows that $d \geq 2$.

We associate with f the map $F = (F_0, F_1, \ldots, F_k)$ from $\mathbb{C}^{k+1}$ to $\mathbb{C}^{k+1}$. If

$$\pi : \mathbb{C}^{k+1} \setminus \{0\} \to \mathbb{P}^k$$

is the canonical projection, we have the commutation property $\pi \circ F = f \circ \pi$. We denote the Jacobian of F by $J(F)$.

The function f is not actually defined everywhere. There is an indeterminacy set $I = \pi(F^{-1}\{0\})$ in $\mathbb{P}^k$. Since the F_j have no common factor, I is an analytic set of codimension ≥ 2.

We call a map f *dominant* if it is generically of rank k (i.e., the Jacobian $J(F)$ is not identically zero); $\mathcal{M}_d$ denotes the space of dominant rational maps of degree d. When $I = \emptyset$, we say that f is holomorphic; $\mathcal{H}_d$ denotes the semigroup of holomorphic maps of degree d from $\mathbb{P}^k$ to $\mathbb{P}^k$. We show that any meromorphic map from $\mathbb{P}^k$ to $\mathbb{P}^k$ is rational. Thus any holomorphic map from $\mathbb{P}^k$ to $\mathbb{P}^k$ is in some $\mathcal{H}_d$.

The space of all maps of degree d,

$$f = [F_0 : F_1 : \ldots : F_k],$$

where the F_j are homogeneous polynomials of the same degree, can be identified with $\mathbb{P}^N$, where $N = (k+1)\frac{(d+k)!}{d!k!} - 1$. Thus

$$\mathcal{H}_d \subset \mathcal{M}_d \subset \mathbb{P}^N.$$

PROPOSITION 1.1.1. *$\mathcal{H}_d$ and $\mathcal{M}_d$ are Zariski open subsets of $\mathbb{P}^N$. In particular, they are connected subsets of $\mathbb{P}^N$.*

PROOF. The complement of $\mathcal{M}_d$ is the analytic subset of functions f on $\mathbb{P}^N$ defined by $J(F, z) \equiv 0$. We show that the complement Σ of $\mathcal{H}_d$ is an analytic subset of $\mathbb{P}^N$. Indeed, Σ is the subset of functions $f \in \mathbb{P}^N$ for which there exists $z \in \mathbb{P}^k$ such that $F(z) = 0$. It is thus the projection to $\mathbb{P}^N$, under the proper map $(f, z) \mapsto f$, of the analytic subset Σ' of $\mathbb{P}^N \times \mathbb{P}^k$ defined by

$$\Sigma' = \{(f, z) \in \mathbb{P}^N \times \mathbb{P}^k : F(z) = 0\}.$$

□

COROLLARY 1.1.2. *If $f \in \mathcal{M}_d$, then the degree of $J(F, z)$ is $(k+1)(d-1)$.*

PROOF. Since the degree of $J(F, z)$ depends continuously on f, it is constant because $\mathcal{M}_d$ is connected. To find its value, it suffices to compute it for the map $f = [z_0^d : z_1^d : \ldots : z_k^d]$. □

EXAMPLES 1.1.3.

(1) Let $k = 2$. The map $f_0 = [z^d : w^d : t^d]$ and, more generally, maps of the form $f_1 = [P(z, w) : Q(z, w) : t^d]$, where P and Q are homogeneous polynomials of degree d such that $\{P = Q = 0\} = \{0\}$, are in $\mathcal{H}_d$.

The map $f_2 = [(z - 2w)^2 : (z - 2t)^2 : z^2]$ is in $\mathcal{H}_2$, and the orbit of its critical set is algebraic.

(2) The map from $\mathbb{C}^2$ to $\mathbb{C}^2$ defined by $f(z, w) = (p(z) + aw, z)$, where p is a polynomial of degree $d \geq 2$ in one variable and $a \neq 0$, is a polynomial automorphism of $\mathbb{C}^2$ called a *Hénon map.* It is useful to consider its meromorphic extension to $\mathbb{P}^2$, which we continue to denote by f by abuse of notation. Then

$$f([z : w : t]) = \left[t^d p\left(\frac{z}{t}\right) + awt^{d-1} : zt^{d-1} : t^d\right],$$

and the inverse automorphism is

$$f^{-1}([z : w : t]) = \left[wt^{d-1} : \frac{1}{a}\left(zt^{d-1} - t^d p\left(\frac{w}{t}\right)\right) : t^d\right].$$

The indeterminacy sets of f and f^{-1}, respectively, are the two points

$$I_+ = [0 : 1 : 0], \qquad I_- = [1 : 0 : 0].$$

The points of indeterminacy are in the hyperplane at infinity, defined by $t = 0$.

(3) We now give an example of a dynamically interesting automorphism of $\mathbb{C}^3$:

$$g([x : y : z : t]) = [\alpha x^2 + \beta y^2 + azt : \gamma x^2 + byt : xt : t^2],$$

where $\alpha, \beta, a, \gamma, b \in \mathbb{C}^*$. We have

$$g^{-1}([x : y : z : t]) = \left[zt^3 : \frac{1}{b}(yt^3 - \gamma z^2 t^2) : \frac{1}{a}\left(xt^3 - \alpha z^2 t^2 - \beta\left(\frac{yt - \gamma z^2}{b}\right)^2\right) : t^4\right].$$

Here $I_+ = [0 : 0 : 1 : 0]$ and $I_- = \{[x : y : 0 : 0]\}$. Note that the degree of g^{-1} is 4, while that of g is 2.

1.2. Critical Set. Blow-up

Let $f \in \mathcal{M}_d$. The Jacobian $J(F)$ of the transformation F on $\mathbb{C}^{k+1}$ associated with f is a homogeneous polynomial of degree $(k+1)(d-1)$. The set $\{J(F) = 0\}$ defines an analytic subset of $\mathbb{P}^k$; this is the critical set C_f of f.

For $f \in \mathcal{M}_d$, we cannot discuss $f(p)$ when p is a point of indeterminacy. We introduce the blow-up $\mathcal{B}_p$ of f at p. We denote by $B(p, \varepsilon)$ the ball with center p and radius ε, for a fixed metric on $\mathbb{P}^k$. We define

$$\mathcal{B}_p = \bigcap_{\varepsilon > 0} \overline{f(B(p, \varepsilon) \setminus I)}.$$

$\mathcal{B}_p$ can also be defined by considering the graph

$$\Gamma_f = \{(z, f(z)) : z \in \mathbb{P}^k \setminus I\},$$

which is an analytic subset of $(\mathbb{P}^k \setminus I) \times \mathbb{P}^k$. Its closure $\overline{\Gamma}_f$ is also an analytic set. Then $\mathcal{B}_p$ is the analytic subset of $\overline{\Gamma}_f$ that projects to p. We sometimes write $\mathcal{B}_p^+$ for the blow-up at p associated with f, and $\mathcal{B}_q^-$ for the blow-up at q associated with the inverse of f.

Clearly $\mathcal{B}_p$ is connected because $(B(p,\varepsilon) \setminus I)$ is connected for every $\varepsilon > 0$. If the dimension of $\mathcal{B}_p$ is 0, then f extends holomorphically to p, and p is not in I.

Recall that a map $f \in \mathcal{M}_d$ is *birational* if there exists $g \in \mathcal{M}_d$ such that $f \circ g = \mathrm{Id}$ and $g \circ f = \mathrm{Id}$ outside a hypersurface $\mathcal{C}$ of $\mathbb{P}^k$. We will write $g = f^{-1}$. We will denote the indeterminacy set of f by I_+ and that of f^{-1} by I_-. Blow-ups have the following property.

PROPOSITION 1.2.1. *Let f be birational on $\mathbb{P}^k$. If p is in $\mathcal{B}_q^-$, then q is in $\mathcal{B}_p^+$. In particular, $f(\mathcal{B}_q^- \setminus I_+) = q$.*

PROOF. Let $p \in \mathcal{B}_q^-$. There exists a sequence $\{z_n\}$, $z_n \notin I_-$, such that $\lim_n z_n = q$, $\lim_n f^{-1}(z_n) = p$, and $f^{-1}(z_n) \notin I_+$. It follows that $f(f^{-1}(z_n)) = z_n$ approaches q, and hence that q is in $\mathcal{B}_p^+$. □

EXAMPLE 1.2.2. Let f be a polynomial automorphism of $\mathbb{C}^2$. Let $[z : w : t]$ denote the coordinates in $\mathbb{P}^2$, and let H be the hyperplane at infinity defined by $t = 0$. Clearly $\mathcal{B}_p^+ = H$ for every point $p \in I_+$ (since $\mathcal{B}_p^+$ must be an analytic subset of dimension 1, and $f^{-1}(\mathcal{B}_q^+ \setminus I_-) = p$ by Proposition 1.2.1). Similarly, $\mathcal{B}_q^- = H$ for every $q \in I_-$. Hence the preceding proposition states that $f(H \setminus I_+) = q$ for every $q \in I_-$. It follows that I_+ and I_- each reduce to a single point.

DEFINITION 1.2.3. Let V be an analytic subset of dimension ℓ of $\mathbb{P}^k$, and let $f \in \mathcal{M}_d$. We say that V is a degeneracy set for f if the rank of the restriction $f|_{V \setminus I}$ is $\leq \ell - 1$. If $k = 2$, this means that $f_{V \setminus I}$ is constant.

EXAMPLE 1.2.4. If $f([z : w : t]) = [z^2 : w^2 : wt]$, then the image under f of the line $(w = 0)$ is the point $[1 : 0 : 0]$.

1.3. Preimages of a Point

Recall the statement of Bézout's theorem:

THEOREM 1.3.1. *Let $P_1, \ldots, P_k$ be homogeneous polynomials on $\mathbb{P}^k$. If the set of their common zeros is discrete, then the number of these zeros, counting multiplicity, equals the product $d_1 \cdots d_k$ of the degrees of the polynomials $P_1, \ldots, P_k$.*

This implies the following corollary:

COROLLARY 1.3.2. *Let $f \in \mathcal{H}_d(\mathbb{P}^k)$. Then for every $a \in \mathbb{P}^k$ the cardinality of $\{f^{-1}(a)\}$, counting multiplicity, is d^k.*

PROOF. Let $f = [F_0 : F_1 : \ldots : F_k]$, where $F_0, F_1, \ldots, F_k$ are homogeneous polynomials of degree d in $z = (z_0, z_1, \ldots, z_k)$. We introduce a new variable t and solve the system

$$F_j(z) - t^d a_j = 0, \quad j = 0, 1, \ldots, k,$$

of homogeneous equations in $\mathbb{P}^{k+1}$. Considering this system in $\mathbb{C}^{k+2}$, we see that the intersection with $t = 0$ reduces to 0. Hence $f = at^d$ defines an analytic subset of $\mathbb{P}^{k+1}$ that does not intersect $t = 0$. It is thus a compact analytic subset of $\mathbb{C}^{k+1}$, i.e. a finite set. Applying Bézout's theorem in $\mathbb{P}^{k+1}$, we see that the set $S = \{f = at^d\}$

has d^{k+1} points, and $\{f^{-1}(a)\}$ can be identified with the projection of S to $\mathbb{P}^k$. But S does not intersect $t = 0$ and does not contain the point $[0 : \dots : 0 : 1]$. Thus there are exactly d distinct points of S over each point in $\{f^{-1}(a)\}$; they are obtained by multiplying t by the d dth roots of unity. It follows that the cardinality of $\{f^{-1}(a)\}$ is d^k. □

The proof of the following theorem is similar.

THEOREM 1.3.3 ([**30**]). *Let f and g be holomorphic maps from $\mathbb{P}^k$ to $\mathbb{P}^k$. Assume that f has degree d and g has degree d', $d' < d$. Then the set $\{f = g\}$ is discrete and, counting multiplicity, its cardinality is* $\dfrac{d^{k+1} - {d'}^{k+1}}{d - d'}$.

COROLLARY 1.3.4. *Let f be a holomorphic map from $\mathbb{P}^k$ to $\mathbb{P}^k$, $f \in \mathcal{H}_d$, where $d \geq 2$. Then the number of periodic points of order n of f, counting multiplicity, is* $\dfrac{d^{n(k+1)} - 1}{d^n - 1}$.

PROOF. Recall that a point p is said to be periodic of period n if $f^n(p) = p$, where we assume, as usual, that n is the smallest such positive integer. The corollary follows from the preceding theorem applied to f^n and $g = \mathrm{Id}$. □

Now recall the following theorem of Shub and Sullivan.

THEOREM 1.3.5 ([**73**]). *Let $F : U \to \mathbb{R}^p$ be a $\mathcal{C}^1$ map defined in a neighborhood $U \subset \mathbb{R}^p$ of 0. Assume that 0 is an isolated fixed point of f^n for every n. Then $\mu(f^n - I, 0)$, the multiplicity of $f^n - I$ at 0, is bounded by a constant independent of n.*

COROLLARY 1.3.6. *Let f be a holomorphic map from $\mathbb{P}^k$ to $\mathbb{P}^k$, $f \in \mathcal{H}_d$. Then f has infinitely many distinct periodic orbits.*

Observe that meromorphic maps do not necessarily have periodic points. For instance, let f be defined by

$$f(x, y) = (\alpha x + P(y), y + \beta),$$

where P is a polynomial of degree d and $\alpha, \beta \neq 0$. Clearly the map has no periodic points in $\mathbb{C}^2$, and passing to homogeneous coordinates gives

$$f[x : y : t] = \left[\alpha x t^{d-1} + t^d P\left(\frac{y}{t}\right) : y t^{d-1} + \beta t^d : t^d\right].$$

In this example, $t = 0$ is mapped to the point of indeterminacy $[1 : 0 : 0]$, so f has no periodic point in $t = 0$.

1.4. Indeterminacy Set. Degrees of Iterates. Algebraically Stable Maps.

Let $f \in \mathcal{M}_d$. Let I be the indeterminacy set of f, and for each $n \in \mathbb{N}$, let I_n be the indeterminacy set of f^n. I_n is the analytic subset of $\mathbb{P}^k$ defined by $(F_n)^{-1}(0)$, where F_n is a lift of f^n.

DEFINITION 1.4.1. We denote by $E = \overline{\bigcup_{n\in\mathbb{N}} I_n}$ the closure of the set of points of indeterminacy of the iterates of f.

PROPOSITION 1.4.2. *Let $f \in \mathcal{M}_d$. If $z \notin E$ and $f(z) \in E$, then z is in a degeneracy set.*

PROOF. Let

$$\Sigma = \{z \in \mathbb{P}^k \setminus E : \dim f^{-1}(f(z)) > 0\}.$$

We know ([**45**], p. 136) that Σ is a proper analytic subset of $\mathbb{P}^k \setminus E$. It is clear that every point z of Σ is in the degeneracy set $f^{-1}(f(z))$. If $z \notin \Sigma$, then there exist arbitrarily small neighborhoods V of z such that $f^n_{|V} : V \to U$ is surjective, proper, and finite. But if each image contains a point in I_ℓ, then V contains a point in $I_{\ell+n}$. Hence $z \in E$. □

Let $f \in \mathcal{M}_d$. The degree d_n of f^n may be less than d^n because the components of F^n may have a common factor. Thus $d_{n+m} \leq d_n \cdot d_m$ in general. However, equality $d_n = d^n$ does hold if there is no hypersurface of $\mathbb{P}^k$ whose image under f^{n-1} is contained in I. More precisely, we have the following result.

PROPOSITION 1.4.3 ([**29**]). *Let $f \in \mathcal{M}_d$, $g \in \mathcal{M}_{d'}$. The algebraic degree of $f \circ g$ is $d \cdot d'$ if and only if there is no hypersurface V such that $g(V \setminus I_g) \subset I_f$.*

PROOF. If $g(V \setminus I_g) \subset I_f$, then all the components of $F \circ G$ have a common factor, which gives the defining equation of V. The degree of $f \circ g$ is thus less than $d \cdot d'$. The converse is also clear. □

DEFINITION 1.4.4. A map $f \in \mathcal{M}_d$ is *algebraically stable* if there does not exist any integer n or hypersurface V such that all the components of F^n are zero on V.

It follows from Proposition 1.4.3 that if f is algebraically stable and has degree d, then f^n has degree d^n.

If f is algebraically stable and I_n denotes the indeterminacy set of f^n, then $I_n \subset I_m$ for $m > n$. Algebraically stable maps are introduced in [**32**], where they are called "generic maps".

EXAMPLES 1.4.5.

(1) Hénon maps (Example 1.1.3(2)) are algebraically stable. Indeed, $\{t = 0\}$ is mapped to the fixed point $[1 : 0 : 0]$.

(2) The automorphism g (Example 1.1.3(3)) is algebraically stable because the image of $\{t = 0\}$ is the set $\{[\alpha x^2 + \beta y^2 : \gamma x^2 : 0 : 0]\}$, which is stable under g and does not intersect I_+.

(3) The map h defined by $h([x : y : t]) = [xt + t^2 : yt + x^2 : t^2]$ is not algebraically stable because $\{t = 0\}$ is sent to the point of indeterminacy $[0 : 1 : 0]$.

EXAMPLES 1.4.6.

(1) Let $f \in \mathcal{M}_d$ be an algebraically stable map. A meromorphic map g that is birationally equivalent to f need not be algebraically stable. It suffices to take

$$f[z : w : t] = [z^d : w^d : t^d], \quad \varphi_a = [az^2 + wt : zt : t^2].$$

We have

$$\varphi_a^{-1}[z : w : t] = [wt : zt - aw^2 : t^2]$$

and

$$g_a[z : w : 1] = \varphi_a^{-1} \circ f \circ \varphi_a[z : w : 1] = (z^d, (az^2 + w)^d - az^{2d}).$$

When $a \neq 0$, the image of the hyperplane $(t = 0)$ is the point of indeterminacy $I = [0 : 1 : 0]$. If a approaches 0, then g_a approaches f. It is clear that the dynamics of g is essentially that of f. Thus we must choose the "good" representation for a map in a given conjugacy class.

(2) We give an example (due to V. Guedj) of an algebraically stable automorphism of $\mathbb{C}^3$ such that f^{-1} is not algebraically stable but f^{-2} is algebraically stable. For $\lambda \in \mathbb{C}^*$, $a \in \mathbb{C}^*$, set

$$f(x, y, z) = (x^2 + \lambda y + az,\ \lambda^{-1}x^2 + y, x).$$

Then $f(t = 0) = [\lambda : 1 : 0 : 0]$, which is a fixed point; thus f is algebraically stable. But

$$f^{-1}(x, y, z) = \left(z, y - \lambda^{-1}z^2, \frac{1}{a}(x - \lambda y) \right),$$

and $f^{-1}(t = 0) = [0 : 1 : 0 : 0]$ is a point of indeterminacy for f^{-1}. One can check that f^{-2} is algebraically stable and even that $I(f^2) \cap I(f^{-2}) = \emptyset$.

(3) Let f be the automorphism of $\mathbb{C}^3$ defined by

$$f(x, y, z) = (x^2 + y,\ z^2 + x,\ z).$$

Then

$$f^{-1}(x, y, z) = (y - z^2,\ x - (y - z^2)^2,\ z).$$

One can check that f is algebraically stable, but f^{-n} is not algebraically stable for any $n \geq 1$ because $f^{-n}(t = 0) = [0 : 1 : 0 : 0] \in I_-$.

If f is not algebraically stable, then $F^n = h_n F_n$ for sufficiently large n, where h_n is a nonconstant homogeneous polynomial and the components of F_n have no common factor. We have $d_n < d^n$, where d_n denotes the degree of F_n. One can check that

$$d_{n+m} \leq d_n \cdot d_m$$

and hence that

$$\lim_{n \to \infty} d_n^{1/n} = \inf_n d_n^{1/n}.$$

Thus we can state the following definition:

DEFINITION 1.4.7. Given $f \in \mathcal{M}_d$, we define the dynamical degree δ_f of f by

$$\delta_f = \lim_{n \to \infty} d_n^{1/n}.$$

If g is birationally equivalent to f, it is easy to check that $\delta_f = \delta_g$. If f is algebraically stable, then $\delta_f = d$. Hence $\delta_g = d$ if g is birationally equivalent to an algebraically stable map $f \in \mathcal{M}_d$.

1.5. Fatou Sets. Julia Sets. Normal Maps.

DEFINITION 1.5.1. Let $f \in \mathcal{M}_d$ be an algebraically stable meromorphic map.

- A point p is in the *Fatou set* of f if there exists a neighborhood U of p such that the family $f^n_{|U}$ is equicontinuous.
- The *Julia set* is the complement of the Fatou set.
- A point p is called *normal* if there exist a neighborhood U of p and a neighborhood V of the indeterminacy set I such that $f^n(U) \cap V = \emptyset$ for every $n \in \mathbb{N}$.

It follows from the definitions above that the Fatou set is open and disjoint from E (see Definition 1.4.1), the Julia set is closed, and the set of normal points is an open subset of $\mathbb{P}^k \setminus E$.

DEFINITION 1.5.2. A meromorphic map is *normal* if the set N of its normal points is $\mathbb{P}^k \setminus E$.

EXAMPLES 1.5.3.

(1) Hénon maps are normal. Indeed, the point of indeterminacy $p_- = [0:1:0]$ is attracting for f^{-1}. Hence, if $f^{n_j}(p_j) \to p_-$, then $p_j = f^{-n_j}(f^{n_j}(p_j)) \to p_-$.

(2) All the maps in $\mathcal{H}_d$ are normal because E is empty.

(3) Let g be the map defined by

$$g([z:w:t]) = [z^d : w^d : t^{d-1}w].$$

Then

$$g^n([z:w:t]) = [z^{d^n} : w^{d^n} : t^{(d-1)^n} w^{d^n-(d-1)^n}],$$

and $E = I = \{[0:0:1]\}$. Let $\Omega = \{[z:w:t] : |z| < |w| < |t|\}$. In the chart $t = 1$, we have, for $[z:w:t] \in \Omega$,

$$g^n(z,w) = \left(\frac{z^{d^n}}{w^{d^n-(d-1)^n}}, w^{(d-1)^n} \right).$$

Hence $g^n_{|\Omega}([z:w:t])$ converges to $[0:0:1]$. The points of Ω are therefore not normal. However, the open set Ω is contained in the Fatou set.

(4) If $f([z:w:t]) = [z^d : w^d : t^d]$, then the Julia set of f is

$$J_f = \{|z| = |w| \geq |t|\} \cup \{|z| = |t| \geq |w|\} \cup \{|w| = |t| \geq |z|\}.$$

1.6. Green's Current Associated with a Meromorphic Map

Let $f \in \mathcal{M}_d$ be a dominant meromorphic map on $\mathbb{P}^k$. We will associate with it a closed positive current T of bidegree $(1,1)$ whose properties are tied to the dynamics of f.

Let ω be the standard Kähler form on $\mathbb{P}^k$. If π is the projection from $\mathbb{C}^{k+1} \setminus \{0\}$ to $\mathbb{P}^k$, the Kähler form is defined by the property that $\pi^*\omega = dd^c \log |z|$, where $z = (z_0, z_1, \ldots, z_n)$. We have seen (see Section A.5) that for every closed positive current S of bidegree $(1,1)$ on $\mathbb{P}^k$, there exists a p.s.h. function u on $\mathbb{C}^{k+1}$ such that

$$u(\lambda z) = c \log |\lambda| + u(z), \quad \forall \lambda \in \mathbb{C}^*, \quad \text{and } \pi^* S = dd^c u.$$

Moreover, $\|S\| = \displaystyle\int S \wedge \omega^{k-1} = c$, and u is unique if we impose the normalization $\displaystyle\int_{\partial B} u = 0$. We call u the potential of S. Weak convergence of currents translates into convergence of potentials in $L^1_{\text{loc}}(\mathbb{C}^{k+1})$.

Let $f \in \mathcal{M}_d$, let S be a closed positive current on $\mathbb{P}^k$, and let u be the potential of S. We can then define $f^*(S)$ by the relation $\pi^*(f^*(S)) = dd^c(u \circ F)$. In other words, $f^*(S)$ is the current of the potential $u \circ F$. Since f is dominant, the image of an open set is not contained in an analytic subset, and the operation $S \mapsto f^*(S)$ is continuous in the topology of currents.

THEOREM 1.6.1. *Let $f \in \mathcal{M}_d$ $(d \geq 2)$ be a meromorphic map on $\mathbb{P}^k$ that is dominant and algebraically stable. The sequence of currents $\left\{\frac{1}{d^n}(f^n)^*(\omega)\right\}$ converges to a closed positive current T of bidegree $(1,1)$ that satisfies*

$$f^*(T) = d \cdot T.$$

If $F = (F_0, F_1, \ldots F_k)$ is a lift of f, then T has a plurisubharmonic potential G on $\mathbb{C}^{k+1}$ satisfying

$$\begin{cases} G(\lambda z) = \log|\lambda| + G(z) \\ G(F(z)) = d \cdot G(z). \end{cases}$$

Moreover, if a function v satisfies the properties above, then $v \leq G$.

PROOF. Since the function F is defined up to a multiplicative constant, we may assume that $\sup_B |F| = 1$ on the unit ball of $\mathbb{C}^{k+1}$. By homogeneity, it follows that

$$|F(z)| \leq |z|^d, \quad \text{so} \quad |F^{n+1}(z)| \leq |F^n(z)|^d.$$

Set $G_n = \frac{1}{d^n} \log|F^n|$. Then

$$G_{n+1}(z) = \frac{1}{d^{n+1}} \log|F^{n+1}(z)| \leq \frac{1}{d^n} \log|F^n(z)| = G_n(z).$$

The sequence $\{G_n\}$ is thus a decreasing sequence of p.s.h. functions. Hence G_n converges to a limit G that is either a p.s.h. function or $\equiv -\infty$. We will show later that $G \equiv -\infty$ does not occur.

Assuming this for now, it follows from the definition of G_n that

$$\frac{1}{d} G(F(z)) = \lim \frac{1}{d} G_n(F(z)) = \lim G_{n+1}(z) = G(z).$$

Hence $G(F(z)) = d \cdot G(z)$.

Similarly, $G(\lambda z) = \log|\lambda| + G(z)$ follows from the same relation satisfied by the G_n. The current T such that $\pi^* T = dd^c G$ thus has the stated properties:

$$f^* T = d \cdot T \quad \text{and} \quad \|T\| = 1.$$

We now show that $G \equiv -\infty$ does not occur. For $N \geq 1$, set

$$\sigma_N = \frac{1}{N} \sum_{n=0}^{N-1} \frac{1}{d^n} (f^n)^*(\omega).$$

Then $\{\sigma_N\}$ is a sequence of closed positive currents such that $\|\sigma_N\| = 1$. Hence it has a convergent subsequence, $\sigma_{N_i} \to \sigma$. It follows easily from the definition of the σ_N's and the continuity of the operation f^* that $\frac{1}{d} f^*(\sigma) = \sigma$. Let h be the potential of σ. The equality above implies that $\frac{1}{d}(h \circ F) = h + c$, where c is a constant.

The function H defined by $H := h + \frac{cd}{d-1}$ is thus a potential for σ such that $H \circ F = d \cdot H$. We show that such a potential is always a lower bound for G, which implies that $G \not\equiv -\infty$.

The relation $H(\lambda z) = \log|\lambda| + H(z)$ gives the upper bound

$$H(z) \leq \log|z| + C,$$

so

$$H(z) = \frac{1}{d^n} H(F^n) \leq \frac{1}{d^n}(\log|F^n| + C).$$

Passing to the limit as n goes to $+\infty$, we obtain $H \leq G$. Thus $G \not\equiv -\infty$. □

REMARKS 1.6.2.

(i) Special cases of Theorem 1.6.1 are proved in [**32**]. Diller [**23**] has shown that $G \not\equiv -\infty$ for algebraically stable birational maps of $\mathbb{P}^2$.

(ii) The proof above also shows that $v \leq G$ for every function v satisfying $v \circ F = d \cdot v$ and $v(z) \leq \log|z| + C$.

PROPOSITION 1.6.3. *Let $\mathcal{G}$ be the set of closed positive currents S of bidegree $(1,1)$ such that $f^*(S) = d \cdot S$ and $\|S\| = 1$. Then the Green's current T is an extreme point of the compact convex set $\mathcal{G}$.*

PROOF. Suppose that

$$T = \frac{T_1 + T_2}{2}, \quad \text{with } f^*(T_j) = d \cdot T_j,\ \|T_j\| = 1,\ j = 1, 2.$$

We can choose potentials G_j for T_j satisfying $G_j(F(z)) = d \cdot G_j(z)$. Remark 6.2 (ii) implies that

$$G_j \leq G \quad \text{and hence that} \quad H = \frac{G_1 + G_2}{2} \leq G.$$

But $H = G$ because $G - H = c$ and $c(F(z)) = d \cdot c$, so $c = 0$. It follows that $G = G_1 = G_2$. □

EXAMPLE 1.6.4. Let f be the map defined by $f = [z^d : w^d : t^d]$. Then we have $F^n = (z^{d^n}, w^{d^n}, t^{d^n})$, and G is the function

$$G = \lim \frac{1}{d^n} \log(\sup\{|z|^{d^n}, |w|^{d^n}, |t|^{d^n}\}) = \sup(\log|z|, \log|w|, \log|t|).$$

The functions $\log|z|$, $\log|w|$, and $\log|t|$ also satisfy the functional equations.

In what follows, except in Section 1.9, we will restrict ourselves to the case of algebraically stable maps; that is, those for which the lift of f^n is F^n.

THEOREM 1.6.5 ([**32**]). *Let $f \in \mathcal{M}_d$ be an algebraically stable map.*

(1) *The support of the current T is contained in the Julia set, which is thus nonempty.*

(2) *If N denotes the set of normal points, then for every compact subset K of N we have*

$$\forall_{j \geq 1},\ |G_{n+j} - G_n| \leq \frac{C_K}{d^n},$$

and $N \cap (\mathbb{P}^k \setminus \operatorname{supp} T)$ is contained in the Fatou set; in particular, the Green's function G is continuous on $\pi^{-1}(N)$.

PROOF. Let $p \in U$, where U is an open set contained in the Fatou set. Shrinking U if necessary, we may assume that a subsequence $\{f^{n_j}\}$ converges in U to a holomorphic map h and that $f^{n_j}(U) \subset \{z_0 = 1,\ |z_j| < 2\}$. We can then write

$$G_{n_j} = \frac{1}{d^{n_j}} \log|F_0^{d^{n_j}}| + \frac{1}{d^{n_j}} \log|(1, A_j^1, \ldots, A_j^k)|.$$

The last term converges uniformly to 0, and the first is pluriharmonic. Hence G is pluriharmonic in U, and U does not intersect the support of T.

Now let U be an open set biholomorphic to a ball, such that $U \subset K \subset N$. We will prove that if $dd^c G = 0$ in $\pi^{-1}(U)$, then the sequence $\{f^n_{|U})$ is equicontinuous. By hypothesis, there exists a constant C_K such that the distance $\delta(f^n(z), I)$ is greater than or equal to C_K. It follows that

$$\left| F\left(\frac{F^n(z)}{|F^n(z)|}\right)\right| \geq C, \quad \text{hence that} \quad |F^{n+1}(z)| \geq C|F^n(z)|^d.$$

Thus

$$\frac{\log C}{d^{n+1}} \leq G_{n+1} - G_n \leq 0.$$

Hence $|G_{n+j} - G_n| \leq \dfrac{c_1}{d^n}$ for every j. Passing to the limit as j goes to $+\infty$, we see that $|G - G_n| \leq \dfrac{c_1}{d^n}$.

By hypothesis, G is pluriharmonic in $\pi^{-1}(U)$. Hence there exists a holomorphic function h such that $G = \log|h|$, and the estimate above can also be written as

$$\left|\frac{1}{d^n}\log\frac{|F^n|}{|h|^{d^n}}\right| = \left|\frac{1}{d^n}\log|F^n| - \frac{1}{d^n}\log|h|^{d^n}\right| \leq \frac{c_1}{d^n}.$$

Thus $e^{-c_1} \leq \dfrac{|F^n|}{|h|^{d^n}} \leq e^{c_1}$. Since $\dfrac{|F^n|}{|h|^{d^n}}$ is a lift of f^n, the family $\{f^n\}$ is equicontinuous in U. □

COROLLARY 1.6.6 ([**32**]). *Let $f \in \mathcal{M}_d$ be algebraically stable. If some subsequence f^{n_i} is equicontinuous in an open set U contained in the set of normality N, then U is in the Fatou set.*

PROOF. The proof of the preceding theorem shows that G_{n_i} converges to a pluriharmonic G in U. Hence U is in the Fatou set. □

COROLLARY 1.6.7 ([**32**]). *If f is normal, then the support of T equals J. The support of T is always connected, and the Fatou set is a Stein manifold.*

PROOF. If f is normal, the preceding theorem implies that the support of T coincides with J.

It is easy to see that the complement of the support of a closed positive current of bidegree $(1,1)$ on a complex manifold is a pseudoconvex set (see [**20, 33**]). In $\mathbb{P}^k$ the Levi problem has a positive solution; that is, every open set in $\mathbb{P}^k$ that is not $\mathbb{P}^k$ and is pseudoconvex is a Stein manifold. We know that if we have a compact subset K of a Stein manifold Ω, then $\Omega \setminus K$ cannot be Stein. Let $\Omega = \mathbb{P}^k \setminus J$. If we had $J = K_1 \cup K_2$, with K_1 and K_2 nonempty disjoint compact sets, then K_1 would be the support of a closed positive current $T_1 = T_{|K_1}$. Then $\mathbb{P}^k \setminus K_1$ would be a Stein manifold, as would $\mathbb{P}^k \setminus (K_1 \cup K_2)$; but this, as we have just recalled, is impossible. □

1.7. Hölder Continuity of the Potential of T in the Domain of Normality

The Hölder continuity of the Green's function was proved for Hénon maps in [**27**], for maps in $\mathcal{H}_d$ by Briend-Duval [**14, 15**], and for some special cases by Kosek [**56**] (also see Carleson-Gamelin [**18**]).

THEOREM 1.7.1. *Let $f \in \mathcal{M}_d$ be an algebraically stable dominant map. Let N be the open set of normal points. Then the Green's function is Hölder-continuous in $\pi^{-1}(N)$.*

PROOF. Let z_1 be such that $\pi(z_1) \in N$. We may assume that $G(z_1) > 0$. Let $U \Subset N$ be an open set containing $\pi(z_1)$, and let $\widetilde{U} = \pi^{-1}(U)$; this is a cone in $\mathbb{C}^{k+1}$. By Theorem 6.5, we may assume that for every $z \in \widetilde{U}$,

$$\frac{1}{d^n}\log|F^n(z)| - \frac{C}{d^n} \le G(z) \le \frac{1}{d^n}\log|F^n(z)|.$$

It follows that if $G(z) = 0$, then $|F^n(z)| \le e^C$. Let $\delta(z)$ denote the distance from z to $G^{-1}(0)$, and let z_0 be the point such that $G(z_0) = 0$ and $\delta(z_1) = |z_1 - z_0|$. Suppose that z_1 is sufficiently close to $\{G = 0\}$ that $z_0 \in \widetilde{U}$. Consider the segment ℓ with endpoints z_0 and z_1. Let m be the first index such that $F^m(\ell)$ is not contained in the ball $B(0, 2e^C)$. Then there exists $z_2 \in \ell$ such that $|F^m(z_2)| \ge 2e^C$. Set $M = \sup_{|z|\le 2e^C} \|F'(z)\|$, where $\|\ \|$ denotes the operator norm. By the mean value theorem,

$$e^C \le |F^m(z_2) - F^m(z_0)| \le M^m|z_2 - z_0| \le M^m\delta(z_1).$$

Set $\alpha = \dfrac{\log d}{\log M}$; then $d^m = M^{m\alpha}$. It follows from the inequalities above that

$$G(z_1) = \frac{1}{d^m}G(F^m(z_1)) \le \sup_{|z|\le 2e^C}\frac{1}{d^m}G(F(z)) \le \frac{A}{d^m} = \frac{A}{M^{m\alpha}} \le C'\delta^\alpha(z_1),$$

where A and C' are suitable constants.

Take z_1 and z_1' in $\widetilde{U}$ close to $\{G = 0\}$, such that $G(z_1)$ and $G(z_1')$ are positive. Let ζ_0 be such that $z_1' = \lambda_0\zeta_0$ and $G(\zeta_0) = 0$, with $|\lambda_0| > 1$. Consider the following function, which is subharmonic in the variable λ:

$$v(\lambda) = G(\lambda\zeta_0 + z_1 - z_1') - C'|z_1 - z_1'|^\alpha.$$

By homogeneity, $G(\lambda\zeta_0) = 0$ still holds for $|\lambda| = 1$. Hence we can apply the estimate above to $z = \lambda\zeta_0 + z_1 - z_1'$. Since

$$\delta(\lambda\zeta_0 + z_1 - z_1') \le |z_1 - z_1'|$$

for $|\lambda| = 1$, it follows that

$$G(\lambda\zeta_0 + z_1 - z_1') \le C'|z_1 - z_1'|^\alpha;$$

that is, $v(\lambda) \le 0$ on $|\lambda| = 1$.

But the subharmonic function $v(\lambda) - \log|\lambda|$ is bounded above at infinity, and less than or equal to zero on $|\lambda| = 1$. By the maximum principle,

$$v(\lambda) \le \log|\lambda| \quad \text{on} \quad |\lambda| \ge 1.$$

But, still by homogeneity, $G(\lambda\zeta_0) = \log|\lambda|$. Thus, taking $\lambda = \lambda_0$, we have

$$G(z_1) - G(z_1') \le C'|z_1 - z_1'|^\alpha.$$

Interchanging the roles of z_1 and z_1', we find that the Green's function is Hölder continuous in a neighborhood of $\{G = 0\}$ in $\pi^{-1}(N)$, hence throughout $\pi^{-1}(N)$ by homogeneity. □

REMARK 1.7.2. If $f \in \mathcal{H}_d$, then G is Hölder continuous with Hölder exponent α for every $\alpha < \alpha_0$, where $\alpha_0 = \dfrac{\log d}{\log \Lambda}$, with $\Lambda = \sup\limits_{G(z)\le 0} \|F'(z)\|$.

The Hölder continuity of the Green's function has consequences for the Hausdorff dimensions of certain invariant subsets associated with f. Classical methods suffice [**48**].

THEOREM 1.7.3. *Let $u_1, \dots, u_\ell$ be p.s.h. functions that are Hölder continuous with exponent $\alpha > 0$ in an open subset of $\mathbb{C}^k$. The current*

$$S_\ell := dd^c u_1 \wedge \cdots \wedge dd^c u_\ell$$

assigns no mass to those sets A whose $[2(k-\ell)+\alpha\ell]$-dimensional Hausdorff measure $\Lambda_{2(k-\ell)+\alpha\ell}(A)$ is zero.

PROOF. We sketch the proof. We may assume that A is compact. Set $\tau = 2(k-\ell) + \alpha\ell$. For a given $\varepsilon > 0$, we can cover A by dyadic cubes Q_i with sides s_i and centers c_i such that $\sum s_i^\tau < \varepsilon$. Let $(\psi_i)_{1 \le i \le N}$ be an associated partition of unity such that

$$\operatorname{supp} \psi_i \subset \frac{3}{2} Q_i, \quad \sum_{i=1}^{N} \psi_i = 1, \quad |D^\alpha \psi_i| \le C_\alpha s_i^{-|\alpha|}.$$

Set $U_\varepsilon = \bigcup_{i=1}^N Q_i$ and $S_{\ell-j} := dd^c u_1 \wedge \cdots \wedge dd^c u_{\ell-j}$. Then

$$\begin{aligned} \int_A S_\ell \wedge \omega^{k-\ell} &\le \int_{U_\varepsilon} S_\ell \wedge \omega^{k-\ell} \\ &\le \sum_i \int_{(3/2)Q_i} S_\ell \cdot \psi_i \wedge \omega^{k-\ell} \\ &= \sum_i \int S_{\ell-1} \wedge u_\ell dd^c \psi_i \wedge \omega^{k-\ell} \\ &= \sum_i \int S_{\ell-1} \wedge (u_\ell(z) - u_\ell(c_i)) dd^c \psi_i \wedge \omega^{k-\ell} \\ &\le C \sum_i s_i^{\alpha-2} \int_{(3/2)Q_i} S_{\ell-1} \wedge \omega^{k-\ell+1} \psi_i. \end{aligned}$$

One then repeats the process. □

COROLLARY 1.7.4. *Let $f \in \mathcal{M}_d$. Let α be a Hölder exponent for G. Then the currents $T^\ell = T \wedge \cdots \wedge T$ (ℓ times) assign no mass to those sets $A \subset N$ such that $\Lambda_{2(k-\ell)+\alpha\ell}(A) = 0$, $\alpha < \alpha_0 = \log d / \log \Lambda$.*

1.8. The Current T Assigns No Mass to Analytic Sets

THEOREM 1.8.1. *Let $f \in \mathcal{M}_d$ be an algebraically stable map. The current T assigns no mass to hypersurfaces.*

PROOF. Suppose that T assigns mass to some hypersurface V_0 of $\mathbb{P}^k$. By Siu's theorem [**76**], we have $T = T_1 + T_2$, where T_1 has no mass on any hypersurface and $T_2 = \sum_j c_j [V_j]$, where the $[V_j]$ are hypersurfaces and the c_j are positive constants. Using the invariance

$$T = \frac{1}{d}(f^* T_1 + f^* T_2),$$

and

$$\frac{f^* T_2}{d} = \frac{1}{d} \sum_j c_j f^*[V_j] \le \sum_j c_j [V_j],$$

and comparing the two sides of the inequality, we see that $T_2 = \frac{1}{d} f^* T_2$.

Since the current T is extremal among the invariant currents, it follows that $T = T_2$; in particular, $G = -\infty$ on $\pi^{-1}(\cup V_j)$.

Let $\alpha = \sup_j \|c_j[V_j]\|$. The supremum is attained because the mass of T is finite and that of each $[V_j]$ is an integer. Suppose, for instance, that $\alpha = \|c_0[V_0]\|$. We have $\left\| \frac{1}{d^j}(f^j)^* c_0[V_0] \right\| = \alpha$. The current T dominates each current $\frac{1}{d^j}(f^j)^* c_0[V_0]$. Since the mass of T is bounded, the components of $f^{-j}(V_0)$ cannot all be distinct. Hence there exists an irreducible submanifold V such that $V \subset f^{-\ell}(V)$. We may assume for simplicity that $\ell = 1$. Then $f(V \setminus I) \subset V$. Now we recursively define a sequence of analytic subsets

$$X_1 = \overline{f(V \setminus I)}, \ldots, X_j = \overline{f(X_{j-1} \setminus I)}.$$

The analytic set $X := \cap_j X_j \neq \emptyset$ because f is algebraically stable. We may consider X to be an irreducible analytic set satisfying $\overline{f(X \setminus I)} = X$; that is, $f_{|X}$ is dominant.

If X reduces to a point x, it is a fixed point and $G(x) \neq -\infty$, which is the desired contradiction. This is the idea we will generalize. Set

$$\sigma_N := \frac{1}{N} \sum_{n=0}^{N-1} \frac{1}{d^n} (f^n)^* \omega_{|X}.$$

We have $\sigma_N - \omega = dd^c v_N$, where $v_N \in L^1(X)$. The masses of the σ_N are bounded. Thus some subsequence converges to a limit σ, which is a current of bidegree $(1,1)$ on X and satisfies

$$\sigma - \omega_{|X} = dd^c v.$$

Hence $\pi^* \sigma$ has a potential $u = \log|z| + v$ on $\widetilde{X} = \pi^{-1}(X)$. Since the current σ is invariant, we proceed as in Theorem 1.6.1; we construct a potential H on $\widetilde{X}$ such that $H(F(z)) = dH(z)$ and show that $H \leq G_{|\widetilde{X}}$. It follows that $G_{|\widetilde{X}}$ is not identically $-\infty$, which completes the proof. □

1.9. The Non-Algebraically Stable Case

Suppose that $f \in \mathcal{M}_d$ is not algebraically stable. Then for n sufficiently large, we have $F^n = h_n F_n$, where h_n is a homogeneous polynomial, F_n is of degree $d_n < d^n$, and the components of F_n have no common factors.

One can verify that $d_{m+n} \leq d_m d_n$ for all m, n. In this case as well, one sets

$$G_n = \frac{1}{d^n} \log|F^n|.$$

As in Theorem 1.6.1, one shows that $G = \lim G_n$ is not identically $-\infty$ and defines the current $T = dd^c G$.

When S is a closed positive current such that $\pi^* S = dd^c u$, the currents S_n and ω_n are defined by the relations

$$\pi^* S_n = \frac{1}{d^n} dd^c u \circ F^n, \quad \pi^* \omega_n = \frac{1}{d^n} dd^c \log|F^n|.$$

THEOREM 1.9.1. *Let $f \in \mathcal{M}_d$ be non-algebraically stable. The following properties hold:*

- *The current T is supported on a union of hypersurfaces.*

– *If* $F^n = h_n F_n$, *then* $T = \lim_n \frac{1}{d^n}[h_n = 0]$, *where* $[h_n = 0]$ *denotes the current of integration on the hypersurface* $(h_n = 0)$.
– $T = \lim_n S_n$ *for every closed positive current* S *of bidegree* $(1,1)$.

PROOF. Since f is not algebraically stable, $F^\ell = h_\ell F_\ell$ for some $\ell \geq 2$, where the degree of F_ℓ is $d_\ell < d^\ell$. The inequality $d_{m+n} \leq d_m d_n$ then implies that $\frac{d_n}{d^n} \to 0$. We have

$$\frac{1}{d^n}\log|F^n| = \frac{1}{d^n}\log|h_n| + \frac{1}{d^n}\log|F_n|.$$

Hence

$$\omega_n = \frac{1}{d^n}[h_n = 0] + \sigma_n,$$

where $\pi^*\sigma_n = \frac{1}{d_n} dd^c \log|F_n|$. We have $\|\sigma_n\| = \frac{d_n}{d^n} \to 0$, and the result follows.

Let S be a closed positive current of bidegree $(1,1)$ and mass 1. Suppose that $\pi^*S = dd^c u$, where u is a p.s.h. function such that $u(\lambda z) = \log|\lambda| + u(z)$.

Then

$$\frac{1}{d^n} u \circ F^n = \frac{1}{d^n}\log|h_n| + \frac{1}{d^n} u \circ F_n.$$

It follows that

$$S_n = \frac{1}{d^n}[h_n = 0] + \sigma'_n,$$

where $\|\sigma'_n\| = \frac{d_n}{d^n} \to 0$, as before.

We now give a description of T. For every n,

$$G \leq G_n = \frac{1}{d^n}\log|h_n| + O(1).$$

The function $G - \frac{1}{d^n}\log|h_n|$ is p.s.h. because it is bounded above near $h_n = 0$. Hence $T - \frac{1}{d^n}[h_n = 0]$ is a closed positive current. The mass of $\frac{1}{d^n}[h_n = 0]$ approaches 1. It follows that all the mass of T is supported on $\bigcup_n [h_n = 0]$. □

REMARKS 1.9.2.

(1) The behavior of the sequence $\{d_n\}$ under a non-algebraically stable map can be fairly diverse. Bonifant [**13**] has studied some examples.
(2) It is natural to study the Cesàro convergence of the currents of the potentials $\frac{1}{d_n}\log|F_n|$. The current T of Theorem 1.9.1 accounts only for the distribution of those hypersurfaces that, under iteration, land on the indeterminacy set.

1.10. Convergence in Mean to T

Let $f \in \mathcal{M}_d$ be an algebraically stable map. Given a closed positive current S of bidegree $(1,1)$, we ask whether the sequence of currents $\frac{1}{d^n}(f^n)^*S$ converges to T. It may not, because there may exist other currents than T that satisfy the equation

$$f^*S = d \cdot S.$$

For instance, if f is the map $f([z:w:t]) = [z^d : w^d : t^d]$, then the currents $[z=0]$ and $[t=0]$ are also solutions. For a given current S, we need hypotheses on the dynamics of f or on the support of S (see Theorem 3.5.1). There is, however, a result on almost sure convergence for a family of currents; its proof goes back to a classical method in value-distribution theory ([**19**, **44**, **61**, **70**, **69**]). Let B_ℓ be the unit ball in $\mathbb{C}^\ell$, with variable denoted by w.

Let $U : \mathbb{C}^{k+1} \times B_\ell \to [-\infty, \infty)$ be a measurable function such that

(1) $U(\lambda z, w) = \log|\lambda| + U(z,w)$;
(2) $z \mapsto U(z,w)$ is p.s.h. for each w;
(3) $w \mapsto U(z,w)$ is p.s.h. for each z;
(4) $U(\zeta, w) \leq 0$ if $|\zeta| = 1$;
(5) $\int_{B_\ell} |U(\zeta,w)| dm(w) \leq C$ a.e. on $|\zeta|=1$, with respect to Lebesgue measure (denoted here by m) on $|\zeta| = 1$.

For each $w \in B_\ell$, let S_w be the closed positive current of bidegree $(1,1)$ on $\mathbb{P}^k$ defined by

$$\pi^*(S_w) = dd^c_z U(z,w).$$

We will call the family S_w an admissible family of currents. Condition (5) ensures that the family is sufficiently large, or else that the potentials are locally bounded.

Let us give some examples:

(1) $U(z,w) = \log|\langle z, w\rangle| \in \mathbb{C}^{k+1} \setminus \{0\}$. We obtain the family of hyperplanes.
(2) $U(z,w) = \log|P(z,w)|$, where $P(\cdot, w)$ ranges over an open set of polynomials of degree d.

The following theorem is a variant of a result of Russakovski-Sodin [**70**] and Russakovski-Shiffman [**69**].

THEOREM 1.10.1. *Let $f \in \mathcal{M}_d$ be an algebraically stable map. Let S_w, $w \in B_\ell$, be an admissible family of closed positive currents of bidegree $(1,1)$. Then there exists a pluripolar set $\mathcal{E} \subset B_\ell$ such that*

$$\frac{1}{d^n}(f^n)^* S_w \longrightarrow T$$

for every $w \in B_\ell \setminus \mathcal{E}$.

We begin by proving the following lemma:

LEMMA 1.10.2. *Let ν be a probability measure on B_ℓ such that*

$$\int |U(\zeta, w)| d\nu(w) \leq C_\nu$$

for almost every $\zeta \in \partial B$ (with respect to Lebesgue measure).

Then $\frac{1}{d^n}(f^n)^* S_w \to T$, ν-almost everywhere, as $n \to \infty$.

PROOF. Let φ be a test form. Set $K(z,w) := \log|z| - U(z,w)$ (≥ 0). Then

$$\begin{aligned} |\langle f^*\omega - f^* S_w, \varphi\rangle| &= |\langle dd^c K(F(z),w), \varphi\rangle| = |\langle K(F(z),w), dd^c\varphi\rangle| \\ &\leq |\varphi|_2 \, |\langle K(F(z),w), \omega_z^k\rangle| =: |\varphi|_2 m_F(w), \end{aligned}$$

where $|\varphi|_2$ denotes the $\mathcal{C}^2$ norm of φ.

Using the property that $K(F(z),w) = -U(F(z)/|F(z)|, w)$ and Fubini's theorem, we have

$$\int K(F(z),w)\omega_z^k d\nu(w) = \int \omega_z^k \int \left| U\left(\frac{F(z)}{|F(z)|}, w\right)\right| d\nu(w) \leq C_\nu.$$

Hence

$$\nu(m_F(w) \geq s) \leq \frac{C_\nu}{s}.$$

Let

$$\mathcal{E}_j^N := \bigcup_{n\geq N} \left\{ \frac{m_{F^n}}{d^n} \geq \frac{1}{j} \right\};$$

then

$$\nu(\mathcal{E}_j^N) \leq \sum_{n\geq N} \frac{jC_\nu}{d^n}.$$

It follows that $\mathcal{E} := \bigcup_j \bigcap_N \mathcal{E}_j^N$ has ν-measure zero. For $w \notin \mathcal{E}$, we have

$$\left| \left\langle \frac{(f^n)^*(\omega)}{d^n} - \frac{(f^n)^*(S_w)}{d^n}, \varphi \right\rangle \right| \to 0.$$

It suffices to observe that $\dfrac{(f^n)^* S_w}{d^n}$ converges to T. □

Proof of the theorem. If the set $\mathcal{E}$ on which $\dfrac{(f^n)^* S_w}{d^n}$ does not converge to T were not pluripolar, then it would contain a nonpluripolar compact set $K \subset B_\ell$. But we know (Appendix, Theorem A.9.2) that for every nonpluripolar compact set there exists a bounded p.s.h. function u_K, defined in B_ℓ, for which $\nu := (dd^c u_K)^\ell$ is a probability measure supported on K. By the Chern-Levine-Nirenberg inequalities (see Proposition A.6.3),

$$\int |U(\zeta,w)|(dd^c u_K)^\ell(w) \leq C\|U_\zeta\|_{L^1(B_\ell)} \leq C_\nu.$$

□

Remark 1.10.3. In the case where the family $\{S_w\}$ consists of hyperplanes, measures ν satisfying the hypotheses of the lemma are studied in [**61**].

Russakovski and Shiffman [**69**] proved a result on the equidistribution of inverse images of subspaces of codimension greater than 1. For each meromorphic map $f : \mathbb{P}^k \to \mathbb{P}^k$, with indeterminacy set I, let

$$\delta_\ell(f) := \int_{\mathbb{P}^k \setminus I} f^*\omega^\ell \wedge \omega^{k-\ell}.$$

$\delta_\ell(f)$ is the generic degree of the variety $f^{-1}(W)$, where $W \in G(\ell,k)$, the Grassmanian of subspaces of codimension ℓ. They prove the following result.

Theorem 1.10.4 ([**69**]). *Let $\{f_n\}$ be a sequence of rational maps from $\mathbb{P}^k$ to $\mathbb{P}^k$. Let $\{a_n\}$ be a sequence of positive numbers such that*

$$\sum_{n=1}^{\infty} \frac{\delta_\ell(f_n)}{a_n} < \infty.$$

There exists a pluripolar set $\mathcal{E} \subset G(\ell, k)$ such that, for $W \in G(\ell, k) \setminus \mathcal{E}$,

$$\frac{1}{a_n}(f_n^*[W] - f_n^*\omega^\ell) \longrightarrow 0.$$

CHAPTER 2

Polynomial Automorphisms of $\mathbb{C}^k$

2.1. Introduction

Let $f = (f_1, f_2, \ldots, f_k)$ be a polynomial automorphism of $\mathbb{C}^k$ of algebraic degree $d \geq 2$. We will denote by $\overline{f}$ the meromorphic extension of f to $\mathbb{P}^k$ (when there is no ambiguity, we write f for $\overline{f}$).

The coordinates in $\mathbb{C}^k$ are denoted by $z = (z_1, z_2, \ldots, z_k)$ and the coordinates in $\mathbb{P}^k$ by $[z_1 : \cdots : z_k : t]$. The equation of the hyperplane at infinity is $t = 0$.

We denote the indeterminacy sets of $\overline{f}$ and $\overline{f^{-1}}$ by I_+ and I_-, respectively. They are analytic sets of codimension at least two in $\mathbb{P}^k$ that are contained in $\{t = 0\}$.

We have seen some examples of automorphisms of $\mathbb{C}^2$ and $\mathbb{C}^3$. We return to the case of $\mathbb{C}^2$. An *elementary* automorphism [**40**] is an automorphism of the form

$$e(z, w) = (\alpha z + p(w), \beta w + \gamma),$$

where p is a polynomial of degree $d \geq 2$. Then

$$\overline{e}([z : w : t]) = \left[\alpha z t^{d-1} + t^d p\left(\frac{w}{t}\right) : \beta w t^{d-1} + \gamma t^d : t^d\right].$$

The automorphism e is not algebraically stable. Indeed, the image of $\{t = 0\}$ is the point of indeterminacy $I_+ = [1 : 0 : 0]$. The dynamics of e is fairly easy to study (see [**40**]). Note that $I_+ = I_-$ in this case.

We also have Hénon automorphisms. These are finite compositions of automorphisms h_j defined by

$$h_j(z, w) = (p_j(z) - a_j w, z),$$

where $a_j \in \mathbb{C}^*$ and p_j is a polynomial of degree $d_j \geq 2$. Each h_j has $I_+ = [0 : 1 : 0]$ as point of indeterminacy; $\{t = 0\}$ is mapped to $I_- = [1 : 0 : 0]$, which is an attracting fixed point. It follows that the elements of the semigroup $\mathcal{H}$ consisting of the compositions $h = h_m \circ \cdots \circ h_1$ are algebraically stable. Note that $I_+ \cap I_- = \emptyset$ for every $h \in \mathcal{H}$.

The following result shows that the only dynamically interesting polynomial automorphisms of $\mathbb{C}^2$ are those that are conjugate to elements of $\mathcal{H}$.

THEOREM 2.1.1 ([**40**]). *Let f be a polynomial automorphism of $\mathbb{C}^2$. Then f is conjugate in the group of polynomial automorphisms to either an elementary map or a Hénon map (i.e., an element of $\mathcal{H}$).*

There is another criterion, due to J.-Ph. Furter:

THEOREM 2.1.2 ([**41**]). *A polynomial automorphism f of $\mathbb{C}^2$ is conjugate to an elementary map if and only if $\deg f^2 \leq \deg f$.*

2.2. Regular Automorphisms of $\mathbb{C}^k$

DEFINITION 2.2.1. Let f be a polynomial automorphism of $\mathbb{C}^k$ of degree d. We say that f is *regular* if the indeterminacy sets I_+ of $\overline{f}$ and I_- of $\overline{f^{-1}}$ are disjoint.

We have seen that a polynomial automorphism of $\mathbb{C}^2$ is conjugate to either an elementary automorphism or a regular automorphism (Hénon map).

Let f be a regular automorphism. If h is an automorphism of $\mathbb{P}^k$ that leaves $\{t=0\}$ stable, then the automorphism $g = h^{-1}fh$ is regular. This is not true when h is a polynomial automorphism of $\mathbb{C}^k$.

EXAMPLE. If $f(z,w) = (z^2+aw, z)$, $a \neq -1$, and $h(z,w) = (w, z+w^2)$, then $g = h^{-1}fh$ is not regular.

Let f be a polynomial automorphism of $\mathbb{C}^k$. Suppose that f and f^{-1} have degrees d and d_-, respectively. Since $d \geq 2$, we also have $d_- \geq 2$. We denote by F and F^{-1} the lifts to $\mathbb{C}^{k+1}$ of $\overline{f}$ and $\overline{f^{-1}}$. Then

$$F \circ F^{-1}(z_1, \dots, z_k, t) = F^{-1} \circ F(z_1, \dots, z_k, t) = t^{d \cdot d_- -1}(z_1, \dots, z_k, t).$$

This implies the inclusions $f(\{t=0\} \setminus I_+) \subset I_-$ and $f^{-1}(\{t=0\} \setminus I_-) \subset I_+$.

Let $E_+ = \overline{\bigcup I_{+,n}}$ and $E_- = \overline{\bigcup I_{-,n}}$ be the closures of the indeterminacy sets of f^n and f^{-n}. The following proposition is a consequence of Definition 2.1.

PROPOSITION 2.2.2. *Let f be a regular polynomial automorphism of $\mathbb{C}^k$. Then*

- *$f(\{t=0\} \setminus I_+) \subset I_-$ and $f^{-1}(\{t=0\} \setminus I_-) \subset I_+$,*
- *$E_+ = I_+$ and $E_- = I_-$,*
- *f is algebraically stable.*

PROOF. The first property always holds. The other two follow from it: since $I_+ \cap I_- = \emptyset$, no point in $\{t=0\} \setminus I_+$ can be sent to I_+. □

PROPOSITION 2.2.3. *Let f be a regular polynomial automorphism of $\mathbb{C}^k$. Then f is normal. In particular, the Green's function $G^+(z) := G(z,1)$ is Hölder continuous in $\mathbb{C}^k$.*

PROOF. Recall that f is normal at $q \in \mathbb{C}^k$ if there exist a neighborhood U of q and a neighborhood V of I_+ such that $f^n(U) \cap V = \emptyset$ for every n.

We have $f^{-1}(\{t=0\} \setminus I_-) \subset I_+$. Moreover, the derivative of f^{-1} in the direction t is zero on $\{t=0\} \setminus I_-$. To see this, observe that the corresponding component is t^d, and d is greater than or equal to 2. It follows that there exists a neighborhood V of I_+ with V disjoint from I_- and $f^{-1}(V) \Subset V$. We use the continuity of f^{-1} on $\{t=0\} \setminus I_-$, the inclusion $f^{-1}(V \cap \{t=0\}) \subset I_+$, and the fact that f^{-1} is contracting in the direction t.

Let $q \in \mathbb{C}^k$. Let V be a neighborhood of I_+ as above. We may assume that $q \notin \overline{V}$. If f were not normal at q, there would exist a sequence $\{q_j\}$ in $\mathbb{C}^k$ such that $q_j \to q$ and $f^{n_j}(q_j) \to I_+$. For sufficiently large j we would then have $f^{n_j}(q_j) \in V$, so $q_j = f^{-n_j} \circ f^{n_j}(q_j) \in V$, contradicting $q \notin \overline{V}$. It is clear that the points in $\{t=0\} \setminus I_+$ are normal. □

We give some examples:

(1) Let f be the automorphism

$$f(z) = (P_1(z_1, \dots, z_{k-1}) + a_k z_k, P_2(z_1, \dots, z_{k-2}) + a_{k-1} z_{k-1}, \dots, P_{k-1}(z_1) + a_2 z_2, z_1),$$

where, for each j, a_j is a nonzero constant and P_j is a polynomial of degree $d \geq 2$ in which the coefficient of z_{k-1}^d is nonzero. It is clear that f is an automorphism.

We have $I_+ = \{[0 : 0 : \ldots : 0 : 1 : 0]\}$ and

$$f(\{t = 0\} \setminus I_+) \subset \{[\zeta_1 : \ldots : \zeta_{k-1} : 0 : 0]\}.$$

Observe that

$$f^{-1}(z) = \left(z_k, \frac{1}{a_2}(z_{k-1} - P_{k-1}(z_k)), \frac{1}{a_3}(z_{k-2} - P_{k-2}(z_k, \frac{1}{a_2}(z_{k-1} - P_{k-1}(z_k)))), \ldots\right).$$

One can check that $\overline{f^{-1}}$ is well defined on I_+.

Thus the automorphism f is regular. It is easy to construct examples of such f for which $\deg f = d$ and $d_- = \deg f^{-1} = d^{k-1}$. If we allow the coefficient of z_{k-j}^d in the polynomial P_j to vanish, then f is still an automorphism, but it is not necessarily regular.

(2) For $a \in \mathbb{C}^*$, let g be the automorphism of $\mathbb{C}^3$ defined by

$$g(z_1, z_2, z_3) = (z_1^2 + az_2, z_1, z_3);$$

then

$$g^{-1}(z_1, z_2, z_3) = \left(z_2, \frac{1}{a}(z_1 - z_2^2), z_3\right).$$

We have

$$I_+ = \{[0 : z_2 : z_3 : 0]\}, \quad I_- = \{[z_1 : 0 : z_3 : 0]\}, \quad I_+ \cap I_- = \{[0 : 0 : 1 : 0]\}.$$

The automorphism g is not regular, but it is dynamically interesting. Its study reduces to the study of a Hénon map on $\mathbb{C}^2$.

Fornaess and Wu [**37**] have described the conjugacy classes of polynomial automorphisms of degree 2 in $\mathbb{C}^3$.

Later we will use the notion of an attractor. We recall the definition:

DEFINITION 2.2.4. Let M be a metric space, and let f be a continuous map from M to M. A compact set $X \subset M$ is said to be attracting for f if there exists a neighborhood U of X such that $f(U) \Subset U$ and $X = \bigcap_{n \geq 0} f^n(U)$. It is said to be an attractor if, for all x, y in X and every $\varepsilon > 0$, there exists an ε-chain from x to y. More precisely, there exist points $x_0 = x, x_1, \ldots, x_k = y$ in X such that $d(f(x_i), x_{i+1}) \leq \varepsilon$ for $0 \leq i \leq k - 1$ (see [**67**]).

DEFINITION 2.2.5. We assign to a polynomial automorphism f of $\mathbb{C}^k$ the sets K^+, K^-, U^+, U^-, and K defined as follows:

$$K^+ = \{z \in \mathbb{C}^k : f^n(z) \text{ is bounded}\}, \quad U^+ = \mathbb{C}^k \setminus K^+,$$

$$K^- = \{z \in \mathbb{C}^k : f^{-n}(z) \text{ is bounded}\}, \quad U^- = \mathbb{C}^k \setminus K^-,$$

$$K = K^+ \cap K^-.$$

We also introduce the notation G^+, G^-, and G_K to denote the functions

$$G^+(z) = G^+(z, 1) = \lim \frac{1}{d^n} \log^+ |f^n(z)|,$$

$$G^-(z) = G^-(z, 1) = \lim \frac{1}{d_-^n} \log^+ |f^{-n}(z)|,$$

and

$$G_K(z) = \sup(G^+(z), G^-(z)).$$

REMARKS.

(1) Bedford and Pambuccian [**5**] have proved the existence of the functions G^+ and G^- and the associated currents for certain (shift-like) automorphisms of $\mathbb{C}^n$.

(2) The sets K^+ and K^- are not closed in general. One can even find examples of non-polynomial automorphisms for which they are everywhere dense ([**34**]).

In the case of regular automorphisms, we nonetheless have the following proposition.

PROPOSITION 2.2.6. *Let f be a regular polynomial automorphism of $\mathbb{C}^k$, of algebraic degree $d \geq 2$. Then the following properties hold:*

- *The functions G^+ and G^- are continuous in $\mathbb{C}^k$, and*

$$K^+ = \{G^+ = 0\}, \quad K^- = \{G^- = 0\}.$$

- *The closures $\overline{K^+}$ and $\overline{K^-}$ of K^+ and K^- in $\mathbb{P}^k$ satisfy*

$$\overline{K^+} \subset K^+ \cup I_+ \quad \textit{and} \quad \overline{K^-} \subset K^- \cup I_-.$$

- *The set I_+ is attracting for f^{-1}, and I_- is attracting for f.*
- *$K = K^+ \cap K^-$ is a compact subset of $\mathbb{C}^k$.*

PROOF. We have seen that the points of $\mathbb{C}^k$, and hence the points of $\mathbb{P}^k \setminus I_+$, are normal. Thus (by Theorem 1.6.5) G is continuous on $\mathbb{C}^{k+1} \setminus \pi^{-1}(I_+)$, and in fact $\exp G$ is continuous on $\mathbb{C}^{k+1}$.

The inclusion $K^+ \subset \{G^+ = 0\}$ follows because f^n is bounded on K^+. Moreover, since the map is normal,

$$|G^+(z) - G_n^+(z)| \leq \frac{C}{d^n}$$

at every point z of $\mathbb{C}^k$. It follows that if $G^+(z) = 0$, then $\log^+ |f^n(z)| \leq C$, and hence $|f^n(z)| \leq e^C$. This gives the opposite inclusion $\{G^+ = 0\} \subset K^+$.

Since the points of $\{t = 0\} \setminus I_+$ are sent to I_-, we certainly have

$$\overline{K^+} \cap \{t = 0\} \subset I_+, \quad \text{whence} \quad \overline{K^+} \subset K^+ \cup I_+.$$

The fact that I_- is attracting for f follows from the property (already used in Proposition 2.2.3) that there exists a neighborhood V of I_- such that $f(V) \Subset V$.

It follows from the hypothesis $I_+ \cap I_- = \emptyset$ that $K = K^+ \cap K^-$ is a compact subset of $\mathbb{C}^k$. $\square$

PROPOSITION 2.2.7. *Let f be a regular automorphism of $\mathbb{C}^k$. Then the distance from $f^n(z)$ to K approaches 0 uniformly on compact subsets of K^+. In particular, the family $\{f^n\}$ is equicontinuous in the interior of K^+.*

PROOF. Let X be a compact subset of K^+. Suppose that $G^- \leq C$ on X. Then, for $z \in X$,

$$G^+(f^n(z)) = 0 \quad \text{and} \quad G^-(f^n(z)) = \frac{1}{d_-^n} G^-(z) \leq \frac{C}{d_-^n}.$$

Hence

$$G_K(f^n(z)) \leq \frac{C}{d_-^n}.$$

The result follows because G_K is continuous and $K = \{G_K = 0\}$. □

DEFINITION 2.2.8. Let f be an algebraically stable polynomial automorphism. Let $\{X_j\}$ be the sequence of analytic sets defined recursively by

$$X_1 = \overline{f(\{t=0\} \setminus I_+)}, \ldots, X_j = \overline{f(X_{j-1} \setminus I_+)}.$$

This sequence is decreasing, hence eventually constant. We define X_+ to be the limit analytic set; that is, $X_+ = X_h$, where h is the first index such that $X_h = X_{h+1}$. The set X_+ is nonempty because f is algebraically stable.

We give an example. Let f be the algebraically stable automorphism of $\mathbb{C}^3$ defined by

$$f(x,y,z) = (x^2 + z^2 + y, z^2 + x, z).$$

We have $I_- = X_1 = \{z = t = 0\}$ and $X_2 = X_+ = \{y = z = t = 0\}$. As can be seen, I_- need not equal X_+.

REMARK 2.2.9. $X_+ = I_-$ if f is a regular automorphism with $\dim I_+ = 0$. Indeed, at every point $q \in I_-$, the blow-up $\mathcal{B}_q^-$ of f^{-1} at q, which has dimension greater than or equal to 1, cannot be contained in I_+. But $f(\mathcal{B}_q^- \setminus I_+) = q$. Hence $f(I_-) = I_-$ and $I_- = X_+$.

PROPOSITION 2.2.10. *Let f be a regular automorphism of $\mathbb{C}^k$. Then X_+ is attracting for f; that is, the distance from $f^n(x)$ to X_+ approaches* 0 *uniformly on compact subsets of U^+. If X_+ has dimension* 0*, then it is an attracting point, U^+ is in the Fatou set, and G^+ is pluriharmonic in U^+. If X_+ has dimension* 0*, then the support in $\mathbb{C}^k$ of the current $T_+ := dd^cG^+$ is $J^+ := \partial K^+$.*

PROOF. We have seen that I_- has a neighborhood V such that $f(V) \Subset V$. Thus a neighborhood of $\{t = 0\} \setminus I_+$ is sent to V. Hence X_+ is attracting.

The set X_+ is always connected. Thus, if it has dimension 0, it reduces to a point, which is necessarily attracting. The family $\{f^n\}$ is then equicontinuous in U^+, so G^+ is pluriharmonic in U^+. Since $K^+ = \{G^+ = 0\}$, it follows from the minimum principle for harmonic functions that G^+ can have no pluriharmonic extension to a neighborhood of a point in J^+. Hence $\operatorname{supp} T \cap \mathbb{C}^k = J^+$. □

REMARK 2.2.11. We denote by $\overline{T}_+$ the current associated with the meromorphic function $\overline{f}$ that extends f. It is a current on $\mathbb{P}^k$. We will sometimes omit the bar because this current assigns no mass to $t = 0$.

THEOREM 2.2.12. *Let f be a regular automorphism of $\mathbb{C}^k$. For every closed positive current S of bidegree $(1,1)$ on $\mathbb{P}^k$ whose potential u in $\mathbb{C}^{k+1}$ is bounded in a neighborhood V of $\pi^{-1}(X_+) \cap \{\|z\| = 1\}$, we have*

$$\lim \frac{1}{d^n}(f^n)^*(S) = \overline{T}_+.$$

Moreover, the convergence is uniform on the set $\mathcal{G}_{V,C}$ of closed positive currents S of bidegree $(1,1)$ and total mass $\|S\| = 1$ whose potentials u are all bounded on the open set V by the same constant C.

PROOF. Let $\{S_j\}$ be a sequence of closed positive currents in $\mathcal{G}_{V,C}$. We may assume that their potentials $\{u_j\}$ are normalized by $\sup_B u_j = 0$ on the unit ball B. We want to show that $\left\{\frac{1}{d^n} u_j \circ F^n\right\}$ converges to G^+, uniformly in j.

Since $u_j \leq \log|z|$, it follows that $\frac{1}{d^n} u_j \circ F^n \leq \frac{1}{d^n} \log|F^n|$. Hence the sequence is locally bounded above. In the cone $\widetilde{V}$ generated by V,

$$\log|z| - C \leq u_j \leq \log|z|.$$

This cone is stable under F because X_+ is attracting. Hence $\{(1/d^n)u_j \circ F^n\}$ converges to G^+ in U^+.

If $\frac{1}{d^n}(f^n)^*(S)$ does not converge uniformly to T_+ with respect to $S \in \mathcal{G}_{V,C}$, then we can find a sequence n_j that goes to infinity and an $S_j \in \mathcal{G}_{V,C}$, with potential u_j, such that

$$\widetilde{v} = \lim \frac{1}{d^{n_j}} u_j \circ F^{n_j} \neq G^+.$$

We know that $\widetilde{v} = G$ on $\pi^{-1}(U^+)$. Now consider the restrictions to $\mathbb{C}^k$. Set $v(z) = \widetilde{v}(z,1)$ and $v_j(z) = \widetilde{u}_j(z,1)$. Then $v = G^+$ on U^+ and $v \leq G^+$ everywhere.

The function G^+ is positive on U^+ and zero on $J^+ = \partial U^+$. The function v is zero on J^+ because it is u.s.c. We must show that $v = 0$ on the interior of K^+ as well. Suppose $v < -2\delta$, where $\delta > 0$, on an open subset W of K^+. Then the Hartogs lemma (see Theorem A.1.2) implies that

$$\frac{1}{d^{n_j}} v_j \circ f^{n_j} < -\delta$$

for sufficiently large j. Let E_j be the set defined by $E_j = \{z \in \mathbb{C}^k : v_j(z) < -\delta d^{n_j}\}$. Proposition A.7.1 of the Appendix and Fubini's theorem imply that

$$\operatorname{vol}(E_j \cap B(0,R)) \leq C_1 e^{-\delta d^{n_j}},$$

where C_1 is a constant independent of j. But the Jacobian of f is constant, with modulus $|a|$. Hence

$$\operatorname{vol}(f^{n_j}(W)) \geq C_2 |a|^{2n_j}.$$

But for sufficiently large n_j, we cannot have $C_2|a|^{2n_j} < C_1 e^{-\delta d^{n_j}}$. Hence $v = G^+$ everywhere. □

COROLLARY 2.2.13. *Let f be a regular automorphism of $\mathbb{C}^k$.*

(1) *The current T_+ is extremal in the cone of closed positive currents of bidegree $(1,1)$.*

(2) *If there exists a nonzero closed positive current S of bidegree $(1,1)$ on $\mathbb{P}^k$ whose support is contained in $\overline{K}^+$, then the support of T_+ is contained in $\overline{K}^+$ and $S = cT_+$, where c is a positive constant.*

PROOF. Let S be a closed positive current of bidegree $(1,1)$, with mass $c > 0$, such that $S \leq T_+$. We must show that $S = cT_+$.

Set $S_n = d^n(f^{-n})^* S$ in $\mathbb{C}^k$, and let $\widetilde{S}_n$ be the trivial extension of S_n to $\mathbb{P}^k$. Now, $\{t = 0\}$ is an analytic set and the mass of S_n is bounded in a neighborhood of $\{t = 0\}$ (see Skoda's theorem, in Appendix A.4, on extending currents that have bounded mass in a neighborhood of an analytic set). Hence $\widetilde{S}_n$ is a closed positive current.

Moreover, $S_n \leq d^n(f^{-n})^*T_+ = T_+$. Indeed, we have $d^nG^+(f^{-n}) = G^+$ in $\mathbb{C}^k$. We also know that $\dfrac{(\overline{f}^n)^*}{d^n}$ preserves the masses of currents in $\mathbb{P}^k$. But $\dfrac{(\overline{f}^n)^*}{d^n}(\widetilde{S}_n) \leq T_+$, and the left-hand side equals S in $\mathbb{C}^k$. Hence equality also holds in $\mathbb{P}^k$. It follows that

$$\|\widetilde{S}_n\| = \|S\| = c.$$

Let v_n and w_n be potentials for $\widetilde{S}_n$ and $T-\widetilde{S}_n$, respectively, with $v_n+w_n = G^+$. We may assume that

$$\int_{\partial B} v_n = 0 \quad \text{and} \quad \int_{\partial B} w_n = \int_{\partial B} G.$$

It follows that v_n and w_n are uniformly bounded above. But G^+ is continuous in a neighborhood of $Y_- := \pi^{-1}(I_-) \cap \partial B$. Hence $|v_n| \leq C$ in a neighborhood of Y_-.

Set $\widetilde{S}_n^1 = \dfrac{S_n}{\|S_n\|}$. By Theorem 2.2.12, $\dfrac{(f^n)^*}{d^n}(\widetilde{S}_n^1)$ converges to T_+. It follows that $\lim \dfrac{S_n}{\|S_n\|} = T_+$, so $S = cT_+$.

Let S be a current supported on $\overline{K}^+$. Suppose $\|S\| = 1$. Let u be its potential. As above, set $S_n = d^n(f^{-n})^*S$, and let $\widetilde{S}_n$ be its extension by 0. The current $\widetilde{S}_n$ is closed by Property (e) of Section A.4.

We show that $S' := \dfrac{(\overline{f}^n)^*}{d^n}(\widetilde{S}_n)$ is equal to S. This is clearly true in $\mathbb{C}^k$, so it suffices to check that S' assigns no mass to $\{t = 0\}$. The currents S_n and $\widetilde{S}_n$ are supported on $\overline{K}^+$. Let $\widetilde{v}_n$ be the potential of $\widetilde{S}_n$. The function $\widetilde{v}_n$ is pluriharmonic in $\pi^{-1}(U^+ \cup \{t=0\} \setminus (I_+))$. Set $\widetilde{U}^+ = (U^+ \cup \{t=0\} \setminus (I_+))$. Since f^n sends $\widetilde{U}^+$ to $\widetilde{U}^+$, the potential of S' is also pluriharmonic on $\pi^{-1}(\widetilde{U}^+)$. Thus we have the desired relation

$$S = \frac{(\overline{f}^n)^*}{d^n}(\widetilde{S}_n).$$

Hence $\|\widetilde{S}_n\| = 1$. Since the potentials $\widetilde{v}_n$ of $\widetilde{S}_n$ are pluriharmonic in a neighborhood of Y_- (where we may choose a point p and normalize by setting $\widetilde{v}_n(p) = 0$), they are bounded above by a fixed constant C in a fixed neighborhood of Y_-. (It suffices to use Harnack's inequalities.) We can therefore apply Theorem 2.2.12 to obtain

$$T_+ = \lim \frac{(\overline{f}^n)^*}{d^n}(\widetilde{S}_n) = \lim \frac{f^n}{d^n} S_n = S.$$

The proof shows that if $\operatorname{supp} T \neq \overline{K}^+$, then there is no nonzero closed positive current of bidegree $(1,1)$ supported on $\overline{K}^+$. □

REMARKS 2.2.14. (1) For Hénon maps on $\mathbb{C}^2$, X_+ reduces to a point. The only closed positive current supported on $\overline{K}^+$ and of mass 1 is T_- ([**29**]). Versions of Theorem 2.2.12 for Hénon maps can be found in [**7, 8, 9, 27, 29**]. We have followed the approach in [**27**].

(2) We return to the example we saw earlier:

$$g([x:y:z:t]) = \left[P(x,y) + \alpha z t^{d-1} : Q(x) + byt^{d-1} : xt^{d-1} : t^d\right],$$

where P and Q are homogeneous polynomials of degree d. We have

$$g(\{t=0\}) = \{z=t=0\} = X_+ \quad \text{and} \quad g^{-1}(\{t=0\}) = \{[0:0:1:0]\}.$$

Hence the only closed positive current supported on $\overline{K^-}$ and of mass 1 is T_-. We can arrange that the support of the current T_+ associated with g is all of $\mathbb{P}^3$. Indeed, if the Jacobian Jg is such that $|Jg| = |ab| > 1$, then K^+ has empty interior.

The polynomials P and Q can be chosen in such a way that the Julia set of

$$g_{|X_+}[x:y] = [P(x,y) : Q(x)]$$

is $\mathbb{P}^1$. Thus the family $\{g^n\}$ is not equicontinuous in U^+. It follows that the Julia set coincides with $\mathbb{P}^3$. Hence the support of T_+ is $\mathbb{P}^3$. So this is an example of an extremal closed positive current of bidegree $(1,1)$ with support $\mathbb{P}^3$. The compact set X_+ is an attractor in this case.

REMARK 2.2.15. In the proof of the theorem we used that $(f^{-1})^*(T_+) = \dfrac{1}{d}T_+$, which follows from $G^+(f^{-1}) = \dfrac{1}{d}G^+$.

Note, however, that

$$(\overline{f^{-1}})^*(\overline{T}_+) = \frac{1}{d}\overline{T}_+ + \left(d_- - \frac{1}{d}\right)[t=0].$$

(The bars are meant to emphasize that we are considering the pullback to projective space.) Indeed, this result follows from

$$d \cdot G(F^{-1}(z)) = G(FF^{-1}(z)) = G((t^{d \cdot d_- - 1})z) = G(z) + (d \cdot d_- - 1)\log|t|.$$

Thus most of the mass of the current $(\overline{f^{-1}})^*(\overline{T}_+)$ is on $\{t=0\}$.

2.3. Relation between the Degrees of f and f^{-1}. Periodic Points. Entropy.

We have seen examples of automorphisms of $\mathbb{C}^k$ of degree d whose inverses were of degree d^{k-1}. We will give a precise relationship between the degree of an automorphism f and that of f^{-1}. We begin by estimating the degree of the inverse (see [**2**]).

PROPOSITION 2.3.1. *Let f be a polynomial automorphism of $\mathbb{C}^k$. Suppose that f has degree d and f^{-1} has degree d_-. Then $d \le (d_-)^{k-1}$.*

PROOF. Let ω be the Kähler form of $\mathbb{P}^k$. Then $\overline{f}^*(\omega)$ has no mass on $\{t=0\}$ because t is not a common factor of the components of f. Hence

$$1 = \frac{1}{d}\int_{\mathbb{C}^k} f^*(\omega)\wedge(\omega)^{k-1} = \frac{1}{d}\int_{\mathbb{C}^k} \omega \wedge ((f^{-1})^*\omega)^{k-1} \le \frac{{d_-}^{k-1}}{d}.$$

$\square$

Now suppose that f is regular. Let ℓ be an integer, $1 \le \ell \le k-1$, such that

$$\dim I_- \le \ell - 1 \quad \text{and} \quad \dim I_+ \le k - \ell - 1.$$

Since $I_+ \cap I_- = \emptyset$, we must have $\dim I_+ + \dim I_- \le k-2$.

PROPOSITION 2.3.2. *Let f be a regular automorphism of $\mathbb{C}^k$. There exists an ℓ such that*

$$d^\ell = (d_-)^{k-\ell}.$$

Moreover, $\dim I_+ + \dim I_- = k-2$ *and* $\dim I_+ = k-\ell-1$.

PROOF. The potential of $(\overline{f})^*\omega$ is locally bounded outside I_+, which has dimension less than or equal to $k-\ell-1$. It follows from the intersection theory of [**75, 33, 22**] (see Section A.6) that we can define the currents $((\overline{f})^*\omega)^j$ for every $j \le \ell+1$, and that they are closed and positive. In particular, the current $((\overline{f})^*\omega)^\ell$, of bidimension $(k-\ell, k-\ell)$, has no mass on I_+, which has dimension $\le k-\ell-1$. Since its potential is locally bounded in $\{t=0\}\setminus I_+$, it has no mass on $\{t=0\}$ (see Theorem A.6.2). We have the same situation for the current $((\overline{f^{-1}})^*\omega)^{k-\ell}$. Hence

$$d^\ell = \int_{\mathbb{P}^k} (\overline{f}^*\omega)^\ell \wedge \omega^{k-\ell} = \int_{\mathbb{C}^k} (f^*\omega)^\ell \wedge \omega^{k-\ell}$$

$$= \int_{\mathbb{C}^k} \omega^\ell \wedge ((f^{-1})^*\omega)^{k-\ell} = \int_{\mathbb{P}^k} \omega^\ell \wedge ((\overline{f^{-1}})^*\omega)^{k-\ell} = d_-^{k-\ell}.$$

Since the result holds for any ℓ satisfying the inequalities of the statement, ℓ is unique, and we must have

$$\dim I_- = \ell - 1 \quad \text{and} \quad \dim I_+ = k-\ell-1.$$

□

REMARK 2.3.3. The proposition above remains true if

$$\dim I_- \le \ell - 1 \quad \text{and} \quad \dim I_+ \le k-\ell-1,$$

even if these sets are not disjoint.

THEOREM 2.3.4. *Let f be a regular automorphism of $\mathbb{C}^k$. For every n, the set*

$$A_n := \{z \in \mathbb{C}^k : f^n(z) = z\}$$

is discrete. The automorphism f has infinitely many distinct periodic orbits. If $N(n)$ denotes the number of periodic points of order n in $\mathbb{C}^k$, counting multiplicity, then $N(kn) = d^{k\ell n}$, where $\dim I_- = \ell - 1$.

PROOF. In $\mathbb{C}^k$, the equation $f^{kn}(z) = z$ is equivalent to

$$f^{\ell n}(z) - f^{-(k-\ell)n}(z) = 0.$$

Passing to homogeneous coordinates, we obtain a system of k polynomial equations, with homogeneous polynomials that all have the same degree $d^{\ell n}$ since $d_-^{k-\ell} = d^\ell$. But I_+ and I_- are disjoint, so there is no solution in $\{t=0\}$. Since the set of solutions is discrete we can apply Bézout's theorem, and we find that $N(kn) = d^{k\ell n}$. It follows that A_n is discrete for all n.

The Shub-Sullivan theorem states that the multiplicity of a periodic point is bounded. Hence there are infinitely many distinct periodic orbits. □

REMARK 2.3.5. Favre ([**24**]) has given precise estimates of the number of periodic points for algebraically stable birational maps of $\mathbb{P}^2$.

REMARK 2.3.6. Let $\mathcal{A}_{k,d}$ be the set of polynomial automorphisms of $\mathbb{C}^k$ with algebraic degree less than or equal to d. More precisely, $f = (f_1, \dots, f_k) \in \mathcal{A}_{k,d}$ if f is an automorphism and $\max_{i\leq k} \deg f_i \leq d$.

$\mathcal{A}_{k,d}$ is clearly an analytic subset of the space of polynomial maps of degree less than or equal to d. This is true because (as we have seen) the degree of the inverse is bounded above by d^{k-1}, so the set $\mathcal{A}_{k,d}$ is the projection, on the space of f's, of the analytic subset defined by

$$(*) \qquad g \circ f = f \circ g = \mathrm{Id},$$

where g is a polynomial map of degree $\leq d^{k-1}$. Since we can think of the maps $(*)$ as equations in $\mathbb{P}^N \times \mathbb{P}^M$ for suitable N and M, we see that $\mathcal{A}_{k,d}$ is an analytic set. One can check that $\mathcal{A}_{k,d}$ is connected. The description of the irreducible components of $\mathcal{A}_{k,d}$ is a delicate problem. For the case where $k = 2$, the question was studied by Friedland-Milnor [**40**].

Let $\widetilde{\mathcal{A}}_{k,d}$ denote the set of automorphisms of $\mathbb{C}^k$ of degree exactly d. It is a Zariski open subset of $\mathcal{A}_{k,d}$. The regular automorphisms of $\widetilde{\mathcal{A}}_{k,d}$ form a Zariski open subset of $\widetilde{\mathcal{A}}_{k,d}$: the condition $I_+ \cap I_- \neq \emptyset$ describes an analytic set that, by the examples of Section 2.2, is not equal to $\widetilde{\mathcal{A}}_{k,d}$.

There are, however, irreducible components that do not contain any regular automorphisms. An example in $\widetilde{\mathcal{A}}_{2,d}$ is the component containing the elementary automorphisms of degree d,

$$e(z, w) = (P(w) + az,\ bw + d),$$

where $\deg P = d$ and $a, b \neq 0$.

The work of Yomdin [**87**] (see also Gromov [**43**]) has shown connections between the entropy of a C^∞ map g from a compact manifold M to itself and the growth of the volume of the image under g of manifolds of dimension s.

In the setting of rational maps of $\mathbb{P}^k$ to $\mathbb{P}^k$, Friedland ([**38**]) introduced the following entropy: let $\rho_j(f)$ be the degree of $f^*[L]$, where L is a generic subspace of dimension $k - j$. Set

$$H(f) = \liminf_{m\to\infty} \max_{1\leq j\leq k} \frac{\log \rho_j(f^m)}{m}.$$

PROPOSITION 2.3.7. *Let f be a regular automorphism of $\mathbb{C}^k$. If $\dim I_- = \ell - 1$, then $H(f) = \log d^\ell$.*

PROOF. Let L be a generic subspace of dimension $k - j$.

- If $j \leq \ell$, then
$$\int_{\mathbb{C}^k} f^*[L] \wedge \omega^{k-j} = d^j$$
because $f^*[L]$ has no mass on I_+, hence none on $\{t = 0\}$.
- If $j > \ell$ and L is not contained in $\{t = 0\}$, then
$$\int_{\mathbb{C}^k} f^*[L] \wedge \omega^{k-j} = \int_{\mathbb{C}^k} [L] \wedge (f^{-1*}\omega)^{k-j} = d_-^{k-j}.$$

Thus

$$\max_{1\leq j\leq k} \log \rho_j(f^m) = d^{\ell m} \quad \text{and} \quad H(f) = \log d^\ell.$$

□

There is also a well-known notion of metric entropy, which quantifies the complexity of the orbits (see [**86**]). We recall the definition. Let (X, d) be a metric space, and let g be a continuous map from X to itself. Set

$$d_n(x, y) = \max_{0 \leq i \leq n-1} d(g^i(x), g^i(y)).$$

Then $d_n(x, y) \geq \varepsilon$ if, before time n, the distance between some pair of points $g^i(x)$ and $g^i(y)$ in the orbits is at least ε. Let $M(n, \varepsilon, X)$ be the minimum number of balls of radius ε needed to cover X when it is equipped with the distance d_n. We define the entropy $h_d(g)$ by the formula

$$h_d(g) = \lim_{\varepsilon \to 0} \limsup_{n \to \infty} \frac{1}{n} \log M(n, \varepsilon, X).$$

When $f : \mathbb{P}^k \to \mathbb{P}^k$ is a meromorphic map , we consider its graph $\Gamma \subset \mathbb{P}^k \times \mathbb{P}^k$. The set of orbits is

$$\Gamma_\infty = \{(x_i)_1^\infty : (x_i, x_{i+1}) \in \Gamma\},$$

and we equip Γ_∞ with the metric

$$\delta((x_i), (y_i)) = \max_{i \geq 1} \frac{d(x_i, y_i)}{2^{i-1}}.$$

We introduce the shift operator σ, where $\sigma(x_i)_1^\infty = (x_i)_2^\infty$, and set

$$h(f) = h_\delta(\sigma).$$

Friedland has shown ([**38, 39**]) that $h(f) \leq H(f)$ always holds. Thus $h(f) \leq \log d^\ell$ in the case we are considering. If we consider the restriction of f to K, then we also have

$$h(f_{|K}) \leq \log d^\ell.$$

In the setting of $\mathcal{C}^\infty$ maps $g : M \to M$ of compact manifolds, Yomdin has shown that the metric entropy $h(g)$ satisfies

$$h(g) \geq \limsup_{m \to \infty} \frac{\log \operatorname{vol}(g^m(V_s))}{m},$$

where V_s denotes a manifold of dimension s.

Smillie [**78**] has proved that Yomdin's argument can be adapted to Hénon maps on $\mathbb{C}^2$. The essential reason is that K is compact. Since $K = K^+ \cap K^-$ is compact in the present case as well, Smillie's arguments can be applied in this setting, and we obtain:

THEOREM 2.3.8. *Let f be a regular automorphism of $\mathbb{C}^k$. Suppose that I_- has dimension $\ell - 1$. Then $h(f_{|K}) = \log d^\ell$.*

2.4. Basins of Attraction. Stable Manifolds

The attracting case. We begin by recalling a result of Sternberg [**79**] on the normal form of a holomorphic map in a neighborhood of an attracting fixed point. ([**79**] treats the more general case of $\mathcal{C}^\infty$ maps.)

Let $f : U \to \mathbb{C}^k$ be a holomorphic map defined in a neighborhood U of 0 and such that $f(0) = 0$. Let $\lambda_1, \ldots, \lambda_k$ be the eigenvalues of $f'(0)$, where we assume

$$0 < |\lambda_k| \leq \cdots \leq |\lambda_1| < 1.$$

We say that 0 is an attracting fixed point. Let

$$R_j = \{I = (i_1, \ldots, i_j) \in \mathbb{N}^j : \lambda_{j+1} = \lambda_1^{i_1} \ldots \lambda_j^{i_j}\},$$

and let $\mathcal{P}_j$ denote the vector space of polynomials in j variables generated by the monomials $z_1^{i_1} \ldots z_j^{i_j}$, where $I \in R_j$.

We will call a map g *triangular*, associated with $\lambda_1, \ldots, \lambda_k$, if it is of the form

$$g = (g_1, \ldots, g_k), \quad \text{where} \quad g_j = \lambda_j z_j + p_j(z_1, \ldots, z_{j-1}), \ p_j \in \mathcal{P}_{j-1}.$$

THEOREM 2.4.1 ([**79**]). *Let $f : U \to \mathbb{C}^k$ be a holomorphic map, $f(0) = 0$. Suppose that the eigenvalues of $f'(0)$ are as above. Then there exist a germ of a biholomorphism h and a triangular map g satisfying*

$$f \circ h = h \circ g, \quad h(0) = 0, \quad h'(0) = \mathrm{Id}.$$

In particular, f is linearizable in a neighborhood of 0 if $R_j = \emptyset$ for every j.

COROLLARY 2.4.2. *Let f be an injective holomorphic map from $\mathbb{C}^k$ to $\mathbb{C}^k$. If 0 is an attracting fixed point, then the basin of attraction of the origin*

$$\Omega = \{z : \lim_n f^n(z) = 0\}$$

is biholomorphic to $\mathbb{C}^k$.

PROOF. Let h and g be the maps whose existence is guaranteed by the theorem. Observe that g is a polynomial automorphism of $\mathbb{C}^k$ and that the basin of attraction of 0 for g is $\mathbb{C}^k$. The map h, which is defined in a ball $B(0, r)$, must be extended to a map defined on $\mathbb{C}^k$. To do this, we define the extension $\widetilde{h}$ by the relation

$$\widetilde{h}(z) = f^{-n} \circ h \circ g^n(z).$$

It is easy to check that for a given z, $\widetilde{h}(z)$ is well defined for sufficiently large n, and that the definition is independent of n. □

We say that $\Omega \subset \mathbb{C}^k$ is a *Fatou-Bieberbach domain* if Ω is biholomorphic to $\mathbb{C}^k$ and $\overline{\Omega} \neq \mathbb{C}^k$.

COROLLARY 2.4.3. *Let f be a regular polynomial automorphism of $\mathbb{C}^k$. The basin of attraction of an attracting fixed point of f is a Fatou-Bieberbach domain contained in the interior of K^+.*

PROOF. We have seen that the set X_+ is nonempty and attracting for f. If p is an attracting fixed point, then its basin of attraction Ω_p is necessarily disjoint from U^+. Hence it is a Fatou-Bieberbach domain. □

THEOREM 2.4.4. *Let f be a regular polynomial automorphism of $\mathbb{C}^k$. Let Ω be a Fatou component associated with a periodic point p of f that has one eigenvalue with modulus less than 1 (all the other eigenvalues have modulus less than or equal to 1). Let S be a closed positive current on $\mathbb{P}^k$, with bidegree $(1,1)$, whose support Σ is disjoint from I_+.*

Then $\Sigma \cap \Omega$ is nonempty, and $\Sigma \cap K^+$ is compact.

PROOF. By Theorem 2.2.12, the hypothesis on the support Σ implies that

$$\frac{1}{d_-^n}(f^{-n})^* S \to T_-.$$

But the support of T_- intersects Ω because the family $\{f^{-n}\}$ cannot be equicontinuous at p. The support of $(f^{-n})^*S$ is contained in $f^n(\Sigma)$, whose intersection with Ω would be empty. Hence $\Sigma \cap \Omega$ is nonempty.

We have seen that $\overline{K}^+ \cap \{t = 0\} \subset I_+$. Hence the compact set $\overline{K}^+ \cap \Sigma$ is contained in $\mathbb{C}^k$. □

REMARKS 2.4.5. (1) If Ω is a Fatou-Bieberbach domain associated with an attracting fixed point of a regular automorphism, it follows that Ω contains no algebraic variety but intersects them all in a bounded set; this was proved for Hénon maps in [**6**].

(2) Some condition on the support of S is clearly necessary; for instance, the support of T_+ does not intersect any Fatou component.

The case of a saddle point. Let $f : U \to \mathbb{C}^k$ be a holomorphic map, $f(0) = 0$. We now consider the case where the eigenvalues of $f'(0)$ satisfy

$$0 < |\lambda_k| \leq \cdots \leq |\lambda_{j+1}| < 1 < |\lambda_j| \leq \cdots \leq |\lambda_1|, \ j < k.$$

In this case, we say that 0 is a *saddle point.*

We may assume that the space $\mathbb{C}^k$ can be split into two complementary subspaces, $\mathbb{C}^k = E_1 \oplus E_2$, where E_2 is the eigenspace associated with $\{\lambda_{j+1}, \ldots, \lambda_k\}$, E_1 is the eigenspace associated with $\{\lambda_1, \ldots, \lambda_j\}$, and the coordinates have been chosen so that

$$E_1 = \{z \in \mathbb{C}^k : z_{j+1} = \cdots = z_k = 0\}, \quad E_2 = \{z \in \mathbb{C}^k : z_1 = \cdots = z_j = 0\}.$$

One shows ([**53**] or [**72**]) that there exists a stable holomorphic manifold

$$W^s(0, r) = \{z \in B(0, r) : \lim_{n \to +\infty} f^n(z) = 0\}.$$

$W^s(0, r)$ is the graph of a holomorphic map φ defined in a neighborhood U_2 of 0 in E_2 by

$$\varphi : U_2 \longrightarrow U, \quad \varphi(z') = (\varphi_1(z'), \ldots, \varphi_j(z'), z'), \quad z' = (z_{j+1}, \ldots, z_k),$$

where $\varphi_1'(0) = \cdots = \varphi_j'(0) = 0$.

When f is defined on $\mathbb{C}^k$, we set

$$W^s(0) = \bigcup_n f^{-n}(W^s(0, r)).$$

This is an immersed manifold.

PROPOSITION 2.4.6. *Let f be an injective holomorphic map from $\mathbb{C}^k$ to $\mathbb{C}^k$. If 0 is a saddle fixed point, with $k - j$ eigenvalues of modulus less than 1, then the manifold $W^s(0)$ is biholomorphic to $\mathbb{C}^{k-j}$.*

PROOF. Consider the map $\varphi^{-1} \circ f \circ \varphi$ in a neighborhood of 0 in $\mathbb{C}^{k-j}$. Sternberg's theorem states that there exist a polynomial automorphism g of $\mathbb{C}^{k-j}$ and a biholomorphism h such that

$$\varphi^{-1} \circ f \circ \varphi = h \circ g \circ h^{-1}.$$

Set $\psi = \varphi \circ h$. Then $f \circ \psi = \psi \circ g$. We are looking for an extension $\widetilde{\psi}$ of ψ to $\mathbb{C}^{k-j}$. It suffices to set

$$\widetilde{\psi}(z) = f^{-n} \circ \psi \circ g^n(z)$$

in $B(0, R)$, for sufficiently large n. □

PROPOSITION 2.4.7. *Let f be a regular automorphism of $\mathbb{C}^k$. Let p be a saddle fixed point of f. Then the stable manifold $W^s(p)$ is contained in J^+.*

PROOF. We have already seen that $W^s(p)$ is biholomorphic to $\mathbb{C}^{k-j}$. Since $W^s(p) = \{z : \lim_{n\to+\infty} f^n(z) = p\}$, we have $W^s(p) \subset K^+$. The family f^n is equicontinuous in the interior of K^+. Let v be a vector such that $f'(p)v = \lambda_1 v$, $|\lambda_1| > 1$. If there were some point q in the intersection of $W^s(p)$ and the interior of K^+, then $(f^n)'(q)v$ would have to be bounded. But this is impossible because f is expanding on $W^s(p)$ in the direction v. □

The case of transformations tangent to the identity. This work was done by Hakim ([**47, 46**]). We summarize the main results here.

Let $f(Z) = Z + P_2(Z) + P_3(Z) + \cdots$ be a holomorphic transformation tangent to the identity, defined in a neighborhood U of 0 in $\mathbb{C}^k$, where, for $h \in \mathbb{N}$, P_h denotes the homogeneous polynomial part of degree h of the power series expansion of f. We show that there exist pieces of invariant manifolds consisting of points of U that are attracted by the origin in a given direction. For $V \in \mathbb{C}^k \setminus \{0\}$, we denote its image in the projective space $\mathbb{P}^{k-1}$ by $[V] = \pi(V)$.

DEFINITION 2.4.8. Let $Z \in U$, and let $V \in \mathbb{C}^k \setminus \{0\}$. We say that Z is attracted by the origin in the direction V if $f^n(Z) \to 0$ (in $\mathbb{C}^k$) and $[f^n(Z)] \to [V]$ (in $\mathbb{P}^{k-1}$).

For simplicity, we consider the case where $P_2(Z)$ is not identically zero. One then proves the following elementary lemma ([**47**]).

LEMMA 2.4.9. *Let $Z \in U$ and $V \in \mathbb{C}^k \setminus \{0\}$. If $f^n(Z) \to 0$ in the direction V, then either $P_2(V) = 0$ or $[P_2(V)] = [V]$.*

DEFINITION 2.4.10. Any direction $V \in \mathbb{C}^k \setminus \{0\}$ such that $P_2(V) \neq 0$ and $[P_2(V)] = [V]$ is called a *nondegenerate characteristic direction.*

Analysis in a neighborhood of a nondegenerate characteristic direction. The characteristic directions are the solutions of a system of $k-1$ homogeneous polynomial equations of degree 3. Thus there are 3^{k-1} such directions in general. Let V be a nondegenerate characteristic direction. We can choose coordinates such that $V = (1, 0, \ldots, 0)$ and $P_2(V) = (-1, 0, \ldots, 0)$. Then, letting (z, w) be the coordinates with $z \in \mathbb{C}$ and $w \in \mathbb{C}^{k-1}$ and taking the blow-up $w = uz$ in a neighborhood of $u = 0$, we can write the transformation f in the coordinates (z, u) as follows:

$$\begin{aligned} z_1 &= z - z^2 + O(\|u\|z^2, z^3), \\ u_1 &= (I - zA)u + O(\|u\|^2 z, z^2), \end{aligned}$$

where I is the identity $(k-1) \times (k-1)$ matrix and A is a $(k-1) \times (k-1)$ matrix associated with V. It is easy to check that the equivalence class of A under the relation of linear change of coordinates is an invariant, and we can choose coordinates so that A is in Jordan canonical form.

The main results are the following ([**46**]):

THEOREM 2.4.11. *Let V be a nondegenerate characteristic direction, and let A be the matrix associated with V. Let d be the sum of the multiplicities of those eigenvalues λ_j of A for which $\Re e\, \lambda_j > 0$ ($d = 0$ if the real parts of all the eigenvalues of A are negative or zero).*

- *Then there exists a stable piece of manifold* W^s *of dimension* $d+1$ *consisting of points attracted by the origin in the direction* V, *such that* 0 *is in* ∂W^s.
- *The piece of manifold* W^s *is the image under an injective holomorphic map (of class* C^1 *up to the boundary) of a sector*

$$S_{\gamma,s,\rho} = \{(z,v) \in \mathbb{C} \times \mathbb{C}^d : |\Im m\, z| \le \gamma \Re e\, z,\ |z| \le s,\ \|v\| < \rho\}.$$

- *The restriction of* f *to* W^s *is conjugate to a transformation on* $S_{\gamma,s,\rho}$:

$$\left\{\frac{1}{z_1} = \frac{1}{z} + 1,\ v_1 = v\right\}.$$

THEOREM 2.4.12. *Let* f *be a global isomorphism of* $(\mathbb{C}^k, 0)$, *let* 0 *be a fixed point of* f *where the transformation is tangent to the identity, and let* V *be a nondegenerate characteristic direction at* 0. *Suppose that the matrix* A *associated with* V *has* d *eigenvalues, counting multiplicity, with positive real part.*

- *If* A *has no eigenvalue with zero real part, let* $Z_n = f^n(z)$ *and*

$$\Omega_{(0,v)} = \{Z \in \mathbb{C}^p : \lim_{n\to\infty} Z_n = 0 \text{ and } \lim_{n\to\infty} [Z_n] = [V]\}.$$

Then $\Omega_{(0,V)}$ *is isomorphic to* $\mathbb{C}^{d+1}$.
- *If there exist eigenvalues with zero real part, then* $\Omega_{(0,V)}$ *is still isomorphic to* $\mathbb{C}^{d+1}$ *if we modify the definition of* $\Omega_{(0,V)}$ *as follows (using the coordinates* (z,u) *introduced above, and assuming* $V = (0,1)$*; the definition is independent of* α*). Let* $\alpha > 0$ *such that the eigenvalues of the positive real part satisfy* $\Re e\, \lambda_j > \alpha$, *and let*

$$\Omega_{(0,V)} = \{Z \in \mathbb{C}^p : \lim_{n\to\infty} Z_n = 0 \quad \text{and} \quad \lim_{n\to\infty} z_n^{-\alpha} u_n = 0\}.$$

In the transformation f considered above, if the quadratic part $P_2(Z)$ is identically 0, let $P_h(Z)$ be the polynomial part of lowest degree that is not identically zero. It is then necessary to introduce corresponding nondegenerate characteristic directions, that is, the directions $V \in \mathbb{C}^k \setminus \{0\}$ that are solutions to $P_h(V) \neq 0$ and $[P_h(V)] = [V]$. The results above generalize easily; the only difference is that for each direction there are h stable pieces of manifolds, analogous to the h petals of the Fatou domains, for the transformations of $(\mathbb{C}, 0)$ tangent to the identity.

REMARK 2.4.13. Let f be a polynomial automorphism such that $f(0) = 0$ and $f'(0) = I$. Then the Jacobian is 1 everywhere. The map is volume-preserving, and the condition of Theorem 2.4.12 cannot be realized for $d = k-1$ because no domain can be attracted by 0. On the other hand, there are examples of stable pieces of manifolds of dimension $d+1$ with $0 \le d \le k-2$.

2.5. The Invariant Currents $\overline{T}^j_+$ and $\overline{T}^{k-j}_-$

DEFINITION 2.5.1. Let f be a regular automorphism of $\mathbb{C}^k$, with $\dim I_- = \ell - 1$. For $j \le \ell + 1$, we write $\overline{J^+_j}$ for the support in $\mathbb{P}^k$ of the current $\overline{T}^j_+$, and call it the Julia set of order j. Its intersection with $\mathbb{C}^k$ is denoted by J^+_j. There are analogous sets for f^{-1}, denoted by $\overline{J^-_j}$ and J^-_j.

We also define $J := J^+_\ell \cap J^-_{k-\ell}$. In particular, J^+_1 is the support of T_+. It may be equal to $\mathbb{C}^k$ and not equal to $J^+ := \partial K^+$.

Since the potential of the current $\overline{T}_+$ is locally bounded outside I_+, which has dimension $k-\ell-1$, we can define the currents $\overline{T}_+^j$ for $j \le \ell+1$ and $\overline{T}_-^j$ for $j \le k-\ell+1$ (see Section A.6).

THEOREM 2.5.2. *Let f be a regular automorphism of $\mathbb{C}^k$. Then it has the following properties.*

(1) *$(dd^cG^+)^\ell = 0$ in U^+. The current $\overline{T}_+^{\ell+1}$ is a cycle supported on I_+. In particular, $(dd^cG^+)^{\ell+1} = 0$ in $\mathbb{C}^k$ and $J_\ell^+ \subset J^+(=\partial K^+)$.*

(2) *$(dd^cG^-)^{k-\ell} = 0$ in U^-. The current $\overline{T}_-^{k-\ell+1}$ is a cycle supported on I_-. In particular, $(dd^cG^-)^{k-\ell+1} = 0$ in $\mathbb{C}^k$ and $J_{k-\ell}^- \subset J^-(=\partial K^-)$.*

(3) *Let $\mu := T_+^\ell \wedge T_-^{k-\ell}$. Then μ is a probability measure on $J = J_\ell^+ \cap J_{k-\ell}^-$. It is invariant under f; that is, $f^*\mu = \mu$.*

PROOF. We first show that $T_+^{\ell+1}$ is zero in $\mathbb{C}^k$.

$$\int_{\mathbb{C}^k} \left(\frac{f^{n*}\omega}{d^n}\right)^{\ell+1} \wedge \omega^{k-\ell-1} = \frac{1}{d^{(\ell+1)n}} \int_{\mathbb{C}^k} \omega^{\ell+1} \wedge (f^{-n*}\omega)^{k-\ell-1}$$
$$\le \left(\frac{(d_-)^{k-\ell-1}}{d^{\ell+1}}\right)^n = \frac{1}{(dd_-)^n} \to 0.$$

Since $T_+^{\ell+1}$ has no mass on $\{t=0\} \setminus I_+$, it is supported on I_+. But it is a closed positive current of bidimension $(k-\ell-1, k-\ell-1)$ that is supported on the analytic set I_+, which has dimension $k-\ell-1$. Hence (A.4, property h) the support of $T_+^{\ell+1}$ is contained in I_+. We will show that its support equals I_+. We know that

$$G \le \frac{1}{d} \log |F|.$$

For $p \in I_+$, let Δ_p be a polydisk with center p and dimension $\ell+1$, transverse to I_+. We have $G_{|\Delta_p} \le c \log |z-p|$, $z \in \Delta_p$, $c > 0$. It follows that $(dd^cG_{|\Delta_p})^{\ell+1} \ge c\varepsilon_p$, where ε_p denotes the Dirac measure at p (see [**22**]). Slicing theory applied to the current $T_+^{\ell+1}$ allows us to conclude that p is in the support of $T_+^{\ell+1}$. The argument for $T_-^{\ell+1}$ is identical.

First, suppose I_- has dimension 0. We may assume that $I_- = [1:0:\cdots:0]$. Let $f_1^n, \ldots, f_k^n$ be the components of f^n. For $z \in U^+$ we have $|f_1^n(z)| \gg |f_j^n(z)|$, $j \ge 2$. It follows that in U^+,

$$G^+(z) = \lim \frac{1}{d^n} \log^+ |f^n(z)| = \lim \frac{1}{d^n} \log^+ |f_1^n(z)|.$$

Hence G^+ is pluriharmonic in U^+ and $dd^cG^+ = 0$ in U^+.

Next, observe that if $g_1, \ldots, g_\ell$ are ℓ holomorphic functions without common zeros in an open set W, then $(dd^c \log(|g_1|^2 + \cdots + |g_\ell|^2))^\ell = 0$. Indeed, if $g_1 \ne 0$ in a neighborhood of a point, we can write

$$\log(|g_1|^2 + \cdots + |g_\ell|^2) = \log(|g_1|^2) + \log\left(1 + \left|\frac{g_2}{g_1}\right|^2 + \cdots + \left|\frac{g_\ell}{g_1}\right|^2\right).$$

But the restriction of the function to the variety of codimension $\ell-1$ defined by $g_2 = a_2 \cdot g_1, \ldots, g_\ell = a_\ell \cdot g_1$ is pluriharmonic. Hence, on any ℓ-dimensional subspace, $dd^c \log(|g|^2)$ has zero as an eigenvalue.

We now consider the general case, where I_- has dimension $\ell - 1$. We can find coordinates in $\{t = 0\}$, identified with $\mathbb{P}^{k-1}$, such that

$$|z_{\ell+1}| + \cdots + |z_k| \leq C(|z_1| + \cdots + |z_\ell|)$$

on I_-. It suffices that the subspace $z_1 = \cdots = z_\ell = 0$ not intersect I_-; this is always possible for dimensional reasons. Since f^n converges to I_- in U^+, in this case we have

$$G^+(z) = \lim \frac{1}{2d^n} \log(|f_1|^2 + \cdots + |f_\ell|^2),$$

where the convergence is locally uniform. The same argument as above shows that $(dd^c G^+)^\ell = 0$ in U^+. Hence $G^+(dd^c G^+)^\ell = 0$ in $\mathbb{C}^k$, so $(dd^c G^+)^{\ell+1} = 0$ in $\mathbb{C}^k$.

We know that $\mu := \overline{T}_+^\ell \wedge \overline{T}_-^{k-\ell}$ is a probability measure. We have seen that $\overline{T}_+^\ell$ is supported on $J^+ = \partial \overline{K^+}$ and that $\overline{T}_-^{k-\ell}$ is supported on $J^- = \partial \overline{K^-}$. Hence the measure μ is supported on $J = J^+ \cap J^-$.

Now we prove that μ is invariant under f:

$$f^*\mu = f^*T_+^\ell \wedge f^*T_-^{k-\ell} = d^\ell T_+^\ell \wedge \frac{1}{(d_-)^{k-\ell}} T_-^{k-\ell} = \mu.$$

□

The following result was obtained with V. Guedj.

PROPOSITION 2.5.3. *Let f be a regular automorphism of $\mathbb{C}^k$. Then we have $f(\{t = 0\} \setminus I_+) = I_-$, so $X_+ = I_-$ and $X_- = I_+$.*

PROOF. We will show that $X_- = I_+$. We have seen (Proposition 2.2.2) that $X_- \subset I_+$. Let $p \in I_+$. We know that if $\mathcal{B}_p^+ \setminus I_-$ is nonempty, then $f^{-1}(\mathcal{B}_p^+ \setminus I_-) = p$ (Proposition 1.2.1). Thus it suffices to verify that $\mathcal{B}_p^+ \not\subset I_-$. If $\mathcal{B}_p^+ \subset I_-$, then $f(B(p, \varepsilon) \setminus I_+)$ would be contained in a neighborhood V of I_- for sufficiently small ε. We may assume that $\overline{T}_+^\ell = 0$ in V. The relation $f^*(\overline{T}_+) = d\overline{T}_+$ implies that $\overline{T}_+^\ell = 0$ in $B(p, \varepsilon) \setminus I_+$. One checks this first in $B(p, \varepsilon) \setminus \{t = 0\}$, then uses the fact that the potential is locally bounded outside I_+.

Since the current $\overline{T}_+^\ell$ has bidimension $(k - \ell, k - \ell)$, it has no mass on I_+, which has dimension $k - \ell - 1$ (Appendix A.4, Property (h)). Hence $\overline{T}_+^\ell = 0$ in $B(p, \varepsilon)$. But the support of $\overline{T}_+^{\ell+1}$ contains p, by Theorem 2.5.2. This is the desired contradiction. It follows that $f^{-1}(\{t = 0\} \setminus I_-) = I_+$. Applying the same result to f^{-2}, we find that $f^{-1}(I_+) = I_+$. Hence $X_- = I_+$. □

PROPOSITION 2.5.4. *Let f be a regular automorphism of $\mathbb{C}^k$. Suppose that $\dim I_- = \ell - 1$. Then there exists a surjective holomorphic map from $\mathbb{P}^{\ell-1}$ onto $I_- = X_+$. In particular, if I_- is smooth, then I_- is isomorphic to $\mathbb{P}^{\ell-1}$ and the restriction f_0 of f to I_- is conjugate to an endomorphism of degree d of $\mathbb{P}^{\ell-1}$.*

PROOF. We have seen that $f(\{t = 0\} \setminus I_+) = I_- = X_+$ (Proposition 2.5.3). The rank of the restriction f_0 of f to $\{t = 0\}$ is therefore $\ell - 1$. Since the set I_+ has dimension $k - \ell - 1$, we can find a subspace $\mathbb{P}^{\ell-1}$ contained in $\{t = 0\}$, disjoint from I_+, and on which the rank is $\ell - 1$. It follows that $f_0(\mathbb{P}^{\ell-1}) = X_+$.

A theorem of Lazarsfeld [**58**] states that the existence of a surjective holomorphic map from $\mathbb{P}^{\ell-1}$ onto a smooth projective variety X of the same dimension implies that X is isomorphic to $\mathbb{P}^{\ell-1}$.

Let G_0 be the p.s.h. function in $\mathbb{C}^{k+1}$ defined by

$$G_0(z,t) = \lim_{n\to\infty} \frac{1}{d^n} \log |F^n(z,0)|.$$

We can associate with G_0 a closed positive current T_0 of bidegree $(1,1)$ on $\mathbb{P}^k$. The potential of T_0 is continuous in a neighborhood of $X_+ = I_-$. If T_1 denotes the restriction of T_0 to the variety I_-, then

$$f_0^* T_1 = dT_1$$

and

$$f_0^*(T_1^{\ell-1}) = d^{\ell-1} T_1^{\ell-1}.$$

Hence f_0 is an endomorphism of topological degree $d^{\ell-1}$. □

REMARK 2.5.5. The results of Chapter 3 on the dynamics of endomorphisms can be applied to the endomorphism f_0. For example, the repelling periodic points of f_0 have stable manifolds of dimension $k-\ell+1$ in U^+. The structure of the currents T_+^j, $j \leq \ell - 1$, is tied to that of the currents T_1^j associated with the endomorphism f_0. It is clear that studying the automorphisms of $\mathbb{C}^k$ leads to considering the endomorphisms of $\mathbb{P}^{\ell-1}$. We will not pursue this connection here.

COROLLARY 2.5.6. *If f is a regular automorphism of $\mathbb{C}^k$, then $\overline{K}^+ = K^+ \cup I_+$.*

PROOF. We have seen (Proposition 2.2.6) that $\overline{K}^+ \subset K^+ \cup I_+$. It suffices to show that the points of I_+ are accumulation points of K^+. But I_+ is contained in the support of $\overline{T}_+^\ell$ if $\dim I_- = \ell - 1$; indeed, we have seen that the support of $\overline{T}_+^{\ell+1}$ is I_+. Moreover, by Theorem 2.5.2, $\overline{T}_+^\ell$ has no mass in $U^+ \cup (\{t=0\} \setminus I_+)$, the basin of attraction of X_+. Hence the set ∂K^+ has all the mass of $\overline{T}_+^\ell$, and it follows that I_+ is contained in the closure of ∂K^+. □

The results in Sections 2.6 and 2.7 are due to Bedford and Smillie ([**7, 8, 9**]) in the setting of Hénon maps on $\mathbb{C}^2$.

2.6. Convergence to $\overline{T}_+$ and Density of Stable Manifolds

PROPOSITION 2.6.1. *Let R be a closed positive current of bidegree (j,j) in a ball B. Let $\psi \geq 0$, $\psi \in \mathcal{C}_0^\infty(B)$. Let f be a regular automorphism of $\mathbb{C}^k$. Set*

$$R_n = \frac{1}{(d_-)^{(k-j)n}} (f^n)^*(\psi R).$$

Every current S that is a limit point of the sequence $\{R_n\}$ is positive and closed, and has mass

$$c = \int \psi R \wedge (dd^c G^-)^{k-j}.$$

Moreover, the masses of ∂R_n and $i\partial\overline{\partial} R_n$ converge to 0.

PROOF. We compute the mass $\|R_n\|$ of R_n and find its limit. We have

$$\langle R_n, \omega^{k-j} \rangle = \left\langle \psi R, \frac{1}{(d_-)^{(k-j)n}} (f^{-n})^* \omega^{k-j} \right\rangle.$$

But $\dfrac{1}{(d_-)^{(k-j)n}}(f^{-n})^*\omega^{k-j} = (dd^cG_n^-)^{k-j}$. The potentials G_n^- are decreasing and have limit G^-, which is continuous. Hence, by Theorem A.6.2,

$$R \wedge (dd^cG_n^-)^{k-j} \longrightarrow R \wedge (dd^cG^-)^{k-j}.$$

Let θ be a test $(0,1)$-form. Set $\phi = \theta \wedge \omega^{k-j-1}$. Let $S(\psi)$ be the support of ψ. Setting $c_n = (d_-)^{(k-j)n}$ and using the Cauchy-Schwarz inequality (Section A.4, Property (h)), we have

$$\begin{aligned}
\left|\int \partial R_n \wedge \phi\right| &= \frac{1}{c_n}\left|\int \partial\psi \wedge R \wedge (f^{-n})^*\theta \wedge (f^{-n})^*\omega^{k-j-1}\right| \\
&\leq \frac{1}{c_n}\left|\int R \wedge \partial\psi \wedge \overline{\partial}\psi \wedge (f^{-n})^*\omega^{k-j-1}\right|^{1/2} \\
&\quad \cdot \left|\int_{S(\psi)} R \wedge (f^{-n})^*\theta \wedge \overline{(f^{-n})^*\theta} \wedge (f^{-n})^*\omega^{k-j-1}\right|^{1/2} \\
&\leq \frac{1}{(d_-)^{n/2}}\left|\int R \wedge \partial\psi \wedge \overline{\partial}\psi \wedge \frac{(f^{-n})^*\omega^{k-j-1}}{(d_-)^{(k-j-1)n}}\right|^{1/2} \\
&\quad \cdot \left|\int \frac{(f^n)^*R}{(d_-)^{(k-j)n}} \wedge \theta \wedge \overline{\theta} \wedge \omega^{k-j-1}\right|^{1/2}.
\end{aligned}$$

Since both integrals are bounded, the right-hand side approaches zero. Hence $\|\partial R_n\|$ approaches zero.

Let α be a $\mathcal{C}^\infty$ function bounded by 1. If $-C\omega \leq i\partial\overline{\partial}\psi \leq C\omega$, then

$$\begin{aligned}
&\left|\int (f^n)^* \frac{R \wedge i\partial\overline{\partial}\psi}{(d_-)^{(k-j)n}} \wedge \omega^{k-j-1}\alpha\right| \\
&\qquad = \frac{1}{(d_-)^n}\left|\int R \wedge i\partial\overline{\partial}\psi \wedge (f^{-n})\frac{\omega^{k-j-1}}{(d_-)^{(k-j-1)n}}\alpha(f^{-n})\right| \\
&\qquad \leq \frac{C}{(d_-)^n}\int R \wedge \omega \wedge (dd^cG_n^-)^{k-j-1}.
\end{aligned}$$

Hence the mass of $i\partial\overline{\partial}R_n$ approaches 0. □

REMARK. If $j < \ell$, then of course $R_n \to 0$ because $(dd^cG^-)^{k-j-1} = 0$.

THEOREM 2.6.2. *Let f be a regular automorphism of $\mathbb{C}^k$ such that* $\dim I_- = 0$ *(that is, $\ell = 1$). Let p be a hyperbolic fixed point of saddle type for f, where we assume that*

$$|\lambda_1| > 1 > |\lambda_2| \geq \cdots \geq |\lambda_k|.$$

If $p \in J_{k-1}^-$, then $W^s(p)$ is dense in $J^+ = \partial K^+$.

PROOF. Let D be a polydisk of dimension $k-1$ passing through p, transverse to the unstable direction, and such that

$$\int_D (dd^cG^-)^{k-1} = c_1 \neq 0.$$

Such polydisks exist. Since we have assumed that $p \in J_{k-1}^- = \operatorname{supp}(dd^cG^-)^{k-1}$, it suffices to apply slicing theory (**[25]**) to the closed positive (hence, in Federer's terminology, locally flat) current T_-^{k-1}.

Let $\psi \geq 0$ be a test function equal to 1 in a neighborhood of p and 0 in a neighborhood of ∂D. The sequence $\frac{1}{d^n}(f^n)^*(\psi[D])$ converges to cT_+, where

$$c = \int_D \psi(dd^c G^-)^{k-1}.$$

Indeed, Proposition 2.6.1 ensures that every limit point of the sequence is a closed positive current of mass c. It is necessarily supported on $\overline{K^+}$. Hence, by Corollary 2.2.12, it is proportional to T_+. Moreover, since p is hyperbolic, $(f^{-n})(D) \cap B(p,r)$ converges to $W_r^s(p)$. It follows that the current T_+ is supported on $\overline{\bigcup f^{-n}(W_r^s(p))}$, which is thus equal to the support of T_+. □

REMARK 2.6.3. In $\mathbb{C}^2$, the hypothesis $p \in J_{k-1}^-$ is always satisfied ([**8**]). For $dd^c G^-_{|W_r^s(p)} = 0$ would imply that the function G^- was harmonic on $W_r^s(p)$. But G^- is nonnegative, and zero at p. By the minimum principle for harmonic functions, G^- would be zero on W^s. But this is impossible because W^s is an image of $\mathbb{C}$, and the set K, where $G_K = 0$, is compact.

Although there is a minimum principle for the solutions of $(dd^c u)^{k-1} = 0$ (for $k \geq 2$), it is not as strong as for harmonic functions. The function u may attain its minimum without necessarily being constant.

THEOREM 2.6.4. *Let f be a regular automorphism of $\mathbb{C}^k$. Suppose the dimension of I_- is 0. Let Ω be a Fatou component associated with a fixed point p that has one eigenvalue with modulus less than 1, and the others with modulus less than or equal to 1. Suppose that $p \in J_{k-1}^-$. Then $\partial\Omega = J^+$.*

PROOF. Let D be a polydisk of dimension $k-1$ passing through p and such that $\int_D (dd^c G^-)^{k-1} = c_1 \neq 0$. As in the preceding theorem, we show that if ψ is a nonnegative test function equal to 1 in a neighborhood of p and 0 in a neighborhood of ∂D, then the sequence $\frac{1}{d^n}(f^n)^*(\psi D)$ converges to cT_+, $c \neq 0$. Hence the boundary $\partial\Omega$ contains J^+. It is clear that $\partial\Omega \subset \partial K^+ = J^+$. □

REMARK 2.6.5. When $k = 2$, the fixed point p must be in K^-, and the support of $dd^c G^-$ is K^-. Of course, it would be interesting to give a dynamical characterization of the set J_{k-1}^-.

2.7. Mixing When $\dim I_- = 0$

We recall a few concepts [**86**]. Let $(X, \mathcal{A}, m)$ be a measure space with probability measure m. Let $f : X \to X$ be a measure-preserving map. We say that f is mixing if

$$\lim_{n\to\infty} m(f^{-n}(A) \cap B) = m(A)m(B)$$

for all $A, B \in \mathcal{A}$. If X is contained in a manifold M and A is the σ-algebra of Borel sets, it suffices to show that

$$\lim_{n\to\infty} \left(\int_X \psi(f^n)\varphi\, dm\right) = \left(\int_X \psi\, dm\right)\left(\int_X \varphi\, dm\right)$$

for arbitrary positive test functions ψ, φ.

THEOREM 2.7.1. *Let f be a regular automorphism of $\mathbb{C}^k$. Suppose that I_- has dimension* 0. *Then the measure $\mu := T_+ \wedge T_-^{k-1}$ is mixing. More precisely, if ψ and φ are test functions, then*

$$\int \psi(f^n)\varphi d\mu \longrightarrow \left(\int \psi d\mu\right)\left(\int \varphi d\mu\right).$$

PROOF. Since $\dim I_- = 0$, we have $\ell = 1$ and $d = (d_-)^{k-1}$. By Proposition 2.6.1,

$$\psi(f^n)(T_+) = \frac{1}{d^n}(f^n)^*(\psi T) \longrightarrow S,$$

where S is closed and positive. But if $0 \leq \psi \leq 1$, as we may assume, then $\psi(f^n)(T_+) \leq T_+$. Thus $S \leq T_+$. Corollary 2.2.13 implies that $S = cT_+$, where $c = \int \psi T_+ \wedge T_-^{k-1} = \int \psi d\mu$.

We will show that $\psi(f^n)T_+ \wedge T_-$ approaches $cT_+ \wedge T_-$. If θ is a test form of bidegree $(k-2, k-2)$, then

$$\langle \psi(f^n)T_+ \wedge T_-, \theta\rangle = \left\langle dd^c\left(\frac{1}{d^n}(f^n)^*(T_+\psi) \wedge \theta\right), G^-\right\rangle.$$

Set $R_n = \dfrac{1}{d^n}(f^n)^*T_+\psi$. Then the right-hand side of the inequality above equals

$$\langle dd^c R_n \wedge \theta + R_n \wedge dd^c\theta + 2dR_n \wedge d^c\theta, G^-\rangle.$$

But we have seen that $\|dR_n\|$ and $\|dd^c R_n\|$ approach 0. Hence, passing to the limit, we have

$$\lim_n \langle \psi(f^n)T_+ \wedge T_-, \theta\rangle = c\langle T_+ \wedge dd^c\theta, G^-\rangle = c\langle T_+ \wedge dd^c G^-, \theta\rangle$$

since G^- is continuous. Then

$$\psi(f^n)T_+ \wedge T_- = (f^n)^* \frac{\psi T_+ \wedge T_-}{(d_-)^{(k-2)n}}$$

because $(f^n)^*T_+ = d^n T_+ = (d_-)^{(k-1)n}T_+$ and $(f^n)^*T_- = \dfrac{1}{(d_-)^n}T_-$. Now we can set $R_n = (f^n)^* \dfrac{\psi T_+ \wedge T_-}{(d_-)^{(k-2)n}}$ and prove by induction on $j \leq k-1$ that

$$\psi(f^n)T_+ \wedge T_-^j \longrightarrow cT_+ \wedge T_-^j,$$

which gives the result. □

COROLLARY 2.7.2. *If f is a regular automorphism of $\mathbb{C}^2$ or $\mathbb{C}^3$, then the measure μ is mixing for f.*

PROOF. The case $\mathbb{C}^2$ is clear. Let f be an automorphism from $\mathbb{C}^3$ to $\mathbb{C}^3$. If $\dim I_- = 0$, we apply the preceding result. If $\dim I_- = 1$, then $\dim I_+ = 0$. We apply the preceding result to f^{-1} and the measure $T_- \wedge (T_+)^2$. □

CHAPTER 3

Holomorphic Endomorphisms of $\mathbb{P}^k$

In this chapter we study the dynamics of holomorphic endomorphisms of $\mathbb{P}^k$. We have denoted by $\mathcal{H}_d$ the semigroup of endomorphisms of algebraic degree d. Recall that $f \in \mathcal{H}_d$ lifts to a map $F = (F_0, \ldots, F_k)$ on $\mathbb{C}^{k+1}$, where the F_j are homogeneous polynomials of degree d and $F^{-1}(0) = 0$. The indeterminacy set considered in Chapter 1 is empty.

We summarize the results obtained so far; some were proved in the setting of dominant meromorphic maps, and can be applied to maps in $\mathcal{H}_d$.

PROPOSITION 3.0.3. *Let f be a holomorphic endomorphism of $\mathbb{P}^k$.*

(i) *For every $a \in \mathbb{P}^k$, the number of points in $f^{-1}(a)$, counting multiplicity, is d^k.*
(ii) *The cardinality of the set of periodic points of order n is $(d^{n(k+1)}-1)/(d-1)$. There are infinitely many distinct periodic orbits.*
(iii) *The critical set C associated with the endomorphism f is an algebraic hypersurface of degree $(d-1)(k+1)$.*

We denote the closure of the orbit of the critical set by $\mathcal{C} = \overline{\bigcup_{n\geq 0} f^n(C)}$. We say that f is *critically finite* if $\mathcal{C}$ is an algebraic variety, and write

$$\mathcal{C}_\infty = \bigcap_{n\geq 0} \overline{\bigcup_{j\geq n} f^j(C)}.$$

3.1. Some Examples

(i) Ueda [**84**] used the following construction to give some examples whose dynamics follow from the dynamics of rational maps of $\mathbb{P}^1$.

Let $\Phi : \mathbb{P}^1 \times \mathbb{P}^1 \to \mathbb{P}^2$ be the two-fold branched covering defined by

$$\Phi([z_0 : z_1], [w_0 : w_1]) = [z_0w_0 : z_1w_1 : z_0w_1 + z_1w_0].$$

The covering is branched over $\Phi(\Delta)$, where Δ denotes the diagonal of $\mathbb{P}^1 \times \mathbb{P}^1$. Note that $\Phi([z_0 : z_1], [w_0 : w_1]) = \Phi([w_0 : w_1], [z_0 : z_1])$. Let h be a rational map of degree $d \geq 2$ on $\mathbb{P}^1$. We define a map $\widehat{h}$ on $\mathbb{P}^2$ such that the following diagram commutes:

$$\begin{array}{ccc} \mathbb{P}^1 \times \mathbb{P}^1 & \xrightarrow{(h(z),h(w))} & \mathbb{P}^1 \times \mathbb{P}^1 \\ \Phi \downarrow & & \downarrow \Phi \\ \mathbb{P}^2 & \xrightarrow{\widehat{h}} & \mathbb{P}^2 \end{array}$$

The dynamical properties of $\widehat{h}$ can easily be derived from those of h. Set $\widetilde{h} = (h, h)$, and let L be the union of all the components of $\widetilde{h}^{-1}(\Delta)$ except Δ.

One can check that $\Phi(L)$ is critical for $\widehat{h}$ and that $\widehat{h}(\Phi(L)) = \Phi(\Delta)$. Moreover, $\widehat{h}(\Phi(\Delta)) = \Phi(\Delta)$.

(a) If (Ω_j) are the Fatou components of h in $\mathbb{P}^1$ and we take $j \neq k$, it is clear that $\widehat{h}$ has a Fatou component isomorphic to $(\Omega_j \times \Omega_k)$. In particular, there may exist Fatou components isomorphic to a product of annuli.

(b) If the Julia set of h is $\mathbb{P}^1$, then the Julia set of $\widehat{h}$ is $\mathbb{P}^2$. One can check that since the repelling points of h are dense in $\mathbb{P}^1$, the repelling points of $\widehat{h}$ are dense in $\mathbb{P}^2$ [**18**].

(c) If h is critically finite, so is $\widehat{h}$. Observe that $\Phi(\Delta)$ is always in the critical set of $\widehat{h}$; we have seen that $\widehat{h}(\Phi(\Delta)) = \Phi(\Delta)$.

(d) Let σ_k denote the group of permutations on $\{1, \ldots, k\}$. The map Φ shows that $\mathbb{P}^2$ is isomorphic to $\mathbb{P}^1 \times \mathbb{P}^1/\sigma_2$. Similarly, $(\mathbb{P}^1)^k/\sigma_k$ is isomorphic to $\mathbb{P}^k$. It suffices to consider the map that assigns to a point $(z_1, \ldots, z_k) \in \mathbb{C}^k$ the symmetric functions of the (z_j) and to extend the map to $(\mathbb{P}^1)^k$.

In the same way as for $\mathbb{P}^2$, one can obtain examples of endomorphisms of $\mathbb{P}^k$ whose dynamics follow easily from the dynamics of the corresponding maps from $(\mathbb{P}^1)^k$ to $(\mathbb{P}^1)^k$.

(ii) We give an example of a critically finite map of $\mathbb{P}^2$ that is not of the type above. We define [**28**]

$$g[z : w : t] = [(z - 2w)^2 : (z - 2t)^2 : z^2].$$

The critical set is $C = \{z = 2w\} \cup \{z = 2t\} \cup \{z = 0\}$. We have

$$(z = 2w) \to (z = 0) \to (t = 0) \to (t = w) \to (z = w) \to (z = t) \to (t = w).$$

Similarly,

$$(z = 2t) \to (w = 0) \to (z = t) \to (t = w).$$

This example is not of the type above because there is no component of the critical set whose image under g is stable under g (see 1.(i).(a)). It is shown in [**28**] that the Julia set of g is $\mathbb{P}^2$. Critically finite maps are studied in [**30**], [**28**], Ueda [**84**], Jonsson [**52**].

3.2. Fatou Components

Let $f \in \mathcal{H}_d$. We know that the Julia set J of f is equal to the support of the current T. Hence an open set U in $\mathbb{P}^k$ is contained in a Fatou component if and only if the function G is pluriharmonic in $\pi^{-1}(U)$.

A Fatou component U for f is a connected component of $\mathbb{P}^k \setminus J$. We say that U is fixed if $f(U) = U$, periodic if $f^n(U) = U$ for some $n \geq 1$, and wandering if the $(f^n(U))$ are pairwise disjoint. By a theorem of Sullivan [**80**], we know that for $k = 1$ there are no wandering components if f is a rational map of $\mathbb{P}^1$. The question is open for $k \geq 2$.

We have seen that for automorphisms of $\mathbb{C}^k$ there may exist Fatou components biholomorphic to $\mathbb{C}^k$. We will see that there are no such components for maps in $\mathcal{H}_d$; the Fatou components are more like bounded domains.

Recall the definition of Kobayashi hyperbolicity. Let M be a complex manifold. For $p \in M$ and $\xi \in T_p(M)$, consider the set of holomorphic maps from the unit disk D with values in M, $\varphi : D \to M$, $\varphi(0) = p$ and $\varphi'(0) = c\xi$.

The Kobayashi-Royden pseudometric is defined by setting

$$K_M(p,\xi) = \inf\left\{\frac{1}{|c|} : \varphi : D \to M,\ \varphi(0) = p,\ \varphi'(0) = c\xi\right\}.$$

If M contains holomorphic images of $\mathbb{C}$ in the direction ξ, then $K_M(p,\xi) = 0$.

DEFINITION 3.2.1. Let Y be an immersed submanifold in a compact Hermitian manifold M. We say that Y is hyperbolically embedded if there exists a constant $C > 0$ such that

$$K_Y(p,\xi) \geq C|\xi| \text{ for every } (p,\xi) \in T_Y,$$

where $|\xi|$ denotes the Hermitian norm of the vector ξ.

We refer the reader to Kobayashi [**55**], Lang [**57**] for more about this concept. Recall, however, that the Kobayashi-Royden metric is contracting for holomorphic maps. If $\Phi : M \to N$ is holomorphic, then

$$K_N(\Phi(p), \Phi'(p)\xi) \leq K_M(p,\xi).$$

THEOREM 3.2.2 ([**85**]). *Let $f \in \mathcal{H}_d$, $d \geq 2$. If Ω is a Fatou component of f, then Ω is hyperbolically embedded.*

PROOF. Set $A = \{z : z \in \mathbb{C}^{k+1},\ G(z) = 0\}$. Here G denotes the Green's function associated with f. We know that G is continuous and grows at infinity like $\log\|z\|$. Hence A is compact. The set of normality of a holomorphic map is $\mathbb{P}^k$; thus G is pluriharmonic on $\pi^{-1}(\Omega)$.

Let $\varphi : D \to \Omega$ be a holomorphic map. We will show that there exists a section σ of π such that $\sigma \circ \varphi$ takes values in in A. Let $q = \varphi(\zeta_0)$. Let s be a holomorphic section of π in a neighborhood of q. Then $G \circ s \circ \varphi = \log|h|$, where h is holomorphic in a neighborhood of ζ_0. It suffices to set $\sigma = s/h$. Then $G(\sigma \circ \varphi) = 0$. Since the disk is simply connected, we construct a holomorphic lift $\widetilde{\varphi}$ of φ with values in the compact set A. It follows that the derivative of $\varphi = \pi \circ \widetilde{\varphi}$ is uniformly bounded at 0, independently of φ. □

REMARKS 3.2.3.

(i) Let S be a closed positive current of bidegree $(1,1)$ on $\mathbb{P}^k$. Suppose that S has a locally bounded potential H. Every immersed manifold V in $\mathbb{P}^k$ on which H is pluriharmonic is hyperbolically embedded. One argues as before. The set $A = \{H = 0\}$ is bounded in $\mathbb{C}^{k+1}$, and one constructs sections as above.

(ii) Ueda's result also holds in the following setting. Let $f \in \mathcal{H}_d$. Let $p \in \mathbb{P}^k$. Suppose p is in a hyperbolic set in the dynamical sense [**67**]. One can then define the stable manifold of p,

$$W^s(p) = \{z : \operatorname{dist}(f^n(z), f^n(p)) \to 0,\ n \to \infty\}.$$

It is an immersed complex manifold (not necessarily connected). Then the stable manifold of p is hyperbolically embedded.

It suffices to check that the Green's function G restricted to $\pi^{-1}(W^s(p))$ is pluriharmonic. Suppose that $f^{n_i}(p) \to [0 : \cdots : 0 : 1]$. We have

$$\frac{1}{d^{n_i}} \log|F^{n_i}|_{|\pi^{-1}(W^s(p))} = \frac{1}{d^{n_i}} \log|F_k^{n_i}| + \frac{O(1)}{d^{n_i}}.$$

Passing to the limit shows that $G_{|\pi^{-1}(W^s(p))}$ is pluriharmonic.

We will prove another result of Ueda [**85**], on the equicontinuity of inverses.

THEOREM 3.2.4 ([**85**]). *Let W be an immersed complex manifold in $\mathbb{P}^k$. If there exist inverses h_{j_n} of f^{j_n} on W, then (h_{j_n}) is locally equicontinuous on W.*

PROOF. We know there exists a constant $C > 0$ such that

$$\frac{1}{C}|z|^d \leq |F(z)| \leq C|z|^d.$$

It follows that there exist r and R, $0 < r < R < \infty$, such that $|F(z)| < r$ whenever $|z| < r$ and $|F(z)| > R$ whenever $|z| > R$. Hence

$$F^{-1}(\{r < |z| < R\}) \subset \{r < |z| < R\}.$$

Note also that F is of the form $t \to ct^d$ on every complex line passing through 0; its only branch point is 0. Thus, in a ball B in W, we can define sections s_n of π such that $F^{-j_n} := s_n \circ h_{j_n} \circ \pi$ take values in $\{r < |z| < R\}$. Hence, if π^{-1} is a section of π over B, the functions

$$h_{j_n} = \pi \circ F^{-j_n} \circ \pi^{-1}$$

are equicontinuous in B. □

COROLLARY 3.2.5. *Let $f \in \mathcal{H}_d$. Let p be a periodic point of order n of f. Let $\lambda_1, \ldots, \lambda_k$ be the eigenvalues of the derivative of f^n at p. Suppose $|\lambda_1| \geq \cdots \geq |\lambda_k|$. If $|\lambda_k| < 1$ or if λ_k is a root of unity, then $p \in \mathcal{C} = \overline{\bigcup_{j\geq 0} f^j(C)}$.*

PROOF. We may assume that p is a fixed point. If $p \notin \mathcal{C}$, then there exists a neighborhood B of p disjoint from $\bigcup_{j\geq 0} f^j(C)$. Hence we can define inverse branches h_j of f^j, $j \geq 0$, with $h_j(p) = p$. The equicontinuity of the h_j implies that the eigenvalues of the derivative have moduli less than or equal to 1, and any that has modulus 1 cannot be a root of unity. □

PROPOSITION 3.2.6. *Let $f \in \mathcal{H}_d$. Let U be a Fatou component containing an attracting fixed point p. Then U intersects the critical set C of f.*

PROOF. Since U contains a fixed point p, we have $f(U) = U$. If $U \cap C = \emptyset$, then $U \cap f^{-j}(C) = \emptyset$ for every $j \geq 1$. In some ball $B(p, r)$ we can define an inverse h_j of f^{-j} such that $h_j(p) = p$. The equicontinuity of the (h_j), which follows from the fact that U is hyperbolically embedded, implies that the derivatives of h_j at p are bounded. But this is impossible because p is attracting. □

REMARKS 3.2.7.

(i) If p is a fixed point of f with some eigenvalues of modulus less than one and others of modulus greater than one, then the stable manifold of p intersects the critical set C. The proof is the same as for a Fatou component.

(ii) Let p be a periodic point of period ℓ, attracting for f. Let $\{p_0, \ldots, p_{\ell-1}\}$ denote the cycle. The basin of attraction U of p is defined by $U = \bigcup_{j=0}^{\ell-1} U_j$, where $U_j = \{z : d(f^{\ell n}(z), p_j) \to 0\}$. We denote by U_j^0 the connected component of U_j containing p_j, $0 \leq j \leq \ell - 1$, and set $U^* := \bigcup_{j=1}^{\ell-1} U_j^0$. It follows from Proposition 3.2.6 that U^* intersects the critical set C. In one variable, this leads to a proof that there exist at most $(2d - 2)$ attracting cycles. Such a conclusion is not possible in $\mathbb{P}^k$ for $k \geq 2$.

Using Newhouse's method [**62**] for constructing infinitely many attracting cycles for diffeomorphisms of surfaces, E. Gavosto [**42**] produced elements of $\mathcal{H}_d(\mathbb{P}^2)$ that have infinitely many attracting cycles. The result has been improved by G. Buzzard [**17**].

THEOREM 3.2.8 ([**17**]). *There exist a positive number d and a nonempty open set $\mathcal{N} \subset \mathcal{H}_d(\mathbb{P}^2)$ containing a dense G_δ set $\mathcal{N}' \subset \mathcal{N}$ such that every $f \in \mathcal{N}'$ has infinitely many attracting cycles.*

DEFINITION 3.2.9. Let $f \in \mathcal{H}_d$. Let U be a Fatou component. We say that U is a Siegel domain if there exists a subsequence (f^{n_j}) that converges to the identity on compact subsets of U.

THEOREM 3.2.10 ([**85**]). *Let $f \in \mathcal{H}_d$ have critical set C. If U is a Siegel domain for f, then $\partial U \subset \mathcal{C} = \overline{\bigcup_{j\geq 0} f^j(C)}$.*

PROOF. Let $p \in \partial U$. Suppose $p \notin \mathcal{C}$. Then there exists a neighborhood W of p that does not intersect $\bigcup_{j\geq 0} f^j(C)$. Let $W_1 \subset W \cap U$. We define an inverse g_{n_j} of f^{n_j} in W_1 such that $g_{n_j}(W_1) \subset U$; this is possible because we may assume that $f^{n_j} \to \mathrm{Id}$. We extend g_{n_j} to W; this is possible because W does not intersect C. By Theorem 3.2.4 we can find a subsequence $g_{n'_j}$ that converges in W. We must have $g_{n'_j} \to \mathrm{Id}$ in W. Hence $(f^{n'_j})$ converges to the identity in W. It follows from Corollary 1.6.6 that $U \cup W$ is in a Fatou component. Hence $U \cup W = U$, which is the desired contradiction. □

Recall that a Fatou component U is recurrent if, for some $p_0 \in U$, there exists a subsequence (n_j) such that $f^{n_j}(p_0) \to q \in U$. A recurrent component is clearly periodic. Hence we can restrict to the case where it is fixed.

Our knowledge of the dynamics in a recurrent component is incomplete. However, we have the following partial result.

THEOREM 3.2.11 ([**31**]). *Let $f \in \mathcal{H}_d(\mathbb{P}^2)$, $d \geq 2$. Let U be a recurrent Fatou component such that $f(U) = U$. Then one of the following properties is satisfied.*

(i) *U is the basin of attraction of a fixed point $p \in U$.*
(ii) *There exists a closed complex submanifold Σ of U of complex dimension one such that $f^n(K) \to \Sigma$ for every compact subset K of U. The Riemann surface Σ is biholomorphic to either a disk D, a punctured disk, or an annulus, and $f_{|\Sigma}$ is conjugate to an irrational rotation.*
(iii) *The domain U is a Siegel domain.*

At the beginning of this section we introduced the notion of Kobayashi hyperbolicity. We will characterize the Brody-hyperbolic attractors.

DEFINITION 3.2.12. Let X be a compact subset of a complex manifold M. We say that X is Brody hyperbolic if every holomorphic map $h : \mathbb{C} \to M$ such that $h(\mathbb{C}) \subset X$ is constant.

The following theorem characterizes Brody-hyperbolic compact sets [**57**, p. 68].

THEOREM 3.2.13. *Let X be a compact subset of a complex manifold M. X is Brody hyperbolic if and only if X has a hyperbolically embedded neighborhood.*

A result of H. Tsuji [**82**] now allows us to characterize the Brody-hyperbolic attractors. More precisely, we have the following theorem:

THEOREM 3.2.14 ([**82**]). *Let Y be a Kobayashi-hyperbolic complex manifold, and let $f : Y \to Y$ be a holomorphic map such that $f(Y) \Subset Y$. Then*

(1) *$N := \bigcap_{k\geq 1} f^k(Y)$ is a nonempty analytic subset of Y;*
(2) *there exists an integer ℓ_0 such that $f^{\ell_0}|_N = \mathrm{Id}_N$.*

COROLLARY 3.2.15. *Let $f \in \mathcal{M}_d$. Suppose that for every ℓ, the periodic points of order ℓ of f form a discrete set. Let X be a Brody-hyperbolic closed subset of $\mathbb{P}^k$. If X is attracting for f, then it is a finite union of attracting periodic orbits.*

PROOF. Since the compact set X is attracting, there is a neighborhood U of X in $\mathbb{P}^k$ such that $f(U) \Subset U$. We have implicitly assumed that U does not intersect the indeterminacy set of f. Moreover, $X = \bigcap_{n\geq 0} f^n(U)$. Since the compact set X is Brody hyperbolic, we may assume that U is Kobayashi hyperbolic. By Theorem 3.2.14, X is an analytic set; by the hypothesis on the periodic points of f, X is finite. So it is certainly a finite union of attracting cycles. □

REMARK 3.2.16. In particular, if $f \in \mathcal{H}_d$, then the nontrivial attractors of f contain images of $\mathbb{C}$. On this topic, see the paper by Fornaess-Weickert [**36**].

3.3. The Currents T^ℓ $(1 \leq \ell \leq k)$ and Their Supports J_ℓ

We have seen that we can assign to a holomorphic map $f \in \mathcal{H}_d$ a closed positive current T of bidegree $(1,1)$ and mass 1. Recall that T is defined by

$$\pi^* T = dd^c G,$$

where G is p.s.h. in $\mathbb{C}^{k+1}$ and defined by

$$G(z) = \lim_n \frac{1}{d^n} \log |F^n(z)|.$$

The function G satisfies the relation $G(F(z)) = dG(z)$. Hence

$$f^* T = dT.$$

We have shown that the function G is locally Hölder continuous, and in particular continuous. By Appendix A.6 we can define the currents $T^\ell := T \wedge \cdots \wedge T$ (ℓ terms). These are closed positive currents of bidegree (ℓ,ℓ) and satisfy the functional equation

$$f^*(T^\ell) = d^\ell T^\ell.$$

DEFINITION 3.3.1. Let $f \in \mathcal{H}_d$. The Julia set of order ℓ associated with f is the support of the current T^ℓ. We denote it by J_ℓ.

We will see that the set J_ℓ has properties similar to those of the Julia set in one variable. We also have $J_1 = J$, the Julia set.

THEOREM 3.3.2 ([**32, 33**]). *Let f be a holomorphic map of degree d on $\mathbb{P}^k$.*

(i) *The support of T is the Julia set of f.*
(ii) *The currents T^ℓ have bidegree (ℓ,ℓ) and mass 1. In particular, the Julia sets J_ℓ are nonempty and totally invariant. Moreover, J_ℓ is connected if $2\ell \leq k$. The open set $\mathbb{P}^k \setminus J_\ell$ is $(k-\ell)$-pseudoconvex.*
(iii) *The measure $\mu := T^k$ is a probability measure that is invariant under f. It satisfies the relation $f^*\mu = d^k \mu$ and maximizes entropy.*

PROOF.

(i) Since the map f is holomorphic, its set of normality is $\mathbb{P}^k$. Hence the property follows from Theorem 1.6.5.

(ii) The mass of T^ℓ is 1. Indeed, T is in the fundamental class of ω and $\|T\| = 1$. We can apply Corollary A.6.5. Since $f^*T^\ell = d^\ell T^\ell$, it is clear that the support of T^ℓ is totally invariant. We refer the reader to [**33**] for the notion of $(k-\ell)$ pseudoconvexity. Recall, however, that the usual pseudoconvexity coincides in this terminology with $(k-1)$-pseudoconvexity. An open set is $(k-\ell)$ pseudoconvex if it satisfies the Kontinuitätssatz for holomorphic disks of dimension ℓ.

(iii) The equation $f^*\mu = d^k\mu$ follows from the functional equation satisfied by the Green's function G. In particular, the measure μ is invariant: if φ is a test function, then

$$\langle f_*\mu, \varphi\rangle = \langle \mu, \varphi\circ f\rangle = \frac{1}{d^k}\langle f^*\mu, \varphi\circ f\rangle = \frac{1}{d^k}\left\langle \mu, \sum_{f(z_i)=z} (\varphi\circ f)(z_i)\right\rangle = \langle \mu, \varphi\rangle.$$

Gromov [**43**] has shown that the metric entropy of a map $f \in \mathcal{H}_d$ equals $\log d^k$. A measure satisfying the equation $f^*\mu = d^k\mu$, with no mass on the critical set, has entropy $\log d^k$ [**64**]. □

THEOREM 3.3.3. *Let $f \in \mathcal{H}_d$. Set $\Lambda = \sup_{G\le 0}\|F'(z)\|$ and*

$$\alpha_0 = \frac{\log d}{\log \Lambda}.$$

(i) *For $1 \le \ell \le k$, the current T^ℓ assigns no mass to sets of Hausdorff dimension less than $2(k-\ell)+\ell\alpha_0$.*

(ii) *Let X be a compact set that is attracting for f. Suppose that $\Lambda_{2\ell}(X) = 0$, where $\Lambda_{2\ell}$ denotes the 2ℓ-dimensional Hausdorff measure. Then T^ℓ is zero on the basin of attraction of X. In particular, an attractor of dimension less than 2 is an attracting cycle.*

PROOF.

(i) We have seen that since f is holomorphic, T^ℓ can be defined for every ℓ. Hence the property follows from Remark 1.7.2 and Theorem 1.7.3.

(ii) Recall that X is an attractor if there exist arbitrarily small neighborhoods V of X such that $f(V) \Subset V$.

If $\Lambda_{2\ell}(X) = 0$, then there exists a complex subspace L of codimension ℓ that does not intersect X [**25, 71**]. We may assume that $L = \{z_0 = \cdots = z_{\ell-1} = 0\}$. Hence there exists a constant C such that

$$X \subset \{|z_0| + \cdots + |z_{\ell-1}| > C(|z_\ell| + \cdots + |z_k|)\} := U.$$

If an open set B is in the basin of attraction of X, then $f^n(B) \subset U$ for sufficiently large n. Hence, for $z \in \pi^{-1}(B)$,

$$G(z) = \lim_{n\to\infty} \frac{1}{d^n}\log(|F_0^n| + \cdots + |F_{\ell-1}^n|),$$

where F_j^n denotes the jth component of F^n. We already pointed out in Section 2.5 that

$$(dd^c(\log|h_0| + \cdots + |h_{\ell-1}|))^\ell = 0$$

if $h_0, \ldots, h_{\ell-1}$ are holomorphic functions without common zeros. Thus $(dd^cG)^\ell = 0$ on $\pi^{-1}(B)$.

If $\Lambda_2(X) = 0$, we find that $T = 0$ on the basin of X, which is thus contained in the Fatou set. Replacing f by an iterate if necessary, we may assume that X is contained in a Fatou component W and that $f(W) \subset W$. But W is hyperbolically embedded. The Kobayashi metric is contracting. Using equicontinuity and the fact that $\Lambda_2(X) = 0$, we see that the limits of convergent subsequences of f^n are constants. To see that these limits are attracting fixed points, observe that if $\{f^{n_j}\}$ converges to p on a compact subset containing p and $f(p)$, then p is fixed, hence necessarily attracting. □

EXAMPLES.

(i) If $f([z : w : t]) = [z^d : w^d : t^d]$, then $J_2 = \{|z| = |w| = |t|\}$, and the measure μ is Lebesgue measure on this two-dimensional torus.

(ii) Let h be a rational map of $\mathbb{P}^1$ whose Julia set is $\mathbb{P}^1$. Let $\widehat{h}$ be the associated map from $\mathbb{P}^2$ to $\mathbb{P}^2$, as in the example in Section 3.1. It is fairly easy to check that $J_1 = J_2 = \mathbb{P}^2$: if ν is the measure associated with h on $\mathbb{P}^1$, then $\mu = \Phi_*(\nu \otimes \nu)$.

(iii) For the example $f([z : w : t]) = [(z - 2w)^2 : (z - 2t)^2 : z^2]$, which is critically finite, Jonsson [**52**] has shown that $J_2 = \mathbb{P}^2$.

REMARK 3.3.4. It follows from the functional equation for G that the Green's function associated with f^n is the same as that associated with f. Hence the Julia sets J_ℓ associated with f^n are the same as those associated with f.

3.4. Dynamic Hyperbolicity

The only totally invariant set J_ℓ on which f can be hyperbolic is J_k. We begin by recalling a few ideas (see Ruelle [**67**]).

Let $f \in \mathcal{H}_d$. Let K be a compact subset of $\mathbb{P}^k$ such that $f(K) = K$. Consider

$$\widehat{K} = \{x = (x_n)_{n \le 0} : \ f(x_n) = x_{n+1}\}.$$

This is a compact subset of $K^{\mathbb{N}}$, which we equip with the product topology. The tangent bundle to $\widehat{K}$, denoted by $T_{\widehat{K}}$, consists of pairs (x, ξ), where $x = (x_n)_{n \le 0}$ and ξ is a tangent vector at x_0. Let $\pi : \widehat{K} \to K$ be defined by $\pi(x) = x_0$. The map f has a lift $\widehat{f}$ such that $\pi \circ \widehat{f} = f \circ \pi$. The derivative Df of f lifts to $D\widehat{f}$. We call $(\widehat{K}, \widehat{f})$ the natural extension of (K, f).

DEFINITION 3.4.1. Let $K \subset \mathbb{P}^k$ be a compact set such that $f(K) = K$. We say that f is hyperbolic on K if there exists a continuous decomposition $E^u \oplus E^s$ of $T_{\widehat{K}}$ such that $D\widehat{f}(E^u) \subset E^u$, $D\widehat{f}(E^s) \subset E^s$, and there exist constants $c > 0$, $\lambda > 1$ such that, for every $n \ge 1$,

$$\begin{cases} |D\widehat{f}^{-n}(\xi)| \le c\lambda^{-n}|\xi| & \text{if } \xi \in E^u, \\ |D\widehat{f}^{n}(\xi)| \le c^{-1}\lambda^{-n}|\xi| & \text{if } \xi \in E^s. \end{cases}$$

The fiber dimension of E^u is called the unstable dimension and that of E^s the stable dimension.

THEOREM 3.4.2 ([**33**]). *Let $f \in \mathcal{H}_d$. The map f cannot be hyperbolic on any of the sets $\mathbb{P}^k, J_1, \ldots, J_{k-1}$. If f is hyperbolic on J_k, then the unstable dimension of J_k is k.*

PROOF. It actually suffices to prove the last assertion because, on the one hand, $J_k \subset J_{k-1} \subset \cdots \subset \mathbb{P}^k$, and, on the other, the critical set C intersects $J_{k-1}, \ldots, \mathbb{P}^k$ (the current $T^\ell \wedge [C]$ is nonzero if $\ell \leq k-1$; see Appendix A.6). So we suppose that f is hyperbolic on J_k and the stable dimension is at least 1. Since $\dim E^s \geq 1$, through every point of J_k there is a disk on which the local potential u of T is harmonic. It follows (see [**33**], Lemma 5.5, or Appendix A.10.3) that $(dd^c u)^k = 0$, which contradicts the fact that μ is a probability measure. □

3.5. Convergence to T

We give a theorem on the convergence to T of the pullbacks of a current that is more precise than in the case of maps in $\mathcal{M}_d$. Let $f \in \mathcal{H}_d$. Recall that the multiplicity of f at $a \in \mathbb{P}^k$ equals the generic number of solutions in a neighborhood of a of the equation $f(z) = w$, for w in a neighborhood of $f(a)$; we denote this number by $\nu(f, a)$.

THEOREM 3.5.1 ([**32**]). *Let $f \in \mathcal{H}_d$. Let P be the set of points of multiplicity $\nu(f,a) \geq d$. Suppose there exists an attractor A such that the set $P \setminus A$ is finite and contains no periodic points. Let S be a closed positive current of mass 1 whose potential is locally bounded in a neighborhood of A. Then the sequence of currents $\left(\frac{(f^n)^* S}{d^n}\right)$ converges to T.*

PROOF. Let u be the potential of S. Let Ω be the basin of attraction of A. In a neighborhood of $\pi^{-1}(A)$, we have

$$\log|z| - C \leq u \leq \log|z| + C.$$

It follows that $u(F^n)/d^n \to G$ on $\pi^{-1}(\Omega)$. We first consider the case where P is finite, with no periodic points, and A is empty. We show that the sequence $u(F^n)/d^n$ does not converge uniformly to $-\infty$. If it did, then for sufficiently large n we would have

$$u\left(\frac{F^n}{|F^n|}\right) \leq -d^n.$$

Then the image under $F^n/|F^n|$ of the unit sphere would be contained in

$$\{u < -d^n\} \cap \{|\xi| = 1\}.$$

But the Lebesgue measure of this set approaches 0, giving a contradiction (Remark A.8.3).

Hence we may assume that $v_{n_i} := \dfrac{1}{d^{n_i}} u(F^{n_i}) \to v$ in L^1_{loc}, and it suffices to show that $v = G$ (Theorem A.1.2). We know that $v \leq G$. Suppose $\{v < G\}$ were nonempty. Since G is continuous, there exist $\delta > 0$ and an open set $U \subset \mathbb{P}^k$ such that $v < G - 2\delta$ on $\pi^{-1}(U)$. By the Hartogs lemma, for sufficiently large n_i we have

$$\frac{1}{d^{n_i}} u\left(t \frac{F^{n_i}}{|F^{n_i}|}\right) < -\delta$$

on $\pi^{-1}(U)$, for all $t \in [1/2, 1]$. Hence $\pi^{-1}(f^{n_i}(U)) \cap \{1/2 \leq |z| \leq 1\}$ is contained in $X := \{u < -\delta d^{n_i}\}$.

Let L be a complex line on which u is not identically $-\infty$. Since u grows logarithmically on L, the logarithmic capacity of $X \cap L$ is at most $e^{-\delta d^{n_i}}$. A classical estimate (Proposition A.8.2) shows that any disk contained in $X \cap L$ has radius of order $e^{-\delta d^{n_i}}$. To reach a contradiction, it suffices to show that $X \cap \{1/2 \leq |z| \leq 1\}$ contains balls of radius $r^{(d-1/2)^{n_i}}$ for sufficiently small r. This follows from a Lojasiewicz-type lemma, details of which are given in [**32**]. □

LEMMA 3.5.2. *Let $g : W \to \mathbb{C}^k$ be a holomorphic map in a neighborhood W of a compact set $K \subset \mathbb{C}^k$. Suppose that the fibers $S_w = f^{-1}(f(w))$, $w \in K$, are discrete and that at every point $w \in K$ the multiplicity is less than or equal to m. Then there exists a constant $C > 0$ such that, for all $w \in f(W)$ and $z \in K$,*

$$|f(z) - w| \geq C \operatorname{dist}(z, S_w)^m.$$

In particular, there exist $a > 0$, $r_0 > 0$ such that for all $z \in K$ and $0 < r < r_0$,

$$f(B(z, r)) \supset B(f(z), ar^m).$$

END OF THE PROOF OF THE THEOREM. The open set $f^n(U)$ does not intersect the basin of attraction of A because $v = G$, as we saw above. Let ℓ be the number of points in $P \setminus A$ where the multiplicity is greater than or equal to d; the multiplicity is less than d^k. We know that $P \setminus A$ has no periodic points. It follows that for sufficiently large N, the local multiplicity of f^N is at most $(d^k)^\ell (d-1)^{N-\ell} \leq (d - 1/2)^N$. Hence, for $B(z, r) \subset U$, the image $f^N(B(z, r))$ contains a ball of radius

$$r^{d^{k\ell}(d-1)^{N-\ell}} \geq r^{(d-1/2)^N}.$$

This is the desired contradiction because $X \cap L$ contains only disks with radius less than or equal to $e^{-\delta d^{n_i}}$. □

COROLLARY 3.5.3. *For $d > 2^k$, there is a Zariski dense open set $\mathcal{U}_d \subset \mathcal{H}_d(\mathbb{P}^k)$ such that $\dfrac{(f^n)^* S}{d^n} \to T$ for every $f \in \mathcal{U}_d$ and every closed positive current S of bidegree $(1,1)$.*

PROOF. Set $Z_d = \{(f, z) : f \in \mathcal{H}_d,\ \nu(f, z) \geq d\}$. Then Z_d is an analytic subset of $\mathcal{H}_d \times \mathbb{P}^k$. Let Y_d denote its projection to $\mathcal{H}_d$. This is an analytic subset of $\mathcal{H}_d$ because projection to $\mathcal{H}_d$ is a proper map. It suffices to check that it is not equal to $\mathcal{H}_d$ for $d > 2^k$. We will then set $\mathcal{U}_d := \mathcal{H}_d \setminus Y_d$, and the result will follow from Theorem 3.5.1.

It is easy to construct an endomorphism f from $\mathbb{P}^1$ to itself of degree d such that the local multiplicity at each point is less than or equal to 2; it suffices that all the zeros of the derivative be simple. Consider the map $\widehat{f}$ that makes the following diagram commutative:

$$\begin{array}{ccc} (\mathbb{P}^1)^k & \xrightarrow{\ \widetilde{f}\ } & (\mathbb{P}^1)^k \\ \pi \big\downarrow & & \big\downarrow \pi \\ (\mathbb{P}^1)^k/\sigma_k & \xrightarrow{\ \widehat{f}\ } & (\mathbb{P}^1)^k/\sigma_k \end{array}$$

We know that $\mathbb{P}^k$ is isomorphic to $(\mathbb{P}^1)^k/\sigma_k$, where σ_k denotes the group of permutations on k elements. We have set $\widetilde{f}(z_1, \dots, z_k) = (f(z_1), \dots, f(z_k))$.

The local multiplicity of $\widehat{f}$ is less than or equal to $2^k < d$. □

3.6. The Measure μ. Mixing

Let $f \in \mathcal{H}_d$. We have seen that the measure $\mu := T^k$ associated with f is an invariant probability measure that maximizes entropy. We will show that f is mixing for μ and derive several consequences.

THEOREM 3.6.1 ([**32, 39**]). *Let $f \in \mathcal{H}_d$. Let $\mu = T^k$ be the associated probability measure. Then*

(i) *the map f is mixing for μ;*
(ii) *there exists a pluripolar set $\mathcal{E}$ such that, for $a \notin \mathcal{E}$,*

$$\mu_n^a := \frac{1}{d^{nk}}(f^n)^*\varepsilon_a \to \mu$$

(where ε_a denotes the Dirac measure at a).

PROOF. Let φ be a test function. Set

$$\lambda_n(a,\varphi) := \left\langle \frac{1}{d^{nk}}(f^n)^*\varepsilon_a, \varphi \right\rangle = \frac{1}{d^{nk}} \sum_{f^n(z_i)=a} \varphi(z_i),$$

where the sum is taken over the preimages of a, counting multiplicity.

Recall that a subset $\mathcal{E}$ of $\mathbb{P}^k$ is pluripolar if, for every $a \in \mathbb{P}^k$, there exist a neighborhood U of a and a p.s.h. function $u \not\equiv -\infty$ such that $\mathcal{E} \cap U \subset \{u = -\infty\}$ (Definition A.2.3).

Given a compact set $K \subset B(0,1) \subset \mathbb{C}^k$, we set (as in [**11**])

$$C(K,B) = \sup_u \left\{ \int_K (dd^c u)^k \;:\; u \text{ is p.s.h. in } B,\ 0 \le u \le 1 \right\}.$$

Bedford and Taylor [**11**] have shown that $C(K,B) = 0$ is equivalent to K being pluripolar (Property (c) of Section A.9).

The Siciak function of K is defined by

$$u_K(z) = \sup\{v : v \text{ is p.s.h. in } \mathbb{C}^k,\ v \le \log^+|z| + O(1),\ v \le 0 \text{ on } K\}.$$

If K is not pluripolar and $m(K) := \sup_{B(0,1)} u_K$, we have the following inequalities:

$$\log^+|z| \le u_K(z) \le m(K) + \log^+|z|. \tag{1}$$

The constant $m(K)$ is related to $C(K,B)$ by the Alexander-Taylor inequality [**1**] (Theorem A.9.3): if $K \subset B(0,r)$, $r < 1$, then

$$m(K) \le \frac{A(r)}{C(K,B)}. \tag{2}$$

□

The theorem will be a consequence of the following lemma.

LEMMA 3.6.2. *There exists a constant M such that, for $s > 0$ and every test function φ,*

$$\mu(|\lambda_n(a,\varphi) - c| \ge s) \le \frac{M|\varphi|_2}{d^n}$$

and

$$C\left((|\lambda_n(a,\varphi) - c| \ge s) \cap B(0,1/2), B\right) \le \frac{M|\varphi|_2}{d^n}.$$

We have set $c := \int \varphi\, d\mu$, and $|\varphi|_2$ is the $\mathcal{C}^2$ norm of φ. For the second inequality, we take B to be the unit ball of a chart $\mathbb{C}^k \subset \mathbb{P}^k$.

PROOF OF THE LEMMA. We work in the chart $z_0 \neq 0$ of $\mathbb{P}^k$ and set

$$K_s = \{a : a \in B(0, 1/2) \subset \mathbb{C}^k,\ \lambda_n(a, \varphi) - c \geq s\}.$$

We may assume that K_s is not pluripolar, since otherwise there is nothing to show.

Let u_s be the Siciak function of K_s. For $z \in \mathbb{C}^{k+1}$, set

$$v_s(z) = \log|z_0| + u_s\left(\frac{z_1}{z_0}, \dots, \frac{z_k}{z_0}\right).$$

The function v_s is p.s.h. in $\mathbb{C}^{k+1}$ and satisfies the usual homogeneity condition. To see this, observe that it is locally bounded above in a neighborhood of $z_0 = 0$, which is pluripolar (Appendix A.2). Hence there is a closed positive current S of bidegree $(1,1)$ defined by $\pi^* S = dd^c v_s$. It has mass 1 and admits a locally bounded potential in $\mathbb{C}^{k+1} \setminus \{0\}$. Thus we can consider the probability measure $\nu_s := S^k$, and since $(f^n)^* \mu = d^{nk} \mu$ and ν_s is supported on K_s, we have

$$\begin{aligned}
s &\leq \int (\lambda_n(a, \varphi) - c) d\nu_s \\
&= \int \lambda_n(a, \varphi) d\nu_s - \int \varphi d\mu \\
&= \int \lambda_n(a, \varphi) d\nu_s - \int \lambda_n(a, \varphi) d\mu \\
&= \lambda_n(a, \varphi)(S^k - T^k) \\
&= \int \lambda_n(a, \varphi)[S - T] \wedge \left[\sum_{i=0}^{k-1} S^i \wedge T^{k-1-i}\right] \\
&= \int \varphi (f^n)^* \frac{[S-T]}{d^n} \wedge \frac{1}{d^{n(k-1)}} (f^n)^* \left[\sum_{i=0}^{k-1} S^i \wedge T^{k-1-i}\right] \\
&:= I.
\end{aligned}$$

But $(f^n)^*(S - T) = dd^c(v_s - G)(f^n)$. It is important to note that $v_s - G$ is a function on $\mathbb{P}^k$. Moreover, the current $(f^n)^* S^i \wedge T^{k-1-i}$ has mass $(d^n)^{k-1}$. Thus

$$I = \int \frac{dd^c \varphi}{d^n} (v_s - G) \circ f^n \wedge \frac{1}{d^{n(k-1)}} \sum_{i=0}^{k-1} (f^n)^* S^i \wedge (f^n)^* T^{k-1-i}.$$

Thus, since there are k terms in the sum and each current in the sum has mass 1 (Corollary A.6.5),

$$I \leq k \frac{|\varphi|_2}{d^n} \sup |v_s - G|. \tag{3}$$

Set

$$m(s) = \sup_B u_s, \quad M = \sup_{|z|=1} G.$$

By the homogeneity of v_s and G,

$$|v_s - G| \leq m(s) + M.$$

By (3) and (2),

$$I \leq k \frac{|\varphi|_2}{d^n} (m(s) + M) \leq k \frac{|\varphi|_2}{d^n} \left(\frac{A(1/2)}{C(K_s, B)} + M\right).$$

It follows that

$$C(K_s, B) \leq \frac{1}{s}|\varphi|_2 \frac{M'}{d^n},$$

where M' is a constant independent of s. This proves the second inequality of the lemma.

Set

$$u = \frac{1}{2}\left(\frac{G(1,z)}{\sup_B |G(1,z)|} + 1\right).$$

Then $0 \leq u \leq 1$ on B and hence, by the definitions of capacity and of μ, there exists a constant $\alpha > 0$ independent of s such that

$$\mu(K_s, B) \leq \alpha C(K_s, B).$$

The lemma follows. □

END OF THE PROOF OF THE THEOREM. Let φ and ψ be test functions. We must estimate

$$I_n := \langle \mu, \psi(f^n)\varphi \rangle - \langle \mu, \psi \rangle \langle \mu, \varphi \rangle.$$

But

$$\langle \mu, \psi(f^n)\varphi \rangle = \frac{1}{d^{nk}} \langle (f^n)^* \mu, \psi(f^n)\varphi \rangle = \langle \mu, \psi \lambda_n(a, \varphi) \rangle.$$

Hence

$$I_n = \langle \mu, \psi[\lambda_n(a, \varphi) - c] \rangle.$$

For fixed p, q with $q > 1$ and $\frac{1}{p} + \frac{1}{q} = 1$,

$$\begin{aligned}
|I_n| &\leq \left(\int \psi^q d\mu\right)^{1/q} \left(\int |\lambda_n(a,\varphi) - c|^p d\mu\right)^{1/p} \\
&\leq \|\psi\|_q \left(\int_0^L p s^{p-1} \mu(|\lambda_n(a,\varphi) - c| \geq s) ds\right)^{1/p} \\
&\leq \|\psi\|_q \left(\int_0^L p s^{p-2} M \frac{|\varphi|_2}{d^n} ds\right)^{1/p} \\
&= O\left(\frac{1}{d^{n/p}}\right) \to 0.
\end{aligned}$$

Thus f is mixing for μ, and we even have an estimate for the rate of convergence of I_n to 0.

For the proof of (ii), we set

$$E_n^\ell = \left\{a \in B\left(0, \frac{1}{2}\right) : \left|\lambda_n(a,\varphi) - \int \varphi \, d\mu\right| > \frac{1}{\ell}\right\}$$

for $n, \ell \in \mathbb{N}$, and define

$$\mathcal{E}(\varphi) := \bigcup_\ell \bigcap_N \bigcup_{n \geq N} E_n^\ell := \bigcup_\ell \mathcal{E}_\ell.$$

For every N, we have

$$C(\mathcal{E}_\ell, B) \leq \sum_{n \geq N} C(E_n^\ell, B) \leq M|\varphi|_2 \sum_{n \geq N} \frac{\ell}{d^n}.$$

Hence $\mathcal{E}_\ell$ has capacity zero in B, and so does $\mathcal{E}(\varphi)$. Taking a dense sequence $\{\varphi_j\}$ and setting $\mathcal{E} = \bigcup_j \mathcal{E}(\varphi_j)$, we see that $\mu_n^a \to \mu$ for $a \in B \setminus \mathcal{E}$. □

Some properties of J_k, the Julia set of order k, can be derived from Theorem 3.6.1. Recall that when $k = 1$, J_1 is the Julia set J.

COROLLARY 3.6.3. *Let $f \in \mathcal{H}_d$. Set $J_k = \operatorname{supp} \mu$.*

(i) *If X is a closed set such that $f^{-1}(X) \subset X$, then either X is pluripolar or $J_k \subset X$.*
(ii) *If U is a union of Fatou components and U is completely invariant, then $J_k \subset \partial U \subset J_1$.*

PROOF. (i) Let $\mathcal{E}$ be the pluripolar set of Theorem 3.6.1. If $a \in X \cap (\mathbb{P}^k \setminus \mathcal{E})$, then $\mu_n^a = \frac{1}{d^{nk}}(f^n)^* \varepsilon_a \to \mu$; hence the support of μ, which is J_k, is contained in X. The other possibility is that $X \subset \mathcal{E}$; it would then be pluripolar.

(ii) The inclusion $\partial U \subset J_1$ is clear because the (f^n) cannot be equicontinuous in a neighborhood of a point in ∂U. Since U is totally invariant, $f^{-1}(\partial U) \subset \partial U$. Thus the result is true unless ∂U is pluripolar. If ∂U is pluripolar, it is contained in the interior of $\overline{U}$. For if a point is a boundary point of U and of the complement of $\overline{U}$, then the boundary cannot be pluripolar at that point because pluripolar sets have connected complements (Properties A.2.4). Let $p \in \partial U$. If ∂U is contained in the interior of $\overline{U}$, then ∂U is open in J_1 because $\overline{U} \cap J_1 = (\overline{U} \setminus U) \cap J_1$. But T has mass on every open subset of J_1 and assigns no mass to pluripolar sets (C.L.N. inequality, Proposition A.6.3). This is the desired contradiction. □

REMARKS 3.6.4.

(i) As we mentioned earlier, Gavosto has shown that there exists $f \in \mathcal{H}_d(\mathbb{P}^2)$ with infinitely many attracting cycles [**42**]. Hence there exist pairwise disjoint, completely invariant open sets $(U_n)_{n \in \mathbb{N}}$, each of which contains J_2 in its boundary.
(ii) There can exist pluripolar closed sets that are completely invariant. The set $X = \{z = 0\}$ is pluripolar, and it is completely invariant for the map $f = [z^d : w^d : t^d]$.

COROLLARY 3.6.5 ([**32**]). *Let $f \in \mathcal{H}_d$. If the interior of J_k is nonempty, then $J_k = \mathbb{P}^k$. More generally, if U is an open set, then $\mathbb{P}^k \setminus \bigcup_{n \geq 0} f^n(U)$ is pluripolar if and only if U intersects J_k.*

PROOF. Suppose $U \cap J_k$ is nonempty. Let $a \notin \bigcup_{n \geq 0} f^n(U)$. The supports of the measures $\mu_n^a = \frac{1}{d^{nk}}(f^n)^* \varepsilon_a$, which are Dirac measures on the preimages of a, do not intersect the open set U. Since μ has mass on U, we must have $a \in \mathcal{E}$, the pluripolar set introduced in Theorem 3.6.1. Hence the closed set $\mathbb{P}^k \setminus \bigcup_{n \geq 0} f^n(U)$ is contained in $\mathcal{E}$. Conversely, the open set $U = \mathbb{P}^k \setminus J_k$ is invariant under f and does not intersect J_k, which is not pluripolar (Proposition A.6.3).

If $U \subset J_k$, then $\cup_{n \geq 0} f^n(U) \subset J_k$. Since the complement of $\cup_n f^n(U)$ is pluripolar and J_k is closed, we have $J_k = \mathbb{P}^k$. □

3.7. Lyapunov Exponents and the Measure μ

Let X be a compact metric space, and let $g : X \to X$ be a continuous map. A point $x \in X$ is wandering if it has a neighborhood V such that $g^n(V) \cap V = \emptyset$ for every $n \geq 1$. Otherwise it is nonwandering. The set $\Omega(g)$ of nonwandering points is closed and invariant; if g is surjective, then $g(\Omega(g)) = \Omega(g)$.

For $f \in \mathcal{H}_d(\mathbb{P}^k)$, it is rather delicate to describe the dynamics of f on $\Omega(f)$, even if we assume that $\Omega(f)$ is compact and hyperbolic. An attempt was made in [**35**], in the case $f \in \mathcal{H}_d(\mathbb{P}^2)$. Moreover, the hypothesis of hyperbolicity is rather hard to verify except when one perturbs very simple examples.

We have seen that the measure μ is mixing for $f \in \mathcal{H}_d(\mathbb{P}^k)$. It follows that $\operatorname{supp}\mu = J_k \subset \Omega(f)$. When J_k is assumed to be hyperbolic, we have seen that the restriction of f to J_k is expanding.

Briend-Duval ([**14, 15**]) have shown that the Lyapunov exponents of f for μ are always positive; that is, the measure μ is hyperbolic. The proof is based on a variant of Pesin's construction [**65**] of stable manifolds and on some estimates from potential theory. We start by recalling the notion of Lyapunov exponents.

Let K be a compact invariant set, $f(K) = K$. We introduced in Section 3.4 the natural extension that conjugates f to a shift on $\widehat{K}$, where

$$\widehat{K} := \{x : x = (x_n)_{n\leq 0},\ f(x_n) = x_{n+1}\}.$$

If $\pi(x) = x_0$, we have a map $\widehat{f} : \widehat{K} \to \widehat{K}$ such that $\pi \circ \widehat{f} = f \circ \pi$. For $(x, \xi) \in T_{\widehat{K}}$, we set $D\widehat{f}(x,\xi) = (\widehat{f}(x), (Df)(x_0)\xi)$.

The measure μ on $K = J_k$ lifts to an invariant measure $\widehat{\mu}$, $\widehat{f}_*\widehat{\mu} = \widehat{\mu}$, which satisfies the relation $\pi_*\widehat{\mu} = \mu$. It is a classical result [**64**] that f is mixing for μ if and only if $\widehat{f}$ is mixing for $\widehat{\mu}$.

Observe that $D\widehat{f}(x)$ is not invertible because $f \in \mathcal{H}_d(\mathbb{P}^k)$ has a critical set. But in Oseledec's theorem [**63**], which we will apply, this is an important hypothesis. In fact, we can easily reduce to this case by means of the C.L.N. inequality (Proposition A.6.3). For if C is the critical set of f, we use the C.L.N. inequality to prove that for the measure μ, which is locally equal to $(dd^c u)^k$ with u a continuous p.s.h. function, $\mu(C) = \mu(f^{-n}(C)) = 0$.

Set $X = J_k \setminus \bigcup_{n\geq 0} f^{-n}(C)$. Then $\mu(X) = 1$. We construct the natural extension $(\widehat{X}, \widehat{f}, \widehat{\mu})$; then $\widehat{\mu}$ is invariant under $\widehat{f}$, and $D\widehat{f}(x,\xi) = (\widehat{f}(x), (Df)(x_0)\xi)$ is invertible.

Let $\|Df(z)\|$ be the norm of the derivative of f at z. One can check that, in a chart, $z \to \log\|Df(z)\|$ is p.s.h. By the C.L.N. inequality, this function is integrable with respect to μ. In this setting, Oseledec's theorem [**63**] reads as follows.

THEOREM 3.7.1. *Let $(\widehat{X}, \widehat{f}, \widehat{\mu})$, $f \in \mathcal{H}_d(\mathbb{P}^k)$ be as above. There exists a Borel set $\widehat{Y} \subset \widehat{X}$ with the following properties.*

(i) $\widehat{\mu}(\widehat{Y}) = 1$.

(ii) *For every $x \in \widehat{Y}$, there is a decomposition of $T_x\mathbb{P}^k \simeq \mathbb{C}^k$ into complex subspaces*

$$\mathbb{C}^k = \bigoplus_{i=1}^{\ell} E_i(x).$$

The decomposition is measurable and is invariant under $D\widehat{f}$.

(iii) *There are real numbers $\lambda_1 > \cdots > \lambda_\ell$ such that, for all $x \in \widehat{Y}$ and $1 \leq i \leq \ell$,*

$$\lim \frac{1}{n} \log |D\widehat{f}^n(x)v| = \lambda_i$$

locally uniformly in $v \in E_i$.

The numbers λ_i are the Lyapunov exponents of f. They are constants because the measure $\widehat{\mu}$ is mixing, hence ergodic. Briend-Duval obtain the following result.

THEOREM 3.7.2 ([**14, 15**]). *Let* $f \in \mathcal{H}_d(\mathbb{P}^k)$. *Let* $\lambda_1 > \cdots > \lambda_\ell$ *be the Lyapunov exponents of* f *with respect to* μ. *Then*

$$\lambda_\ell \geq \frac{\log d}{2}.$$

We refer the reader to Briend's thesis [**14**] and to [**15**] for the proof. We restrict ourselves to showing that $\lambda_\ell \geq 0$ and sketching the idea of the proof.

SKETCH OF THE PROOF. Suppose $\lambda_\ell < 0$. It follows from Pesin theory that for μ-a.e. $z_0 \in J_k$ there exists a holomorphic stable manifold $W^s(z_0)$ passing through z_0. For $z \in W^s(z_0)$, we have $\operatorname{dist}(f^n(z), f^n(z_0)) \to 0$; hence the sequence f^n is equicontinuous, and if u is a local potential for the current T, then $u_{|W^s(z_0)}$ is harmonic. It follows from Corollary A.10.3 that μ assigns no mass to these points. This is the desired contradiction.

Thus we may suppose that $\lambda_\ell \geq 0$. The idea is to construct, for every n and for every point z of a set $E \subset J_k$ satisfying $\mu(E) \geq 1/2$, an approximating stable manifold Δ_z^n on which $\int_{\Delta_z} dd^c u \leq C/d^n$. One estimates the mass of $(dd^c u)^k$ in a Pesin box and counts the number of boxes needed to cover E; this is related to the value of λ_ℓ. The estimate for λ_ℓ follows. The computation requires Hölder continuity of the function G to estimate the mass of μ in a Pesin box. □

As a consequence of Theorem 3.7.2, Briend-Duval obtain the following result.

THEOREM 3.7.3 ([**14, 15**]). *Let* $f \in \mathcal{H}_d(\mathbb{P}^k)$. *Let* μ_n *be the measure*

$$\mu_n := \frac{1}{d^{nk}} \sum_{f^n(y)=y,\ y \text{ repelling}} \varepsilon_y,$$

where ε_y *denotes the Dirac measure at* y. *Then the sequence* (μ_n) *converges weakly to* μ.

APPENDIX

In this appendix we recall the main properties of plurisubharmonic functions and currents that are used throughout this *Panorama*.

A.1. Plurisubharmonic Functions. A Convergence Theorem. Jensen's Formula

DEFINITION A.1.1. Let Ω be an open set in $\mathbb{R}^n$. A function $u : \Omega \to [-\infty, +\infty)$ is subharmonic if

(i) u is not identically $-\infty$ in any component of Ω,
(ii) u is upper semicontinuous (u.s.c.),
(iii) u satisfies the sub-mean value property.

More precisely, for $x_0 \in \Omega$ such that $B(x_0, r) \Subset \Omega$,

$$u(x_0) \leq M(x_0, r) = \int_{|\zeta|=1} u(x_0 + r\zeta) \frac{d\sigma(\zeta)}{c_n},$$

where $c_n = \int_{|\zeta|=1} d\sigma(\zeta)$ and σ is Lebesgue measure on the sphere.

It can be shown that if u is subharmonic, then $u \in L^1_{\text{loc}}(\Omega)$ and $\Delta u \geq 0$ in the sense of distributions. If $v \in L^1_{\text{loc}}(\Omega)$ and $\Delta v \geq 0$, then v is equal almost everywhere to a subharmonic function.

Property (iii) is equivalent to

(iii)′ For every $x_0 \in \Omega$ and every ball $B(x_0, r) \Subset \Omega$,

$$u(x_0) \leq \frac{1}{\tau_n r^n} \int_{B(x_0,r)} u(x) dm(x),$$

where m denotes Lebesgue measure on $\mathbb{R}^n$ and τ_n is the volume of the unit ball in $\mathbb{R}^n$.

THEOREM A.1.2 ([**49**, p. 94]). *Let (v_j) be a sequence of subharmonic functions on a domain $\Omega \subset \mathbb{R}^n$. Suppose that the sequence (v_j) is bounded above on every compact subset of Ω.*

(i) *If (v_j) does not converge to $-\infty$ on compact subsets of Ω, then there is a subsequence (v_{j_k}) that converges in $L^1_{\text{loc}}(\Omega)$ to a subharmonic function.*

(ii) *If v is subharmonic and $v_j \to v$ in L^1_{loc}, then*

$$\limsup_{j\to\infty} \sup_K (v_j - f) \leq \sup_K (v - f)$$

for every compact set $K \subset \Omega$ and every function f that is continuous on K.

Property (ii) is known as the "Hartogs lemma".

The compactness theorem above replaces, in some sense, the normal families argument in the iteration theory of rational functions of one variable.

The Poisson-Jensen formula relates the growth of a subharmonic function to the growth of its Laplacian.

THEOREM A.1.3 (Poisson-Jensen formula). *Let u be a function that is $\mathbb{R}^n$-subharmonic in a neighborhood of the closed ball $\overline{B}(r)$. Then*

$$\int_{r_0}^{r} \frac{dt}{t^{n-1}} \int_{B(t)} \Delta u = \frac{1}{r^{n-1}} \int_{\partial B(r)} u \, d\sigma - \frac{1}{r_0^{n-1}} \int_{\partial B(r_0)} u \, d\sigma.$$

A.2. Plurisubharmonic (p.s.h.) Functions. Pluripolar Sets

DEFINITION A.2.1. Let Ω be an open subset of $\mathbb{C}^n$. Let $u : \Omega \to [-\infty, +\infty)$ be an u.s.c. function. Suppose that u is not identically $-\infty$ in any component of Ω. We say that u is p.s.h. if

$$u(z_0) \leq \frac{1}{2\pi} \int_0^{2\pi} u(z_0 + we^{i\theta}) d\theta$$

for every $w \in \mathbb{C}^n$ such that the disk $z_0 + wD \subset \Omega$, where D denotes the unit disk in $\mathbb{C}$.

We denote the convex cone of p.s.h. functions in Ω by $\mathrm{Psh}(\Omega)$.

P.s.h. functions are $\mathbb{R}^{2n}$-subharmonic. Thus the convergence theorem A.1.2 holds for sequences of p.s.h. functions.

P.s.h. functions have the following properties.

PROPERTIES A.2.2.

(a) If f is holomorphic in Ω, then $\log|f|$ is p.s.h. in Ω.
(b) A function $v \in L^1_{\mathrm{loc}}$ is equal almost everywhere to a p.s.h. function if and only if

$$\sum_{j,k} \frac{\partial^2 v}{\partial z_j \partial \overline{z}_k} w_j \overline{w}_k \geq 0$$

for every $w \in \mathbb{C}^n$. This means that the left-hand side defines a positive measure.
(c) If $g : \Omega \to \Omega'$ is a holomorphic map between two open sets in $\mathbb{C}^n$ and u is p.s.h. in Ω', then $u \circ g$ is either p.s.h. in Ω or $-\infty$. This property allows us to use charts to define p.s.h. functions on holomorphic manifolds.
(d) Let B be the unit ball in $\mathbb{C}^n$. Let $\alpha \in C_0^\infty(B)$ be a function that depends only on $|z|$ and satisfies $\alpha \geq 0$, $\int \alpha = 1$. Set $\alpha_\varepsilon(z) = \dfrac{1}{\varepsilon^{2n}} \alpha(z/\varepsilon)$. If u is p.s.h. in Ω, then the functions $u_\varepsilon = u * \alpha_\varepsilon$ are $\mathcal{C}^\infty$ p.s.h. in $\{z : d(z, \Omega) > \varepsilon\}$. Moreover, $u_\varepsilon \downarrow u$, and $u_\varepsilon \to u$ in $L^1_{\mathrm{loc}}(\Omega)$.
(e) A function u is pluriharmonic if $dd^c u = 0$. It is locally the real part of a holomorphic function. One can also write $u = \log|h|$, where h is a nonvanishing holomorphic function.

DEFINITION A.2.3. A set $\mathcal{E} \subset \Omega \subset \mathbb{C}^n$ is pluripolar if, for every point $z \in \Omega$, there exist a connected neighborhood U of z and $u \in \mathrm{Psh}(U)$ such that $\mathcal{E} \cap U$ is contained in $\{z \in U : u(z) = -\infty\}$.

PROPERTIES A.2.4.

(a) Pluripolar sets have Lebesgue measure zero.
(b) The notion of pluripolar set carries over to holomorphic manifolds.
(c) If $\mathcal{E} \subset \Omega$ is a pluripolar closed set and $v \in \mathrm{Psh}(\Omega \setminus \mathcal{E})$ is bounded above in a neighborhood of every point of $\mathcal{E}$, then v has a unique extension $\widetilde{v} \in \mathrm{Psh}(\Omega)$.
(d) It follows easily from the preceding property that if $\mathcal{E}$ is a pluripolar closed subset of a connected set Ω, then $\Omega \setminus \mathcal{E}$ is connected.

A.3. Currents

See [**66**]. Let M be a $\mathcal{C}^\infty$ manifold of dimension m. We denote by $\mathcal{D}^p(M)$ the space of compactly supported smooth forms of degree p on M. A sequence φ_j approaches 0 in $\mathcal{D}^p(M)$ if the supports of the φ_j lie in a fixed compact set K and the functions φ_j and all their derivatives approach 0 uniformly.

We denote by $\mathcal{D}'_p(M)$ the space of continuous linear forms on $\mathcal{D}^p(M)$. An element S of $\mathcal{D}'_p(M)$ is a current of dimension p (it acts on forms of degree p). It is also said to be of degree $m - p$. Indeed, in a chart we can write

$$S = \sum_{|I|=m-p} S_I dx^I,$$

where $I = (i_1, \dots, i_{m-p})$, $i_1 < \dots < i_{m-p}$, and $dx^I := dx_{i_1} \wedge \dots \wedge dx_{m-p}$.

If $J = (j_1, \dots, j_p)$, $j_1 < \dots < j_p$, then S_I is the distribution defined by

$$(-1)^{\sigma(I,J)} \langle S_I, \varphi_J \rangle = \langle S, \varphi_J dx^J \rangle,$$

where $I \cup J = \{1, \dots, m\}$ and $\sigma(I, J)$ is the signature of the permutation

$$i_1, \dots, i_{m-p}, j_1, \dots, j_p.$$

EXAMPLES AND DEFINITIONS.

(a) Let $g : Y \to M$ be a proper smooth map from a manifold Y of dimension p to M. Then

$$\varphi \longmapsto \int_Y g^* \varphi$$

is a current of dimension p on M.

(b) If S is a current of dimension p on M and α is a smooth form of degree k, we define $S \wedge \alpha$ by the relation

$$\langle S \wedge \alpha, \varphi \rangle := \langle S, \alpha \wedge \varphi \rangle.$$

$S \wedge \alpha$ is a current of dimension $p - k$.

(c) Let α be a form of degree $m-p$ with coefficients in L^1_{loc}. It defines a current of dimension p if we set

$$\langle \alpha, \varphi \rangle := \int_M \alpha \wedge \varphi.$$

(d) Let (S_j) be a sequence in $\mathcal{D}'_p(M)$. We say that $S_j \to S$ if

$$\langle S_j, \varphi \rangle \longrightarrow \langle S, \varphi \rangle$$

for every $\varphi \in \mathcal{D}^p(M)$.

(e) Given a current S of degree q, we define a current dS of degree $q + 1$ by setting

$$\langle dS, \varphi \rangle = (-1)^{q+1} \langle S, d\varphi \rangle, \quad \varphi \in \mathcal{D}^{m-q-1}(M).$$

This is a continuous operation on currents.

(f) Let M and N be manifolds, $\dim M = m$, $\dim N = n$. Let $f : M \to N$ be a smooth map. Let $S \in \mathcal{D}'_p(M)$. Suppose that f restricted to the support of S is a proper map (if X is a compact subset of N, then $f^{-1}(X) \cap \operatorname{supp} S$ is compact). The pushforward $f_* S$ of S is defined by setting

$$\langle f_* S, \varphi \rangle = \langle S, f^* \varphi \rangle, \quad \varphi \in \mathcal{D}^p(N).$$

Then $f_* S \in \mathcal{D}'_p(N)$, and the operation f_* preserves the dimensions of currents. It has the following properties.

(i) $\operatorname{supp} f_* S \subset f(\operatorname{supp} S)$.

(ii) If ψ is a smooth form, then

$$f_*(S \wedge f^* \psi) = f_* S \wedge \psi.$$

(iii) $d(f_* S) = (-1)^{m+n} f_*(dS)$.

Hence d commutes with f_* when $m + n$ is even.

(g) *Pullback of a current under a submersion*

Let $f : M \to N$ be a $\mathcal{C}^\infty$ submersion. Let ψ be a form of class $\mathcal{C}^k$ on M (resp. with coefficients in L^1_{loc}). If f is proper on the support of ψ, we can consider the pushforward $f_* \psi$ of ψ under f. This is a form of class $\mathcal{C}^k$ (resp. in L^1_{loc}) obtained by integrating ψ along the fibers of f. The map $f_* : \mathcal{D}^{m-p}(M) \to \mathcal{D}^{n-p}(N)$ thus constructed is continuous, and

$$\langle f_*\psi, \varphi\rangle = \langle \psi, f^*\varphi\rangle, \quad \varphi \in \mathcal{D}^p(N).$$

When S is a current, f^*S is defined by the formula

$$\langle f^*S, \varphi\rangle = \langle S, f_*\varphi\rangle.$$

When S is a smooth form, f^*S is the usual pullback of the form S. The pullback operation has the following properties.

(1) $\deg f^*S = \deg S$.
(2) If ψ is a smooth form, then $f^*(S \wedge \psi) = f^*S \wedge f^*\psi$.
(3) $d(f^*S) = f^*(dS)$.
(4) $\operatorname{supp}(f^*S) \subset f^{-1}(\operatorname{supp} S)$.
(5) If $S_j \to S$, then $f^*S_j \to f^*S$.

(h) *Currents representable by integration*

Let Ω be an open set in $\mathbb{R}^m$. A current $S = \sum S_I dx^I$ is said to be representable by integration if the distributions S_I are regular measures on Ω. The currents representable by integration extend to continuous linear forms on the space $C_p(\Omega)$ of differentiable forms of degree p that are continuous on Ω. For every open $U \Subset \Omega$, we set

$$M_U(S) = \sup\{|\langle S, \varphi\rangle| : \varphi \in \mathcal{D}_p(U),\ |\varphi(x)| \leq 1,\ x \in U\}.$$

If K is a compact subset of Ω, we set

$$M_K(S) = \liminf_U \{M_U(S) : U \text{ is an open set containing } K\}.$$

The notion of currents representable by integration can be generalized to manifolds by taking charts.

A.4. Positive Currents on Complex Manifolds

See [**54, 59, 22, 77, 75**]. Let Ω be an open set in $\mathbb{C}^m$. We denote by $\mathcal{D}^{p,q}(\Omega)$ the space of compactly supported smooth forms of bidegree (p,q). They can be written

$$\varphi = \sum_{|I|=p,\,|J|=q} \varphi_{IJ} dz_I \wedge d\overline{z}_J,$$

where $dz_I = dz_{i_1} \wedge \cdots \wedge dz_{i_p}$, $d\overline{z}_J = d\overline{z}_{j_1} \wedge \cdots \wedge d\overline{z}_{j_q}$.

We denote by $\mathcal{D}'_{p,q}(\Omega)$ the space of currents of bidimension (p,q), that is, the dual of $\mathcal{D}^{p,q}(\Omega)$. A current $S \in \mathcal{D}'_{p,q}(\Omega)$ can be represented as a differential form of bidegree $(m-p, m-q)$ with distributional coefficients:

$$S = \sum_{\substack{|I'|=m-p \\ |J'|=m-q}} S_{I',J'} dz_{I'} \wedge d\overline{z}_{J'}.$$

The Poincaré d operator can be decomposed into $d = \partial + \overline{\partial}$, where

$$\partial\varphi = \sum \frac{\partial \varphi_{I,J}}{\partial z_k} dz_k \wedge dz_I \wedge d\overline{z}_J, \quad \overline{\partial}\varphi = \sum \frac{\partial \varphi_{I,J}}{\partial \overline{z}_k} d\overline{z}_k \wedge dz_I \wedge d\overline{z}_J.$$

The operator d^c is defined by $d^c := \frac{i}{2\pi}(\overline{\partial} - \partial)$. It is a real operator, in the sense that $\overline{d^c u} = d^c \overline{u}$. We have $dd^c = \frac{1}{\pi}\partial\overline{\partial}$. If S is a current of bidimension (p,p), then

$$\begin{aligned} \langle dS, \varphi \rangle &= -\langle S, d\varphi \rangle \\ \langle d^c S, \varphi \rangle &= -\langle S, d^c \varphi \rangle \\ \langle dd^c S, \varphi \rangle &= \langle S, dd^c \varphi \rangle. \end{aligned}$$

Positive currents. Let S be a current of bidimension (p,p). We say that S is positive if $\langle S, \varphi \rangle \geq 0$ for every test form

$$\varphi = i\alpha_1 \wedge \overline{\alpha}_1 \wedge \cdots \wedge i\alpha_p \wedge \overline{\alpha}_p, \quad \alpha_j \in \mathcal{D}^{1,0}(\Omega).$$

Let S be a current of bidegree $(1,1)$. Then it can be written in the form

$$S = i \sum S_{jk} dz_j \wedge d\overline{z}_k.$$

The current S is positive if for any $w \in \mathbb{C}^n$ the distribution

$$\sum S_{jk} w_j \overline{w}_k \geq 0.$$

It follows that the S_{jk} are measures. In particular, a positive current is representable by integration. A function $u \in L^1_{\text{loc}}(\Omega)$ is equal a.e. to a p.s.h. function if and only if

$$dd^c u = \frac{i}{\pi} \sum \frac{\partial^2 u}{\partial z_i \partial \overline{z}_j} dz_i \wedge d\overline{z}_j \geq 0.$$

PROPERTIES A.4.1.

(a) Every positive current of bidimension (p,p) is representable by integration.

(b) If S is a closed positive current of bidegree $(1,1)$, then for every $z_0 \in \Omega$ there exist a neighborhood U of z_0 and $u \in \mathrm{Psh}(U)$ such that $S = dd^c u$ in U. If $\Omega = \mathbb{C}^m$, then $S = dd^c u$, where $u \in \mathrm{Psh}(\mathbb{C}^m)$. If u_1 and u_2 are potentials of S in U, then $u_1 - u_2$ is a pluriharmonic function.

(c) Let Z be an analytic subset of Ω of pure dimension p. We denote the regular points of Z by $\mathrm{Reg}\, Z$ and set

$$\langle [Z], \varphi \rangle = \int_{\mathrm{Reg}\, Z} \varphi, \quad \varphi \in \mathcal{D}^{p,q}(\Omega).$$

Lelong has shown that $[Z]$ is a closed positive current on Ω, of bidimension (p,p).

(d) Let $\beta = i \sum_j dz_j \wedge d\overline{z}_j$ denote the canonical $(1,1)$ form on $\mathbb{C}^m$. For each positive current $S \in \mathcal{D}'_{p,p}(\Omega)$, we set

$$\sigma_S = \frac{1}{p!} S \wedge \beta^p.$$

This is the trace measure of the current S. It dominates the measures $|S_{I,J}|$.

For the current associated with an analytic set, the trace measure of a region in Z is equal to its volume.

(e) Let E be a closed subset of Ω. Let S be a positive current of bidimension (p,p) on $\Omega \setminus E$. Suppose S has bounded mass in a neighborhood of any point in E. The trivial extension $\widetilde{S}$ of S is the current obtained by extending S by 0 on E. Since the coefficients of S are measures, we simply extend the measures by 0 on E. In general, the current $\widetilde{S}$ need not be closed when S is closed. But Skoda has shown that if E is an analytic set, then $\widetilde{S}$ is a closed positive current.

(f) Let f be a nonzero meromorphic function in Ω. Let $\sum m_j[Z_j]$ be the associated divisor. Then $\log|f| \in L^1_{\text{loc}}(\Omega)$, and we have the Lelong-Poincaré equation

$$dd^c \log|f| = \sum m_j[Z_j].$$

(g) All these concepts carry over to complex-analytic manifolds. Let M be a Kähler manifold with Kähler form ω. If $S \in \mathcal{D}'_{p,p}(M)$ is a positive current and K is a compact subset of M, we set

$$|S|_K := \int_K S \wedge \frac{\omega^p}{p!}.$$

It can be shown that there exists a constant C_K such that

$$\frac{1}{C_K} M_K(S) \leq |S|_K \leq C_K M_K(S)$$

for every positive current S. When M is compact, we set

$$\|S\| = \int S \wedge \frac{\omega^p}{p!}.$$

We do this when $M = \mathbb{P}^k$, for instance.

If a sequence of positive currents on a compact manifold has bounded mass, it has a convergent subsequence.

(h) The following result is a special case of Federer's support theorem [**25**]. Let Λ_α denote the α-dimensional Hausdorff measure Let S be a closed positive current of bidimension (p, p) on $\Omega \subset \mathbb{C}^m$. Let $A = \operatorname{supp} S$.

(i) If $\Lambda_{2p}(A) = 0$, then $S = 0$.
(ii) If A is a complex-analytic subset of dimension p, then $S = \sum c_j[A_j]$, where the A_j are the irreducible components of A and the c_j are positive constants.

(i) *Cauchy-Schwarz inequality*

Let S be a positive current of bidimension $(1,1)$. Let $\varphi, \psi \in \mathcal{D}^{1,0}(\Omega)$. By definition of the positivity of S, $\langle S, i\varphi \wedge \overline{\varphi}\rangle \geq 0$; this gives a Cauchy-Schwarz inequality for the scalar product $(\varphi, \psi) := \langle S, i\varphi \wedge \overline{\psi}\rangle$. We obtain

$$|\langle S, i\varphi \wedge \overline{\psi}\rangle| \leq \langle S, i\varphi \wedge \overline{\varphi}\rangle^{1/2} \langle S, i\psi \wedge \overline{\psi}\rangle^{1/2}.$$

A.5. Currents of Bidegree $(1,1)$ on $\mathbb{P}^k$

See [**32**]. Let $\mathbb{P}^k$ be complex projective space. Let $\pi : \mathbb{C}^{k+1} \setminus \{0\} \to \mathbb{P}^k$ be the canonical projection. We consider $\mathcal{P}$, the convex cone of functions u that are p.s.h. in $\mathbb{C}^{k+1}$ and satisfy, for some $c > 0$,

$$u(\lambda z) = c \log|\lambda| + u(z), \quad z \in \mathbb{C}^{k+1}. \tag{1}$$

Then $u(z) \leq c \log\|z\| + O(1)$. We can normalize the functions in $\mathcal{P}$ by assuming, as we do from now on, that $\int_B u\, dm = 0$, where m denotes Lebesgue measure on the unit ball B. We assign a closed positive current R of bidegree $(1,1)$ to a function $u \in \mathcal{P}$ as follows: let s be a holomorphic section of π on an open set U, and define the current R_s on U by setting $R_s = dd^c(u \circ s)$. This is a closed positive current on U. If s' is another nonzero holomorphic section of π, then $s' = hs$ for some holomorphic h, and

$$\begin{aligned} R_{s'} &= dd^c(u \circ s') = dd^c(u(h \cdot s)) \\ &= cdd^c \log |h| + dd^c u \circ s = R_s. \end{aligned}$$

Thus we have defined an operator $L : \mathcal{P} \to \mathcal{D}'_{k-1,k-1}(\mathbb{P}^k)$, $L(u) = R$, where R is the current that coincides with R_s on U. For instance, the Kähler form ω of $\mathbb{P}^k$ is associated with the function $u(z) = \log |z|$ on $\mathbb{C}^{k+1}$.

Conversely, let R be a closed positive current on $\mathbb{P}^k$, of bidegree $(1,1)$. Since the map $\pi : \mathbb{C}^{k+1} \setminus \{0\} \to \mathbb{P}^k$ is a submersion, we can consider the closed positive current $\pi^* R$ and assign to it its trivial extension to $\mathbb{C}^{k+1}$. Thus there exists a p.s.h. function on $\mathbb{C}^{k+1}$ such that

$$dd^c v = \pi^* R.$$

Define

$$u(z) = \frac{1}{2\pi} \int_0^{2\pi} v(e^{i\theta} z) d\theta;$$

one can check that $dd^c u = \pi^* R$ and that there exists a constant $c > 0$ such that

$$u(\lambda z) = c \log |\lambda| + u(z), \quad \lambda \in \mathbb{C}.$$

The function $v = u - c \log |z|$ is well defined on $\mathbb{P}^k$. Hence

$$\|R\| = \int R \wedge \omega^{k-1} = c \int \omega^k + \int dd^c v \wedge \omega^{k-1} = c \int \omega^k.$$

The correspondence $R \to u$ is continuous and open (see [**32**] for more details); u is called the potential of R. We have proved the following result.

THEOREM A.5.1. *The map L from $\mathcal{P}$ to the space of positive currents of bidegree $(1,1)$ on $\mathbb{P}^k$ is an isomorphism. For every R, there exists a unique $u \in \mathcal{P}$ satisfying $\pi^* R = dd^c u$, $u(\lambda z) = c \log |\lambda| + u(z)$, and $\|R\| = c$.*

EXAMPLE. Let $V = \{P(z) = 0\}$, where P is a homogeneous polynomial of degree d on $\mathbb{C}^{k+1}$. We know that $\pi^*[V] = dd^c \log |P|$. We find that $\|V\| = d$ because

$$\log |P(\lambda z)| = d \log |\lambda| + \log |P(z)|.$$

PROPOSITION A.5.2. *Every closed positive current of bidimension (p,p) on $\mathbb{P}^k$ is a limit, in the sense of currents, of closed positive smooth currents.*

PROOF. Let R be a closed positive current of bidimension (p,p). Let ρ_ε be a smooth approximate identity on the group $U(k)$, which acts transitively on $\mathbb{P}^k$. Let ν denote the Haar measure on $U(k)$. It suffices to set

$$R_\varepsilon = \int_{U(k)} \rho_\varepsilon(g)(g_* R) d\nu(g).$$

One can check that the R_ε have the required properties. □

A.6. Exterior Product of Currents

For this section, we refer the reader to [**11, 22, 75, 33**].

Let S be a closed positive current of bidimension (p,p) in an open set $\Omega \subset \mathbb{C}^m$. Set $\beta = dd^c|z|^2$. We denote by $|S|$ the positive measure defined by

$$|S| = S \wedge \beta^p.$$

If $u \in \operatorname{Psh}(\Omega) \cap L^1_{\text{loc}}(|S|)$, then the product uS is a current on Ω. We define the current $dd^c u \wedge S$ by setting $dd^c u \wedge S := dd^c(uS)$. In particular, when u is bounded, the definition makes sense for all S [**11**].

Proposition A.6.1 ([**33**]). *Let $u \in \operatorname{Psh}(\Omega) \cap L^1_{\text{loc}}(|S|)$. The current $dd^c u \wedge S$ is closed and positive in Ω. If $u_j \to u$ in $L^1_{\text{loc}}(|S|)$, then $dd^c u_j \wedge S \to dd^c u \wedge S$ in the sense of currents.*

Let $u_1, \dots, u_q \in \operatorname{Psh}(\Omega) \cap L^\infty_{\text{loc}}(\Omega)$. The current

$$dd^c u_1 \wedge dd^c u_2 \wedge \cdots \wedge dd^c u_q \wedge S := dd^c(u_1 dd^c u_2 \wedge \cdots \wedge dd^c u_q \wedge S)$$

is defined by recursion. One can show that it is symmetric with respect to $u_1, \dots, u_q$.

Theorem A.6.2 ([**11**]). *Let $u_1, \dots, u_q \in \operatorname{Psh}(\Omega) \cap L^\infty_{\text{loc}}(\Omega)$. Let $u_1^j, \dots, u_q^j$ be decreasing sequences of p.s.h. functions that converge pointwise to $u_1, \dots, u_q$. Then*

(a) $u_1^j dd^c u_2^j \wedge \cdots \wedge dd^c u_q^j \wedge S \longrightarrow u_1 dd^c u_2 \wedge \cdots \wedge dd^c u_q \wedge S$;

(b) $dd^c u_1^j \wedge dd^c u_2^j \wedge \cdots \wedge dd^c u_q^j \wedge S \longrightarrow dd^c u_1 \wedge \cdots \wedge dd^c u_q \wedge S$.

The following variant of the Chern-Levine-Nirenberg inequality [**21**] is quite useful.

Proposition A.6.3 ([**22**]). *Let K, L be compact subsets of Ω, with $L \Subset K$. Let*

$$u_1, \dots, u_q \in \operatorname{Psh}(\Omega) \cap L^\infty(\Omega) \quad \textit{and} \quad V \in \operatorname{Psh}(\Omega).$$

Then

$$\|V dd^c u_1 \wedge \cdots \wedge dd^c u_q\|_L \le C_{K,L} \|V\|_{L^1(K)} \|u_1\|_{L^\infty(K)} \cdots \|u_q\|_{L^\infty(K)}.$$

In particular, the current $dd^c u_1 \wedge \cdots \wedge dd^c u_q$ assigns no mass to pluripolar sets because we can take V to be a function that equals $-\infty$ on the pluripolar set.

It is useful to generalize the definition of the exterior product of closed positive currents in $\mathcal{D}'_{k-1,k-1}(\Omega)$ to the case where the potentials are unbounded (see [**22, 33, 75**]).

For $u \in \operatorname{Psh}(\Omega)$, we define

$$M(u) := \{q : q \in \Omega,\ u \text{ is unbounded in a neighborhood of } q\}.$$

We denote the α-dimensional Hausdorff measure by Λ_α.

Theorem A.6.4 ([**22, 33**]). *Let $u_1, \dots, u_q$ be p.s.h. functions on an open set $\Omega \subset \mathbb{C}^m$. Suppose that*

$$\Lambda_{2(m-\ell+1)}(M(u_{j_1}) \cap \cdots \cap M(u_{j_\ell})) = 0$$

for every choice of indices $j_1 > \cdots > j_\ell$ in $\{1, \dots, q\}$. Then u_1 is locally integrable with respect to the trace measure of $dd^c u_2 \wedge \cdots \wedge dd^c u_q$, and we can set

$$dd^c u_1 \wedge \cdots \wedge dd^c u_q = dd^c(u_1 dd^c u_2 \wedge \cdots \wedge dd^c u_q).$$

If, for each $1 \le k \le q$, there is a sequence u_k^j of functions in $\mathrm{Psh}(\Omega)$ *such that $u_k^j \longrightarrow u_k$ in $L^1_{\mathrm{loc}}(\Omega)$ and $u_k^j \ge u_k$ for all j, then*

$$u_1^j dd^c u_2^j \wedge \cdots \wedge dd^c u_q^j \longrightarrow u_1 dd^c u_2 \wedge \cdots \wedge dd^c u_q$$

in the sense of currents, and

$$dd^c u_1^j \wedge dd^c u_2^j \wedge \cdots \wedge dd^c u_q^j \longrightarrow dd^c u_1 \wedge dd^c u_2 \wedge \cdots \wedge dd^c u_q.$$

The theorem allows us to define the exterior product when the sets $M(u_j)$ are contained in analytic sets A_j satisfying $\mathrm{codim}(A_{j_1} \cap \cdots \cap A_{j_\ell}) \ge \ell$ for every choice of indices. This is a case considered in [**22**].

Since the results are local, they hold for closed positive currents $S_1, \ldots, S_q$ of bidegree $(1,1)$ on a complex manifold M. We can consider the product $S_1 \wedge \cdots \wedge S_q$ provided that the local potentials $u_1, \ldots, u_q$ of $S_1, \ldots, S_q$ satisfy the hypothesis of Theorem A.6.4.

COROLLARY A.6.5 ([**33**]). *Let $S_1, \ldots, S_q$ be closed positive currents of bidegree $(1,1)$ on $\mathbb{P}^k$. Suppose that the local potentials $u_1, \ldots, u_q$ satisfy the hypothesis of Theorem A.6.4. Then $S_1 \wedge \cdots \wedge S_q$ is well defined. It is a closed positive current. Moreover,*

$$\|S_1 \wedge \cdots \wedge S_q\| = \|S_1\| \cdots \|S_q\|.$$

It suffices to observe that $S_j = c_j\omega + dd^c v_j$, where v_j is defined on $\mathbb{P}^k$ and $c_j = \|S_j\|$.

The computation above is a generalization of Bézout's theorem. If $S_j = [V_j]$, where V_j is a hypersurface in $\mathbb{P}^k$, and $\mathrm{codim}(V_{j_1} \cap \cdots \cap V_{j_\ell}) \ge \ell$ for all $j_1 < \cdots < j_m$, then

$$\|\,[V_1 \wedge \cdots \wedge V_q]\,\| = \|V_1\| \ldots \|V_q\|.$$

If $q = k$, we find that the number of points of intersection, counting multiplicity, equals the product of the degrees.

If $u \in \mathrm{Psh}(\Omega)$ and $\Lambda_{2(m-\ell+1)}(M(u)) = 0$, we can define the current $(dd^c u)^\ell$; this means that u is locally integrable with respect to $(dd^c u)^{\ell-1}$ and that $(dd^c u)^\ell$ is a closed positive current. For example, consider $u(z) = \log|z|$ in $\mathbb{C}^m$. One can check that $(dd^c u)^m = 0$ away from the origin. Hence $(dd^c u)^m = c_m \varepsilon_0$, where ε_0 is the Dirac measure at 0 and c_m is a positive constant that can be computed.

A.7. Pullbacks of Closed Positive Currents

See [**60, 29**]. We saw in Section A.3 that we can define the pullback of a current under a submersion. The hypothesis of surjectivity of the differential is imposed because we want the operation to be continuous; that is, we want $S_j \to S$ to imply $f^*S_j \to f^*S$. When we restrict our attention to the class of closed positive currents of a certain type, we can take pullbacks under maps that are not submersions, and even under meromorphic maps. We begin with an immediate consequence of Theorem A.1.2.

PROPOSITION A.7.1. *Let Ω, Ω' be domains in $\mathbb{C}^m$, and let $f : \Omega \to \Omega'$ be a holomorphic map. Suppose that the Jacobian of f is not identically zero. Let $u_j \in L^1_{\mathrm{loc}}(\Omega') \cap \mathrm{Psh}(\Omega')$. If $u_j \to u$ in $L^1_{\mathrm{loc}}(\Omega')$, then $u_j \circ f \to u \circ f$ in $L^1_{\mathrm{loc}}(\Omega)$. More generally, let $f_j : \Omega \to \Omega'$ be a sequence of holomorphic maps that converge to f uniformly on compact sets. Suppose that the Jacobian of f is not identically zero. Then $u_j \circ f_j \to u \circ f$ in $L^1_{\mathrm{loc}}(\Omega)$.*

PROOF. The sequence (u_j) is locally bounded above because it converges in L^1_{loc}, and we can apply the mean-value inequality. Set $v_j = u_j \circ f$. The v_j are locally bounded above in Ω. We want to show that $v_j \to u \circ f$. This follows from Theorem A.1.2. Passing to a subsequence if necessary, we may assume that $v_j \to v$ in L^1_{loc} and a.e. Since f is generically of rank n, we conclude that $v = u \circ f$ a.e., hence everywhere. We may assume that $u_j \circ f_j$ converges to v in L^1_{loc}. After passing to a subsequence, we may assume that u_j converges to u uniformly on sets of arbitrarily large measure. To show that $v = u \circ f$, it suffices to write

$$u_j \circ f_j - u \circ f = (u_j \circ f_j - u \circ f_j) + (u \circ f_j - u \circ f);$$

the first term is small because u_j converges to u uniformly on sets of large measure. □

Let $f : \Omega \to \Omega'$ be holomorphic, of generic rank $m = \dim(\Omega) = \dim(\Omega')$. Let S be a closed positive current of bidegree $(1,1)$ on Ω'. We define the closed positive current f^*S as follows. Given $z_0 \in \Omega$, let $w_0 = f(z_0)$ and $B(w_0, r) \subset \Omega'$. In $B(w_0, r)$ we have $S = dd^c u$ for some p.s.h. u. There exists a positive number r_1 such that $f(B(z_0, r_1)) \subset B(w_0, r)$. We set $f^*S_{|B(z_0,r)} := dd^c u \circ f$. The definition is independent of the potential u because if u_1 and u_2 are two potentials of S, then $dd^c(u_1 - u_2) \circ f = 0$. If $S_j \to S$ in the sense of currents, then we can choose potentials $u_j \to u$ in $L^1_{\text{loc}}(B(w_0, r))$ and apply the preceding proposition.

Pullback under a dominant meromorphic map on $\mathbb{P}^k$. Let S be a closed positive current of bidegree $(1,1)$ on $\mathbb{P}^k$. We know that there exists $u \in \mathcal{P}$ such that $\pi^*S = dd^c u$. Let $f \in \mathcal{M}_d(\mathbb{P}^k)$. Then $f = [F_0 : \cdots : F_k]$, where the F_j are homogeneous polynomials of degree k, and we set $F = (F_0, \ldots, F_k)$. Since f is dominant, the Jacobian of F is not identically zero. We define f^*S by setting

$$\pi^*(f^*S) = dd^c(u \circ F).$$

Then f^*S is a closed positive current on $\mathbb{P}^k$, with potential $u \circ F$. We have seen that if $S_j \to S$, we can choose the potentials u_j and u in such a way that $u_j \to u$ in $L^1_{\text{loc}}(\mathbb{C}^{k+1})$. Hence $f^*S_j \to f^*S$, as desired. If

$$u(\lambda z) = c \log |\lambda| + u(z),$$

then

$$u \circ F(\lambda z) = cd \log |\lambda| + u(F(z)).$$

Thus we have a simple relation between the masses:

$$\|f^*S\| = d\|S\|.$$

The case of currents of bidegree (ℓ, ℓ), $\ell > 1$. The pullback of a closed positive current of bidegree (ℓ, ℓ), $\ell > 1$, was studied by Meo [**60**]; hypotheses on f are necessary. Here we consider the pullbacks of currents of the form $dd^c u_1 \wedge \cdots \wedge dd^c u_q$, where $u_1, \ldots, u_q$ satisfy the hypotheses of Theorem A.6.4.

THEOREM A.7.2. *Let* Ω, Ω' *be open sets in* $\mathbb{C}^m$. *Let* $f : \Omega \to \Omega'$ *be a holomorphic map with discrete fibers. Let* $u_1, \ldots, u_q$ *be functions in* $\mathrm{Psh}(\Omega')$. *Suppose*

$$\Lambda_{2(m-\ell+1)}(M(u_{j_1}) \cap \cdots \cap M(u_{j_\ell})) = 0$$

for every choice of indices $j_1 < \cdots < j_\ell$ *in* $\{1, \ldots, q\}$. *Set*

$$f^*[dd^c u_1 \wedge \cdots \wedge dd^c u_q] := dd^c(u_1 \circ f) \wedge \cdots \wedge dd^c(u_q \circ f).$$

If $u_k^j \to u_k$ *in* $L^1_{\text{loc}}(\Omega')$*, where* $u_k^j \in \text{Psh}(\Omega')$ *and* $u_k^j \geq u_k$ *for all* j*, then*

$$f^* \left[dd^c u_1^j \wedge \cdots \wedge dd^c u_q^j\right] \longrightarrow f^* \left[dd^c u_1 \wedge \cdots \wedge dd^c u_q\right].$$

PROOF. It suffices to observe that

$$\Lambda_{2(m-\ell+1)}(M(u_{j_1} \circ f) \cap \cdots \cap M(u_{j_\ell} \circ f)) = 0.$$

We can then apply Proposition A.7.1 and Theorem A.6.4. □

THEOREM A.7.3. *Let* Ω*,* Ω' *be open sets in* $\mathbb{C}^m$*. Let* $f : \Omega \to \Omega'$ *be a holomorphic map that is generically of rank* m*. For* $u_1, \ldots, u_q \in \text{Psh}(\Omega') \cap L^\infty_{\text{loc}}(\Omega')$*, set*

$$f^*[dd^c u_1 \wedge \cdots \wedge dd^c u_q] := dd^c(u_1 \circ f) \wedge \cdots \wedge dd^c(u_q \circ f).$$

If $u_k^j \to u_k$ *in* $L^1_{\text{loc}}(\Omega')$*, where* $u_k^j \in \text{Psh}(\Omega')$ *and* $u_k^j \geq u_k$ *for all* j*, then*

$$f^* \left[dd^c u_1^j \wedge \cdots \wedge dd^c u_q^j\right] \longrightarrow f^* \left[dd^c u_1 \wedge \cdots \wedge dd^c u_q\right].$$

PROOF. The functions $u_\ell \circ f$ are bounded and p.s.h. It suffices to apply Theorem A.6.2 and Proposition A.7.1. □

A.8. Logarithmic Capacity in $\mathbb{C}$

See [**83**]. We denote by $\mathcal{L}(\mathbb{C})$ the convex set of subharmonic functions u in $\mathbb{C}$ that satisfy

$$u(z) \leq \log^+ |z| + O(1).$$

For every compact set $K \subset \mathbb{C}$, we define the Green's function of K by

$$g(z) = (\sup\{u(z) : u \leq 0 \text{ on } K,\ u \in \mathcal{L}(\mathbb{C})\})^*.$$

The asterisk $*$ denotes the u.s.c. regularization.

When g is not identically $+\infty$, it is harmonic on $\mathbb{C} \setminus K$. At infinity, it has the expansion

$$g(z) = \log |z| + V + O(1).$$

The constant V is called the Robin constant of K.

The Green's function of the disk $D(0, R)$ is $g(z) = \log^+ |z|/R$. We define the logarithmic capacity of K by setting

$$\text{cap}(K) := e^{-V}.$$

If E is a Borel set, we define

$$\text{cap}(E) = \sup\{\text{cap}(K) : K \subset E,\ K \text{ is compact}\}.$$

The following proposition lets us estimate the area of the set where a function in $\mathcal{L}(\mathbb{C})$ is small.

PROPOSITION A.8.1. *Let* $v \in \mathcal{L}(\mathbb{C})$*. Suppose*

$$\limsup_{|z| \to \infty} (v(z) - \log |z|) = A > -\infty.$$

For $m > 0$*, set*

$$E_m = \{z : |z| \leq 1,\ v(z) < -m\}.$$

Then

$$\text{cap}(E_m) \leq e^{-A} e^{-m}$$

and

$$\text{area}(E_m) \leq \pi e (\text{cap}(E_m))^2 \leq \pi e^{-2A} e^{-2m}.$$

PROOF. We have $v + m \leq 0$ on E_m. Hence, by the definition of g,

$$v(z) + m \leq \log|z| + V_m + O(1)$$

for sufficiently large $|z|$, where V_m denotes the Robin constant of E_m. Subtracting $\log|z|$, we obtain $A + m \leq V_m$ and $\text{cap}(E_m) = e^{-V_m} \leq e^{-A}e^{-m}$.

The area estimate via the capacity is given in Tsuji [**83**, p. 85]. □

PROPOSITION A.8.2 ([**83**, p. 85]). *Let U be a continuum in $\mathbb{C}$, with diameter $\delta(E)$. Then*

$$\delta(E) \leq 4 \cdot \text{cap}(E).$$

Thus the sets where a function in $\mathcal{L}(\mathbb{C})$ is small cannot contain disks that are too large.

REMARK A.8.3. If $u \in \text{Psh}(\mathbb{C}^k)$ is of logarithmic growth, one can estimate the volume of $\{z : |z| \leq 1,\ u(z) \leq -m\}$ by applying Proposition A.8.2 to each slice, then using Fubini's theorem.

A.9. Bedford-Taylor Capacity and the Siciak Extremal Function

See [**11, 54**]. In [**11**], Bedford-Taylor introduced a capacity related to plurisubharmonic functions. Let Ω be a bounded open set in $\mathbb{C}^m$.

DEFINITION A.9.1 ([**11**]). For each Borel set $E \subset \Omega$, we define

$$C(E, \Omega) = \sup\left\{\int_E (dd^c u)^k : u \in \text{Psh}(\Omega),\ 0 \leq u \leq 1\right\}.$$

The C.L.N. inequality (Proposition A.6.3) shows that $C(E, \Omega) < \infty$ if $E \Subset \Omega$ and that $C(E, \Omega) = 0$ if E is pluripolar. It is clear that C has the following properties.

(i) If $E_1 \subset E_2$, then $C(E_1, \Omega) \leq C(E_2, \Omega)$.
(ii) $C(\cup E_j, \Omega) \leq \sum_j C(E_j, \Omega)$.
(iii) If $E_j \nearrow C$, then $C(E_j, \Omega) \nearrow C(E, \Omega)$.
(iv) $C(E, \Omega) = \sup\{C(K, \Omega) : K \subset \Omega,\ K \text{ is compact}\}$.

The capacity C also has the following properties, whose proofs are less immediate [**11**].

(a) If $\Omega_1 \subset \Omega_2 \Subset \mathbb{C}^m$, then $C(E, \Omega_1) \geq C(E, \Omega_2)$ for every Borel set $E \subset \Omega_1$.
(b) Let $\omega \Subset \Omega_1$. There exists a constant $A > 0$ such that $C(E, \Omega_1) \leq AC(E, \Omega_2)$ for every Borel set $E \subset \omega$.
(c) Let E be a Borel subset of Ω. If $C(E, \Omega) = 0$, then E is pluripolar.
(d) Let $K \Subset B$, the unit ball in $\mathbb{C}^m$. Set

$$u_{K,B} = \sup\{v : v \in \text{Psh}(B),\ v \leq 0 \text{ on } K,\ 0 \leq v \leq 1\}.$$

Let $u^*_{K,B}$ be the upper semicontinuous regularization of $u_{K,B}$. Then

$$C(K, B) = \int_B (dd^c u^*_{K,B})^m.$$

Siciak (see [**54**]) introduced a generalization to $\mathbb{C}^m$ of the Green's function. Set $\mathcal{L}(\mathbb{C}^m) = \{u \in \text{Psh}(\mathbb{C}^m) : u(z) \leq \log^+|z| + O(1)\}$. If E is a bounded subset of $\mathbb{C}^m$, set

$$U_E(z) = \sup\{u(z) : u \in \mathcal{L}(\mathbb{C}^m),\ u \leq 0 \text{ on } E\}.$$

Let U_E^* denote the u.s.c. regularization of U_E. This is the Siciak function of E. If U_E^* is not identically $+\infty$, then $U_E^* \in \mathcal{L}$ and, at infinity, satisfies

$$U_E^*(z) = \log|z| + O(1).$$

Moreover, $(dd^c U_E^*)^m = 0$ on $\mathbb{C}^m \setminus \overline{E}$.

If $E = B(0,r)$, then $U_E^*(z) = \log^+(|z|/r)$.

THEOREM A.9.2. *The function U_E^* is not identically $+\infty$ if and only if E is not pluripolar. Then $(dd^c u_E^*)^m$ is a probability measure with support in $\overline{E}$.*

Suppose E is not pluripolar and $E \subset B$, the unit ball. Set $m(E) = \sup_B U_B^*$. One can check that

$$\log^+|z| \le U_E^*(z) \le m(E) + \log^+|z|.$$

Alexander and Taylor proved a relationship between $m(E)$ and $C(E,B)$.

THEOREM A.9.3 ([**1**]). *For $r > 1$, there exists a constant $A(r)$ such that*

$$C(K,B)^{-1/k} \le m(K) \le \frac{A(r)}{C(K,B)}$$

for every compact set $K \subset B(0,r)$.

A.10. The Dirichlet Problem for the Monge-Ampère Equation

Let B be the unit ball in $\mathbb{C}^n$. Let $u \in \mathcal{C}(\partial B)$. We define

$$\widehat{u} = \sup\{v : v \in \mathrm{Psh}(B) \cap \mathcal{C}(\overline{B}),\ v \le u \text{ on } \partial B\}. \tag{1}$$

The function $\widehat{u}$ is p.s.h. in B, and it is easy to check that $\widehat{u}_{|\partial B} = u$. This is a construction that generalizes Perron's method of solving the Dirichlet problem. If $m = 1$, then $\widehat{u}$ is the solution of the Dirichlet problem for the Laplacian.

THEOREM A.10.1 ([**10**]). *Let $u \in \mathcal{C}(\partial B)$. The function $\widehat{u} \in \mathrm{Psh}(B) \cap \mathcal{C}(\overline{B})$; it satisfies the Monge-Ampère equation $(dd^c\widehat{u})^m = 0$ in B. If u is Hölder continuous of order $\alpha > 0$, then $\widehat{u}$ is Hölder continuous of order $\alpha/2$.*

Given two function u_1, $u_2 \in \mathrm{Psh}(B)$ satisfying $u_1 \le u_2$, it is useful to be able to estimate $(dd^c u_1)^m$ on certain sets; the following result of Briend-Duval ([**14, 15**]) makes this possible.

THEOREM A.10.2 ([**14, 15**]). *Let u_1, $u_2 \in \mathrm{Psh}(B) \cap L^\infty(B)$. Suppose $u_1 \le u_2$. For $\varepsilon > 0$, set*

$$\Sigma_\varepsilon := \{z : |z| \le 1/2,\ u_2(z) \le u_1(z) + \varepsilon\}.$$

There exists a constant C (independent of u_1 and u_2) such that

$$(dd^c u_1)^m(\Sigma_\varepsilon) \le \int_B (dd^c u_2)^m + C\varepsilon(\|u_2\|_\infty + 1)^{m-2} \cdot \|u_2\|_\infty.$$

In particular, if $(dd^c u_2)^m = 0$, then

$$(dd^c u_1)^m(\Sigma_\varepsilon) \le C\varepsilon(\|u_2\|_\infty + 1)^{m-2}\|u_2\|_\infty.$$

PROOF. Set $u_3 = \dfrac{\log^+(2|z|)}{\log(3/2)}$. The function u_3 is p.s.h., $u_3 = 0$ if $|z| = 1/2$, and $u_3(z) = 1$ if $|z| = 3/4$.

Define $v := u_2 + 3\varepsilon u_3 - 2\varepsilon$. We have $\Sigma_\varepsilon \subset \{v < u_1\} \Subset B$. Indeed, if $z \in \Sigma_\varepsilon$, then

$$v(z) \leq u_1(z) + \varepsilon - 2\varepsilon = u_1(z) - \varepsilon.$$

Moreover, $v \geq u_2 + \varepsilon > u_1$ for $|z| \geq 3/4$.

Let $\chi \in \mathcal{C}_0^\infty(B)$, with $0 \leq \chi \leq 1$ and $\chi = 1$ in a neighborhood of $|z| \leq 3/4$. Set $\theta = \max(u_1, v)$. Observe that $\theta = u_1$ near Σ_ε and $\theta = v$ for $|z| \geq 3/4$, in particular on the support of $d\chi$. Integrating by parts, we have

$$\begin{aligned} I_\varepsilon := \int_{\Sigma_\varepsilon} (dd^c u_1)^m \leq \int \chi (dd^c \theta)^m &= -\int d\chi \wedge d^c\theta \wedge (dd^c\theta)^{m-1} \\ &= -\int d\chi \wedge d^c v \wedge (dd^c v)^{m-1} \\ &= \int \chi (dd^c v)^m. \end{aligned}$$

Expanding $(dd^c v)^m$, we have

$$I_\varepsilon \leq \int \chi \left[(dd^c u_2)^m + \varepsilon C_1 \sum_{j=1}^{m-1} (dd^c u_2)^j \wedge (dd^c u_3)^{m-j} \right].$$

Since χ has compact support in B, applying the C.L.N. inequality gives

$$\begin{aligned} I_\varepsilon &\leq \int_B (dd^c u_2)^m + C_2 \varepsilon \sum_{j=1}^{m-1} \|u_2\|_\infty^j \|u_3\|_\infty^{m-j} \\ &\leq \int_B (dd^c u_2)^m + C\varepsilon \|u_2\|_\infty (1 + \|u_2\|_\infty)^{m-2}. \end{aligned}$$

□

For an u.s.c. function u, we consider the set $E(u)$ of points through which there passes a holomorphic disk on which u is harmonic. More precisely, $E(u)$ is the set of points $z_0 \in B$ for which there exists a nonconstant holomorphic $h : D \to B$ such that $h(0) = z_0$ and $u \circ h$ is harmonic on D.

COROLLARY A.10.3. *Let $u \in \mathrm{Psh}(B) \cap \mathcal{C}(\overline{B})$. Then*

$$(dd^c u)^m (E(u)) = 0.$$

PROOF. Observe that the estimate of Theorem A.10.1 is invariant under homotheties. If the functions u_j are defined in $B(0, r)$, it suffices to apply the theorem to the functions $z \to u_j(rz)$ in B.

Let $E_{a,r}$ denote the set of points $z_0 \in B(a, r)$ for which there exists a nonconstant holomorphic function $h : D \to B(a, r)$ such that $h(0) = z_0$, $h(e^{i\theta}) \in \partial B(a, r)$, and $u \circ h$ is harmonic in D.

If $v \leq u$ on $\partial B(a, r)$ and $v \in \mathrm{Psh}(B(a, r))$, then $v \circ h \leq u \circ h$. Hence $\widehat{u}_{a,r}$, the solution of the Dirichlet problem with boundary data u, satisfies $\widehat{u}_{a,r}(z_0) = u(z_0)$. Taking $u_1 = u$ in Theorem A.10.2 and $u_2 = \widehat{u}_{a,r}$ gives $(dd^c u)^m (E_{a,r}) = 0$.

Choosing a with rational coordinates and r rational, we conclude that

$$(dd^c u)^m \left[\bigcup_{a,r} E_{a,r}\right] = 0.$$

But for any $z_0 \in E(u)$, we can find a, r such that $z_0 \in E_{a,r}$. $\square$

Bibliography

[1] H. J. Alexander and B. A. Taylor, *Comparison of two capacities in* $\mathbb{C}^n$, Math. Z. **186** (1984), 407–417.

[2] H. Bass, E. H. Connell, and D. Wright, *The Jacobian conjecture: reduction of degree and formal expansion of the inverse*, Bull. Amer. Math. Soc. (N.S.) **7** (1982), 287–330.

[3] E. Bedford, M. Lyubich, and J. Smillie, *Distribution of periodic points of polynomial diffeomorphisms of* $\mathbb{C}^2$, Invent. Math. **114** (1993), 277–288.

[4] ———, *Polynomial diffeomorphisms of* $\mathbb{C}^2$*. IV. The measure of maximal entropy and laminar currents*, Invent. Math. **112** (1993), 77–125.

[5] E. Bedford and V. Pambuccian, *Dynamics of shift-like polynomial diffeomorphisms of* $\mathbb{C}^N$, Conform. Geom. Dyn. **2** (1998), 45–55 (electronic).

[6] E. Bedford and J. Smillie, *Fatou-Bieberbach domains arising from polynomial automorphisms*, Indiana Univ. Math. J. **40** (1991), 789–792.

[7] ———, *Polynomial diffeomorphisms of* $\mathbb{C}^2$*: currents, equilibrium measure and hyperbolicity*, Invent. Math. **103** (1991), 69–99.

[8] ———, *Polynomial diffeomorphisms of* $\mathbb{C}^2$*. II. Stable manifolds and recurrence*, J. Amer. Math. Soc. **4** (1991), 657–679.

[9] ———, *Polynomial diffeomorphisms of* $\mathbb{C}^2$*. III. Ergodicity, exponents and entropy of the equilibrium measure*, Math. Ann. **294** (1992), 395–420.

[10] E. Bedford and B. A. Taylor, *The Dirichlet problem for a complex Monge-Ampère equation*, Invent. Math. **37** (1976), 1–44.

[11] ———, *A new capacity for plurisubharmonic functions*, Acta Math. **149** (1982), 1–40.

[12] M. Benedicks and L. Carleson, *The dynamics of the Hénon map*, Ann. of Math. (2) **133** (1991), 73–169.

[13] A. Bonifant, *Degrees of nonlinearity in higher dimensional complex dynamics*, Thesis, Mexico, 1996.

[14] J.-Y. Briend, *Exposants de Liapounoff et points périodiques d'endomorphismes holomorphes de* $\mathbb{CP}^k$, Thesis, Toulouse, 1997.

[15] J.-Y. Briend and J. Duval, *Exposants de Liapounoff et distribution des points périodiques d'un endomorphisme de* $\mathbb{CP}^k$, Acta Math. **182** (1999), 143–157.

[16] H. Brolin, *Invariant sets under iteration of analytic functions*, Ark. Math. **6** (1965), 103–144.

[17] G. T. Buzzard, *Infinitely many periodic attractors for holomorphic maps of* 2 *variables*, Ann. of Math. (2) **145** (1997), 389–417.

[18] L. Carleson and T. W. Gamelin, *Complex dynamics*, Springer-Verlag, New York, 1993.

[19] J. A. Carlson, *A moving lemma for the transcendental Bezout problem*, Ann. of Math. (2) **103** (1976), 305–330.

[20] U. Cegrell, *Removable singularities for plurisubharmonic functions and related problems*, Proc. London Math. Soc. (3) **36** (1978), 310–336.

[21] S. S. Chern, H. I. Levine, and L. Nirenberg, *Intrinsic norms on a complex manifold*, Global Analysis (Papers in Honor of K. Kodaira), Univ. Tokyo Press, Tokyo, 1969, pp. 119–139.

[22] J. -P. Demailly, *Monge-Ampère operators, Lelong numbers and intersection theory*, Complex analysis and geometry, Plenum Press, New York, 1993, pp. 115–193.

[23] J. Diller, *Dynamical Green's functions for rational and birational maps of* $\mathbb{P}^2$, preprint, 1995.

[24] C. Favre, *Points périodiques d'applications birationnelles de* $\mathbb{P}^2$, Ann. Inst. Fourier (Grenoble) **48** (1998), 999–1023.

[25] H. Federer, *Geometric measure theory*, Grundlehren der mathematischen Wissenschaften, Band 153, Springer-Verlag New York Inc., New York, 1969.

[26] J. E. Fornæss, *Dynamics in several complex variables*, CBMS Regional Conference Series in Mathematics, vol. 87, Amer. Math. Soc., Providence, RI, 1996.

[27] J. E. Fornæss and N. Sibony, *Complex Hénon mappings in* $\mathbb{C}^2$ *and Fatou-Bieberbach domains*, Duke Math. J. **65** (1992), 345–380.

[28] ———, *Critically finite rational maps on* $\mathbb{P}^2$, The Madison Symposium on Complex Analysis (Madison, WI, 1991), Contemp. Math., vol. 137, Amer. Math. Soc., Providence, RI, 1992, pp. 245–260.

[29] ———, *Complex dynamics in higher dimensions*, Complex potential theory (Montreal, PQ, 1993), NATO Adv. Sci. Inst. Ser. C Math. Phys. Sci., vol. 439, Kluwer Acad. Publ., Dordrecht, 1994, pp. 131–186.

[30] ———, *Complex dynamics in higher dimension. I*, Complex analytic methods in dynamical systems (Rio de Janeiro, 1992), Astérisque, Soc. Math. France, 1994, pp. 201–231.

[31] ———, *Classification of recurrent domains for some holomorphic maps*, Math. Ann. **301** (1995), 813–820.

[32] J. E. Fornaess and N. Sibony, *Complex dynamics in higher dimension. II*, Modern methods in complex analysis (Princeton, NJ, 1992), Ann. of Math. Stud., vol. 137, Princeton Univ. Press, Princeton, NJ, 1995, pp. 135–182.

[33] J. E. Fornæss and N. Sibony, *Oka's inequality for currents and applications*, Math. Ann. **301** (1995), 399–419.

[34] ———, *Holomorphic symplectomorphisms in* $\mathbb{C}^{2p}$, Duke Math. J. **82** (1996), 309–317.

[35] ———, *Hyperbolic maps on* $\mathbb{P}^2$, Math. Ann. **311** (1998), 305–333.

[36] J. E. Fornæss and B. Weickert, *Attractors in* $\mathbb{P}^2$, Several complex variables (Berkeley, CA, 1995–1996), MSRI Publ., vol. 37, Cambridge Univ. Press, Cambridge, 1999, pp. 297–307.

[37] J. E. Fornæss and H. Wu, *Classification of degree* 2 *polynomial automorphisms of* $\mathbb{C}^3$, Publ. Mat. **42** (1998), 195–210.

[38] S. Friedland, *Entropy of polynomial and rational maps*, Ann. of Math. (2) **133** (1991), 359–368.

[39] ———, *Entropy of algebraic maps*, Proceedings of the Conference in Honor of Jean-Pierre Kahane (Orsay, 1993). J. Fourier Anal. Appl. 1995, Special Issue, 215–228.

[40] S. Friedland and J. Milnor, *Dynamical properties of plane polynomial automorphisms*, Ergodic Theory Dynam. Systems **9** (1989), 67–99.

[41] J. -Ph. Furter, *On the degree of iterates of automorphisms of the affine plane*, Manuscripta Math. **98** (1999), 183–193.

[42] E. A. Gavosto, *Attracting basins in* $\mathbb{P}^2$, J. Geom. Anal. **8** (1998), 433–440.

[43] M. Gromov, *Entropy, homology and semi-algebraic geometry*, Séminaire Bourbaki, vol. 1985/86, Astérisque, vol. 145-146, Soc. Math. France, 1987, pp. 225–240.

[44] L. Gruman, *La géométrie globale des ensembles analytiques dans* $\mathbb{C}^n$, Séminaire Lelong-Skoda, Lecture Notes in Math., vol. 822, Springer, Berlin, 1980, pp. 90–99.

[45] R. C. Gunning, *Introduction to holomorphic functions of several variables. Vol. II*, Wadsworth & Brooks/Cole Advanced Books & Software, Monterey, CA, 1990.

[46] M. Hakim, *Transformations tangent to the identity. stable pieces of manifolds*, preprint, Orsay, 1997.

[47] ———, *Analytic transformations of* $(\mathbb{C}^p, 0)$ *tangent to the identity*, Duke Math. J. **92** (1998), 403–428.

[48] R. Harvey and J. Polking, *Extending analytic objects*, Comm. Pure Appl. Math. **28** (1975), 701–727.

[49] L. Hörmander, *The analysis of linear partial differential operators. I*, Grundlehren der mathematischen Wissenschaften, vol. 256, Springer-Verlag, Berlin, 1983.

[50] J. H. Hubbard, *The Hénon mapping in the complex domain*, Chaotic dynamics and fractals (Atlanta, Ga., 1985), Academic Press, Orlando, FL, 1986, pp. 101–111.

[51] J. H. Hubbard and Ralph W. Oberste-Vorth, *Hénon mappings in the complex domain. I. The global topology of dynamical space*, Inst. Hautes Études Sci. Publ. Math. **79** (1994), 5–46.

[52] M. Jonsson, *Some properties of* 2*-critically finite holomorphic maps of* $\mathbb{P}^2$, Ergodic Theory Dynam. Systems **18** (1998), 171–187.

[53] A. Katok and B. Hasselblatt, *Introduction to the modern theory of dynamical systems*, Cambridge University Press, Cambridge, 1995.

[54] M. Klimek, *Pluripotential theory*, Oxford University Press, New York, 1991.

[55] S. Kobayashi, *Hyperbolic manifolds and holomorphic mappings*, Marcel Dekker Inc., New York, 1970.
[56] M. Kosek, *Hölder continuity property of filled-in Julia sets in* $\mathbb{C}^n$, Proc. Amer. Math. Soc. **125** (1997), 2029–2032.
[57] S. Lang, *Introduction to complex hyperbolic spaces*, Springer-Verlag, New York, 1987.
[58] R. Lazarsfeld, *Some applications of the theory of positive vector bundles*, Complete intersections (Acireale, 1983), Lecture Notes in Math., vol. 1092, Springer, Berlin, 1984, pp. 29–61.
[59] P. Lelong, *Fonctions plurisousharmoniques et formes différentielles positives*, Gordon & Breach, Paris, 1968.
[60] M. Meo, *Transformations intégrales pour les courants positifs fermés et théorie de l'intersection*, Thesis, Grenoble, 1996.
[61] R. E. Molzon, B. Shiffman, and N. Sibony, *Average growth estimates for hyperplane sections of entire analytic sets*, Math. Ann. **257** (1981), 43–59.
[62] S. E. Newhouse, *Diffeomorphisms with infinitely many sinks*, Topology **13** (1974), 9–18.
[63] V. I. Oseledec, *A multiplicative ergodic theorem. Ljapunov characteristic numbers for dynamical systems*, Trans. Moscow Math. Soc. **19** (1968), 197–231.
[64] W. Parry, *Entropy and generators in ergodic theory*, W. A. Benjamin, Inc., New York-Amsterdam, 1969.
[65] Y. Pesin, *Characteristic Ljapunov exponents and smooth ergodic theory*, Russian Math. Surveys **32** (1977), 55–114.
[66] G. de Rham, *Variétés différentiables. Formes, courants, formes harmoniques*, Hermann et Cie, Paris, 1955.
[67] D. Ruelle, *Elements of differentiable dynamics and bifurcation theory*, Academic Press Inc., Boston, MA, 1989.
[68] D. Ruelle and D. Sullivan, *Currents, flows and diffeomorphisms*, Topology **14** (1975), 319–327.
[69] A. Russakovskii and B. Shiffman, *Value distribution for sequences of rational mappings and complex dynamics*, Indiana Univ. Math. J. **46** (1997), 897–932.
[70] A. Russakovskii and M. Sodin, *Equidistribution for sequences of polynomial mappings*, Indiana Univ. Math. J. **44** (1995), 851–882.
[71] B. Shiffman, *On the removal of singularities of analytic sets*, Michigan Math. J. **15** (1968), 111–120.
[72] M. Shub, A. Fathi, and R. Langevin, *Global stability of dynamical systems*, Springer-Verlag, New York, 1987, Translated from the French by J. Christy.
[73] M. Shub and D. Sullivan, *A remark on the Lefschetz fixed point formula for differentiable maps*, Topology **13** (1974), 189–191.
[74] N. Sibony, Talks at Orsay, unpublished (1981), Course at UCLA (1984).
[75] ———, *Quelques problèmes de prolongement de courants en analyse complexe*, Duke Math. J. **52** (1985), 157–197.
[76] Y. T. Siu, *Analyticity of sets associated to Lelong numbers and the extension of closed positive currents*, Invent. Math. **27** (1974), 53–156.
[77] H. Skoda, *Prolongement des courants, positifs, fermés de masse finie*, Invent. Math. **66** (1982), 361–376.
[78] J. Smillie, *The entropy of polynomial diffeomorphisms of* $\mathbb{C}^2$, Ergodic Theory Dynam. Systems **10** (1990), 823–827.
[79] S. Sternberg, *Local contractions and a theorem of Poincaré*, Amer. J. Math. **79** (1957), 809–824.
[80] D. Sullivan, *Quasiconformal homeomorphisms and dynamics. I. Solution of the Fatou-Julia problem on wandering domains*, Ann. of Math. (2) **122** (1985), 401–418.
[81] P. Tortrat, *Aspects potentialistes de l'itération des polynômes*, Séminaire de Théorie du Potentiel, Paris, Lecture Notes in Math., vol. 1235, Springer, Berlin, 1987, pp. 195–209.
[82] H. Tsuji, *A generalization of Schwarz lemma*, Math. Ann. **256** (1981), 387–390.
[83] M. Tsuji, *Potential theory in modern function theory*, Chelsea Publishing Co., New York, 1975, Reprinting of the 1959 original.
[84] T. Ueda, *Fatou sets in complex dynamics on projective spaces*, J. Math. Soc. Japan **46** (1994), 545–555.
[85] ———, *Critical orbits of holomorphic maps on projective spaces*, J. Geom. Anal. **8** (1998), 319–334.

[86] P. Walters, *An introduction to ergodic theory*, Springer-Verlag, New York, 1982.
[87] Y. Yomdin, *Volume growth and entropy*, Israel J. Math. **57** (1987), 285–300.

DYNAMICS OF QUADRATIC POLYNOMIALS

Jean-Christophe Yoccoz

Notes by Marguerite Flexor

Abstract: The study of the dynamical system defined by a holomorphic function consists of analyzing the behavior of the orbits under iteration of the function and describing their distribution in the plane. The simplest nontrivial case is that of quadratic polynomials in one complex variable, which are the subject of this article. The article is divided into three sections. The first is an overview of the elementary theory of the dynamical plane of a polynomial and the parameter space of the family of quadratic polynomials. The second is devoted to hyperbolic aspects and is centered around Jakobson's theorem. The third describes quasiperiodic aspects, which are related to problems of small divisors.

Introduction

The study of the dynamical system defined by a holomorphic function consists of analyzing the behavior of the orbits under iteration of the function and describing their distribution in the plane. This program was initiated in 1919 by G. Julia and P. Fatou. After a long period of dormancy, it underwent a revival of activity in 1942 with C. Siegel and again during the past twenty years.

The simplest nontrivial case is that of quadratic polynomials in one complex variable, which are the subject of this article. Although the function considered is particularly simple, the associated dynamics can turn out to be quite complicated, and can be drastically altered by a slight perturbation of the function.

Moreover, the results obtained can be reinvested in the study of the dynamics of polynomials of higher degree, rational maps, and, more generally, holomorphic functions with simple critical points.

This article consists of three sections. The first is a very general overview of the elementary theory in the dynamical plane, for the analysis of a single polynomial, and the elementary theory in the parameter space, for the analysis of the family of all quadratic polynomials.

Figure 0 represents the subset of the parameter plane that is classically called the *Mandelbrot set.* (Here we have a single complex parameter, and the parameter plane is $\mathbb{C}$.)

The Mandelbrot set consists of those parameters for which the set of points with bounded orbit is connected. The figure shows some of the phenomena that we will touch on in this section, corresponding to various values of the parameter. We

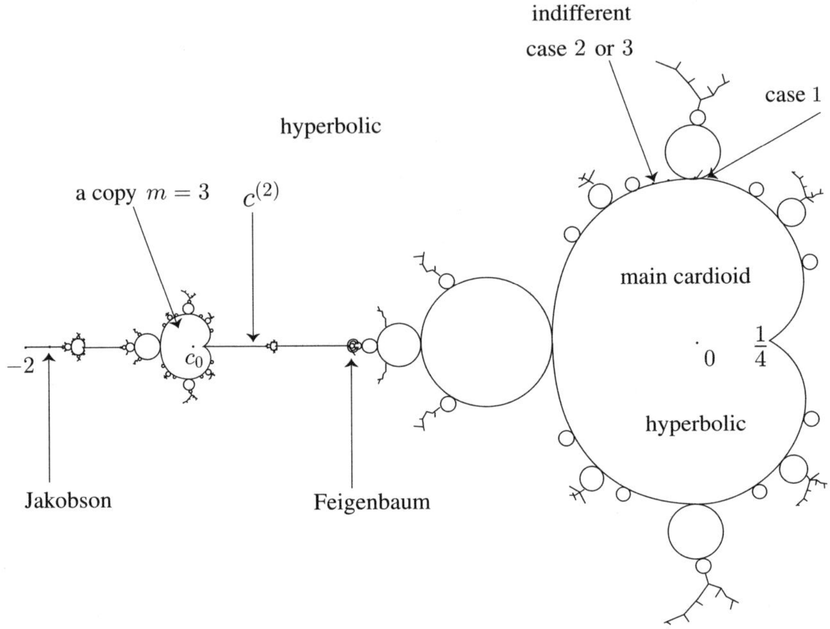

A schematic drawing of the Mandelbrot set

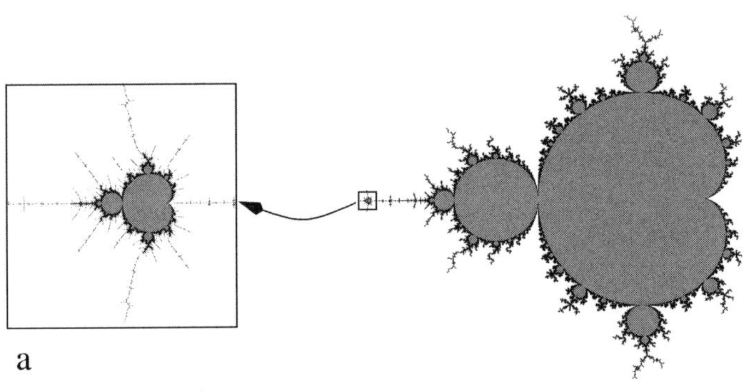

FIGURE 0. The Mandelbrot set M

have deliberately distorted parts of the drawing for clarity. This first section is also an introduction to the results and open questions that will be elaborated on later.

The second section is devoted to hyperbolic aspects and is centered around Jakobson's theorem. The third section describes quasiperiodic aspects, which are related to problems of small divisors.

The point of view adopted here emphasizes the dynamical rather than the geometric or analytic aspects, and thus ignores some important questions. Our goal is to concentrate on some ideas that can be extended to a broader context.

We give no proofs of the deep theorems that are mentioned—at most, for some of them, an indication of the approach followed. References for their proofs are given in the bibliography.

Let us also state explicitly that some of the figures are merely suggestive of the phenomena they illustrate, and are meant to help the reader understand a few real patterns.

1. General Overview

1.1. The dynamical plane and the parameter space. Every quadratic complex polynomial is conjugate under an affine map to a polynomial of the form

$$P_c(z) = z^2 + c.$$

Hence it suffices to study the family of polynomials (P_c), for $c \in \mathbb{C}$.

For $c \in \mathbb{C}$, consider the *filled Julia set*

$$K_c = \{z : \sup_{n \geq 0} |P_c^n(z)| < +\infty\},$$

where $P_c^n = P_c \circ \cdots \circ P_c$, n times. Our first description of K is the following.

Let $R = (1 + \sqrt{1 + 4|c|})/2$. Since $|P_c(z)| > |z|$ for $|z| > R$, we have

$$K_c = \bigcap_{n \geq 0} P_c^{-n}\overline{\mathbb{D}(0, R)}.$$

In particular:

- K_c is an intersection of compact subsets, hence compact.
- K_c is nonempty because it contains all the periodic points (i.e., all the points z for which there exists an integer $n > 0$ such that $P_c^n(z) = z$).
- K_c is totally invariant; i.e., $P_c(K_c) = K_c = P_c^{-1}(K_c)$.
- K_c is full; i.e., $\mathbb{C} \setminus K_c$ is connected. This is true because, by the maximum principle, every bounded open set with boundary in K_c must itself lie completely within K_c.

The boundary $J_c = \partial K_c$ is the *Julia set*. By a theorem obtained independently by Julia and Fatou, J_c is also the closure of the set of repelling fixed points. This is the unstable set of the dynamics.

EXAMPLE. For $c = 0$, $P_0(z) = z^2$; $K_0 = \overline{\mathbb{D}}$ and $J_0 = \mathcal{S}^1 = \{z : |z| = 1\}$.

1.1.1. *The Green's function of K_c.* To study such a compact set, we consider the potential G that it defines. For Julia sets, G is very easy to compute. Since P_c behaves at infinity like $z \mapsto z^2$, the function

$$G_c(z) = \lim_{n \to +\infty} 2^{-n} \log^+ |P_c^n(z)|$$

(where $\log^+ = \max\{\log, 0\}$) is well defined and has the following properties.

(1) $G_c : \mathbb{C} \to \mathbb{R}^+$ is continuous;
(2) $G_c(P_c(z)) = 2G_c(z)$;
(3) $K_c = \{z \in \mathbb{C} : G_c(z) = 0\}$;
(4) G_c is harmonic in $\mathbb{C} \setminus K_c$: it is the uniform limit of a sequence of functions each of which (as the logarithm of the modulus of a holomorphic function) is harmonic;
(5) $G(z) = G_c(z) = \log|z| + O(|z|^{-2})$, $|z| \to +\infty$.

1.1.2. *Critical points of G_c in $\mathbb{C} \setminus K_c$.* The critical point 0 plays a special role in the structure of K_c. By a theorem of Fatou dating from 1919, we have:

Case 1. If $0 \in K_c$, then G_c has no critical points in $\mathbb{C} \setminus K_c$ and K_c is connected.

Case 2. If $0 \in \mathbb{C} \setminus K_c$, then the critical points of G_c are 0 and all its preimages (i.e. the points in $\cup_{n \geq 0} P^{-n}(0)$); $G_c(0)$ is the largest critical value; and K_c is not connected. K_c is actually a Cantor set, as we will see in 1.2.

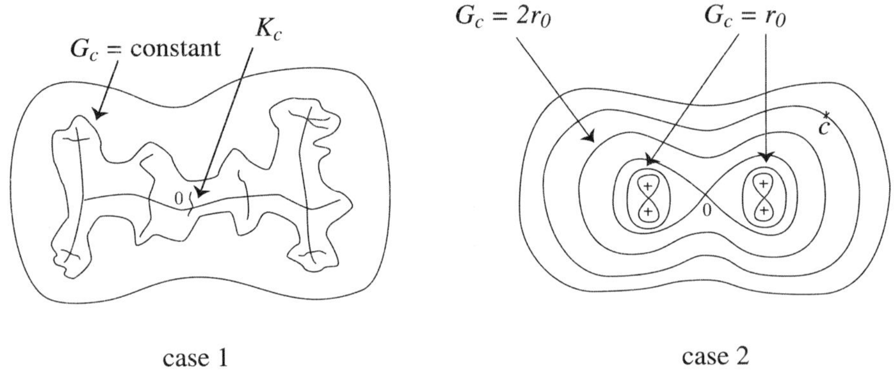

FIGURE 1

Set $g_c(z) = \lim_{n \to +\infty} (P_c^n(z))^{2^{-n}}$. For very large $|z|$, this function is well defined and satisfies the functional equation

$$g_c(P_c(z)) = (g_c(z))^2.$$

The function $g_c(z)$ is defined on $\{z : G_c(z) > G_c(0)\}$ and satisfies

(1) $\log |g_c(z)| = G_c(z)$;
(2) $g_c : \{z : G_c(z) > G_c(0)\} \to \{z : |z| > \exp G_c(0)\}$ is a conformal representation;
(3) $g_c(z) = z + O(z^{-1})$, $|z| \to +\infty$.

In particular, the map

$$\varphi_c = g_c^{-1} : \mathbb{C} \setminus \overline{\mathbb{D}} \to \mathbb{C} \setminus K_c$$

is a conformal representation if K_c is connected.

In the parameter space, consider the *Mandelbrot set*

$$M = \{c \in \mathbb{C} : K_c \text{ is connected}\}.$$

If $|c| > 2$ and $n > 0$, then $|P_c^n(0)| \geq |c|(|c| - 1)^{2^{n-1}}$ and hence $P_c^n(0) \to \infty$; it follows that c is not in M. In fact, $M = \{c : |P_c^n(0)| \leq 2,\ n > 0\}$ and is therefore compact. $\mathbb{C} \setminus M$ has no bounded component by the maximum principle, so it is connected. Hence M is full. Moreover, M is symmetric with respect to the real axis, which it intersects in the interval $[-2, 1/4]$.

K_c is not connected for c in $\mathbb{C} \setminus M$, but $G_c(c) > G_c(0) > 0$ and $g_c(c)$ is well defined. Douady and Hubbard [**5**] showed that the map

$$\varphi : \mathbb{C} \setminus M \to \mathbb{C} \setminus \overline{\mathbb{D}}$$

defined by $\varphi(c) = g_c(c)$ is a conformal representation.

In the remainder of this section, we consider where the parameter c lies in the plane $\mathbb{C}$ and describe the dynamics of P_c accordingly.

1.2. Dynamics for $c \notin M$**.** Let $c \notin M$; let $r_0 > 0$ satisfy $r_0 < G_c(0) < 2r_0$. Let $L = \{z : G_c(z) < 2r_0\}$. The set $P_c^{-1}(L) = \{z : G_c(z) < r_0\}$ has two connected components, L_0 and L_1, which satisfy $P_c(L_j) = L$ for $j = 0, 1$ and do not contain the critical point 0. L_0 and L_1 each contain some point in K_c (otherwise G_c would be harmonic and positive on L_j, hence constant). Similarly, $P_c^{-1}(L_j)$ has two connected components, and so on (see Figure 1, case 2).

Since c is not in L, the map $f_j = P^{-1}_{c|L_j} : L \to L_j$ is well defined and contracting with respect to the Poincaré metric on L (Schwarz's lemma), and

$$K_c = \bigcap_{n \geq 0} P_c^{-n}(L_0 \cup L_1) = J_c$$

is a Cantor set.

We point out two important properties in this case:

– *The dynamics for* $c \notin M$ *is of shift type.*
To see this, let $\Sigma^+ = \{0,1\}^{\mathbb{N}}$ (a compact metric space), and consider the shift $\sigma : \Sigma^+ \to \Sigma^+$ defined by

$$\sigma(\theta)(n) = \theta(n+1).$$

The map $H : K_c \to \Sigma^+$ defined by

$$H(z) = (H(z)(n)), \quad \text{where } H(z)(n) = j \text{ if } P_c^n(z) \in L_j,\ j \in \{0,1\},$$

is a homeomorphism that conjugates $P_{c|K_c}$ to σ. Intuitively, $H(z)$ is the address of the point z; the inverse map consists of taking the unique point with the given address.

– P_c *is hyperbolic*; i.e., there exist $\lambda > 1$, $C > 0$ such that $|D_z P_c^n| \geq C\lambda^n$ for every $z \in K_c$ and $n \geq 0$.
This is an application of the following criterion for $c \in \mathbb{C}$: P_c *is hyperbolic on* J_c *if and only if the closure* T *of the orbit of the critical point* 0 *does not intersect* J_c. To see this, set $S = \mathbb{C} \setminus T$; S is connected, contains J_c, and satisfies $P_c^{-1}(S) \subset S$. If R is a component of $P_c^{-1}(S)$, then the map $P_c^{-1} : S \to R \subset S$ is injective and locally contracting with respect to the Poincaré metric on S. Since J_c is compact, there exists $\lambda > 1$ such that for every $z \in J_c$ and every tangent vector v at z,

$$\|D_z P_c(v)\|_P \geq \lambda \|v\|_P,$$

where $\| \ \|_P$ denotes the norm in the Poincaré metric on S.

1.3. Dynamics for $c \in M$**.** Let $c \in M$, and let z_0 be a periodic point of P_c with period m and associated cycle

$$O(z_0) = \{P_c^i(z_0),\ 0 \leq i < m\}.$$

Let $\mu = (P_c^m)'(z_0)$ be the multiplier of the cycle. There are three cases:

(1) $|\mu| < 1$: $O(z_0)$ is an attracting cycle, superattracting if $\mu = 0$;

(2) $|\mu| = 1$: $O(z_0)$ is an indifferent cycle;
(3) $|\mu| > 1$: $O(z_0)$ is a repelling cycle

The existence (or nonexistence) of a periodic point of type (1) or (2) is a determining factor for the structure of K_c. Douady showed in 1982 that in any case P_c has at most one cycle of type (1) or (2).

1.3.1. *Attracting periodic orbits and hyperbolic components of M.* Let $c \in M$ be such that P_c has an attracting cycle $O(z_0)$ of period m. The set

$$W = W(O(z_0)) = \{z : \lim_{n\to+\infty} d(P_c^n(z), O(z_0)) = 0\}$$

is called the basin of attraction of $O(z_0)$. It is open and totally invariant.

To complete the description of K_c in this case, we mention the following two properties:

- Let U be the component of W containing z_0; U is also a component of $P_c^{-m}(U)$. Since $P^m_{c|U}$ is not an isometry, U contains a critical point of P_c^m and there exists $0 \le k < m$ such that $U_0 = P_c^k(U)$ contains 0.
- W is the interior of K_c. Since the closure of the critical orbit does not intersect J_c, we see as above that $P_{c|J_c}$ is hyperbolic.

In the parameter plane, if P_{c_0} has an attracting periodic orbit $O(z_0)$, then c_0 is in the interior of M (this follows from the reasoning above).

Let V_0 be the connected component of the interior of M containing c_0. One can follow the attracting periodic point $c \mapsto z_0(c)$, $z_0(c_0) = z_0$ in V_0; the cycle $(O(z_0(c))$ is still attracting and has the same period. Moreover, the map $c \mapsto \mu(c)$, $V_0 \to \mathbb{D}$, where $\mu(c)$ is the multiplier of the cycle $O(z_0(c))$, is a conformal representation (Douady-Hubbard, Sullivan) [**5**].

Such a component is called a *hyperbolic component* of M.

EXAMPLE. The set of c such that P_c has an attracting fixed point is the interior of a cardioid, called the *main cardioid*, which contains 0 (see Figure 0). For such c, the interior of K_c has only one component, and J_c is a quasicircle (Sullivan); i.e., it is the image of $\mathcal{S}^1$ under a quasiconformal homeomorphism (defined below) ϕ such that $P_c(\phi(z)) = \phi(z^2)$ for z in a neighborhood of J_c.

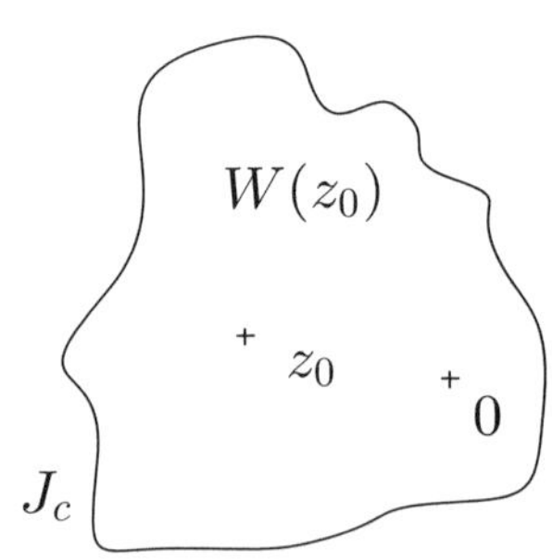

FIGURE 2

DEFINITION. A homeomorphism $\phi : \mathbb{C} \to \mathbb{C}$ is quasiconformal if ϕ satisfies

$$\sup_{z\in\mathbb{C},\, t>0} \frac{\sup_{d(y,z)=t} d(\phi(z), \phi(y))}{\inf_{d(y,z)=t} d(\phi(z), \phi(y))} < \infty.$$

(Intuitively, $D\phi^{-1}$ exists almost everywhere and turns the field of infinitesimal circles into a field of infinitesimal ellipses with bounded eccentricity.)

We now consider the interior of M.

- *The interior of M is dense in M.* More precisely, every point on the boundary of M is approximated by parameter values each of which has a superattracting cycle.

  To see this, let U be an open disk that does not contain 0 and intersects the boundary of M. If no point c in U has a superattracting cycle, then the family of functions $g_n : c \mapsto P_c^n(0)/\sqrt{-c}$, which is well defined on U, does not assume the values $0, 1, \infty$. By Montel's theorem, (g_n) is a normal family. But for c_0 in the boundary of M and c near c_0, either $g_n(c) \to \infty$ or $g_n(c)$ remains bounded (both cases can occur), and (g_n) cannot be a normal family.
- *The interior of M contains all the hyperbolic components.*

The question, of course, is whether the interior of M has other components. Douady and Hubbard conjectured in 1982 that it does not.

CONJECTURE (Hyperbolicity). *All the components of the interior of M are hyperbolic.*

Partial results in this direction have been obtained. We mention the following two important theorems. Theorem 1.1 is due to R. Mañé, P. Sad, and D. Sullivan [**13**]; Theorem 1.2 is due to G. Swiatek [**7**].

THEOREM 1.1 (Stability). *Every interior point c of M is J-stable; i.e., if f is a holomorphic perturbation of P_c in a neighborhood of J_c, then f and P_c are quasiconformally conjugate in a neighborhood of J_c.*

COROLLARY. *If c and c' are in the same component of the interior of M, then P_c and $P_{c'}$ are quasiconformally conjugate in a neighborhood of J_c.*

If P_c is hyperbolic, then it is J-stable. To prove the hyperbolicity conjecture one must show that the converse is true. There is a partial result:

THEOREM 1.2. *The intersection of the hyperbolic components with*

$$M \cap \mathbb{R} = [-2, 1/4]$$

is a dense set.

This result admits an extension due to M. Lyubich [**12**]. However, M. Jakobson's theorem, the subject of Section 2, states that if E is the set of $c \in M \cap \mathbb{R}$ that have an attracting cycle, then $(M \cap \mathbb{R}) \setminus E$ has positive measure.

The proof of Theorem 1.2 is difficult and uses a number of arguments that appear in the proof of Jakobson's theorem.

1.3.2. *Indifferent periodic points and the boundaries of the hyperbolic components.* Let $c \in M$ be such that P_c has an indifferent periodic point z_0; i.e., z_0 has multiplier $\mu = \exp 2i\pi\alpha$, $\alpha \in \mathbb{R}$. Let m denote its period. We distinguish three cases, depending on the properties of α:

(1) The case where $\alpha \in \mathbb{Q}$ (parabolic case). In this case, 0 is in the interior int K_c of K_c. Let U be the component of int K_c containing 0. Then U also contains the set $\{P^{km}(0) : k \geq 0\}$. There is a z_i in $O(z_0)$ such that z_i is in the boundary of

U and the sequence $(P^{km}(0))_k$ converges to z_i. Furthermore, $O(z_0) \subset J_c$ and the sequence of iterates of each point in $\operatorname{int} K_c$ converges to a point in $O(z_0)$.

Since $z_0 \in J$, $P_{|J}$ is not hyperbolic.

EXAMPLE. For $c = 1/4$, the point $z_0 = 1/2$ is a parabolic fixed point and $\operatorname{int} K_c$ has only one component. P_c is topologically conjugate on J_c to the map $z \mapsto z^2$.

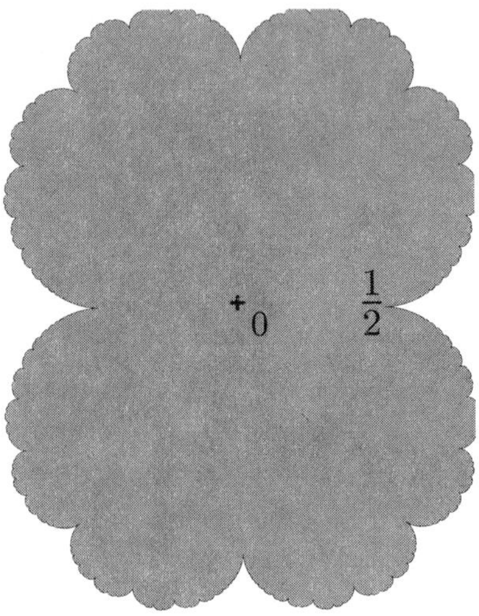

FIGURE 3. The cauliflower $c = 1/4$

(2) The case where α is irrational and satisfies the Brjuno condition. Recall that a number α satisfies the Brjuno condition if

$$\sum_{n \geq 0} q_n^{-1} \log q_{n+1} < +\infty,$$

where the fraction p_n/q_n is the nth convergent to α in its continued-fraction expansion.

In this case, $z_0 \in \operatorname{int} K_c$ and $0 \in J_c$; in particular, it follows that P_c is not hyperbolic. A. Brjuno [**2, 3**] showed that P_c is linearizable; i.e., if U is the component of z_0 in $\operatorname{int} K_c$, then there exists a conformal representation $h : (U, z_0) \to (\mathbb{D}, 0)$ that conjugates $P^m_{|U}$ to the rotation $z \to \lambda z$. The set U is called a *Siegel disk.*

The converse is also true, and was proved by Yoccoz [**22**]: if P_c is linearizable, then α satisfies the Brjuno condition.

(3) The case where α is irrational and does not satisfy the Brjuno condition. In this case, P_c is not linearizable in any neighborhood of z_0, and K_c has empty interior.

We will return to cases (2) and (3) in Section 3.

In the parameter plane, any point $c_0 \in M$ that has an indifferent cycle is in the boundary of some hyperbolic component. This is true because the multiplier of a cycle is an analytic function with modulus that is not locally constant, so we can perturb the polynomial P_{c_0} to a polynomial P_c that has an attracting cycle. (This argument must be very slightly modified if the multiplier equals 1.)

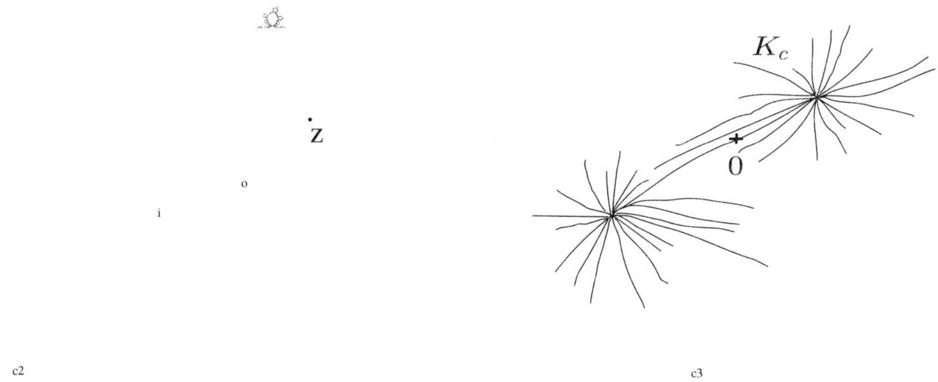

FIGURE 4. Behavior of K_c

1.3.3. *The interior of K_c.* The following theorem completes the description of the interior of K_c.

THEOREM 1.3. *Let U be a component of* $\operatorname{int} K_c$.

(a) *U is preperiodic: there exist $n \geq 0$, $k > 0$ such that $P_c^n(P_c^k(U)) = P_c^k(U)$.*
(b) *If U is periodic, then exactly one of the following is true:*
  - *U has an attracting periodic point.*
  - *The boundary of U contains a parabolic periodic point.*
  - *U is a Siegel disk.*

The classification of the periodic components (assertion (b) of the theorem) is due to Fatou. Assertion (a), also called the nonwandering theorem, is due to Sullivan [**20**].

COROLLARY. *If all the cycles of P_c are repelling, then K_c has empty interior.*

1.3.4. *Quadratic-Like Maps. Renormalization. Results of Douady and Hubbard. Copies of M.* Douady and Hubbard introduced the basic notion of quadratic-like maps, which provides the right setting for studying certain fine properties of quadratic polynomials.

More precisely, a map $f : V \to U$, where V and U are simply connected open sets in $\mathbb{C}$, is *quadratic-like* if V is relatively compact in U and f is a 2-fold branched covering. We continue to denote the critical point of f by 0.

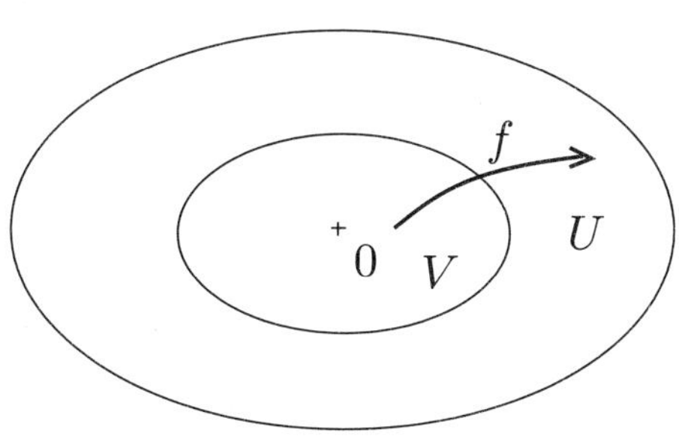

FIGURE 5

We define

$$K_f = \bigcap_{n \geq 0} f^{-n}(V),$$

the filled Julia set of f. It is compact, nonempty, totally invariant, and full. Moreover, K_f is connected if and only if $0 \in K_f$.

EXAMPLE. For $f = P_c$, we can take $U = \mathbb{D}(0, R)$, where R is sufficiently large, and $V = P_c^{-1}(U)$.

Douady and Hubbard [**6**] proved the following theorem:

THEOREM 1.4 (Straightening). *Let $f : V \to U$ be a quadratic-like map. There exists c (unique if K_f is connected) such that f and P_c are conjugate in a neighborhood of K_f under a quasiconformal homeomorphism h satisfying $\overline{\partial} h = 0$ a.e. on K_f.*

EXAMPLE. For $|\mu| = 1$, there exists $\varepsilon_0 > 0$ such that the quadratic polynomial $P(z) = \mu z + z^2$ and the cubic polynomial $Q(z) = \mu z + z^2 + \varepsilon z^3$ are conjugate in a neighborhood of the origin for $|\varepsilon| < \varepsilon_0$. In particular, this occurs for μ that do not satisfy the Brjuno condition; P is not linearizable in this case and has complicated dynamics.

The proof of this straightening theorem is based on *quasiconformal surgery*, a procedure that allows one to start with two dynamical systems acting on two distinct subsets of $\mathbb{C}$ and construct a new dynamical system on $\mathbb{C}$ that respects the initial systems. In the case that interests us, one of the dynamical systems is of course $f : V \to U$. The other is $g = \phi \circ P_0 \circ \phi^{-1} : \mathbb{C} \setminus U \to \mathbb{C}$, where $P_0(z) = z^2$ and $\phi : \mathbb{C} \setminus \mathbb{D} \to \mathbb{C} \setminus U$ is a conformal representation. The existence of the dynamical system $P_c : \mathbb{C} \to \mathbb{C}$ constructed from these two systems is based on the Ahlfors-Bers integrability theorem [**1**], which has no equivalent in real dynamics.

For every pair (m, c_0), $m \geq 2$, satisfying $P_{c_0}^m(0) = 0$, there exists a homeomorphism

$$\Phi_{c_0} = \Phi : M \to M_{m,c_0} \subset M$$

such that $\Phi(0) = c_0$ and, for every $c \in M$, there are simply connected neighborhoods V and U of 0, with V relatively compact in U, such that $P^m_{\Phi(c)|V} : V \to U$ has the following properties: it is quadratic-like; its Julia set is connected; and it can be straightened to P_c (cf. Theorem 1.4).

The parameters $\widehat{c} \in M_{m,c_0}$ are called m-renormalizable, and M_{m,c_0} is called a copy of M. Two copies are either disjoint or nested. Indeed, if $m \leq m'$, then $M_{m',c_0'} \subset M_{m,c_0}$ if and only if m' is a multiple of m and c_0' is in M_{m,c_0}.

A parameter is infinitely renormalizable if it is contained in an infinite sequence of nested copies. For example, the Feigenbaum point $c = -1.401155\ldots$, the only parameter c in $M \cap \mathbb{R}$ for which the real periodic points of the polynomial P_c are of order 2^n for every $n > 0$, is in the intersection of the 2^n copies $(n \geq 0)$ centered on the real axis.

We are very far from understanding all the complexity of M. In 1982 Douady and Hubbard [**6**] proposed the following conjecture.

CONJECTURE (MLC). *M is locally connected.*

An important step in this direction was taken by Yoccoz [**10**] in 1991.

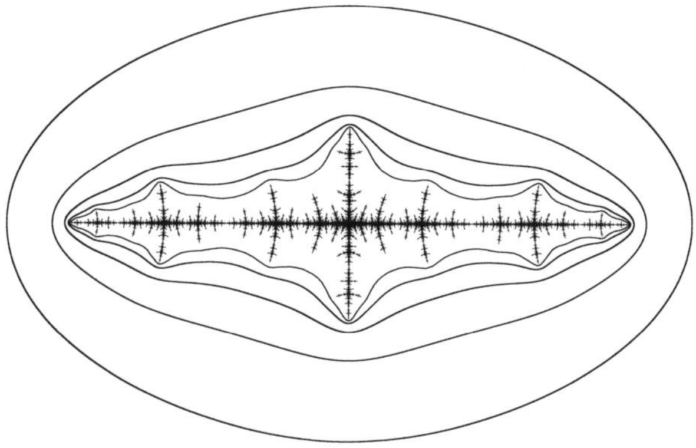

FIGURE 6. Julia set of the Feigenbaum point

THEOREM 1.5. *If c is a point in M that is not infinitely renormalizable, then M is locally connected at c. Moreover, K_c is locally connected for such c.*

This theorem is proved in two steps: the first when c is in the closure of a hyperbolic component, the second when all the periodic orbits are repelling, with the second step broken into two cases according to whether the orbit of the critical point 0 accumulates at 0 (recurrent case) or does not accumulate at 0 (non-recurrent case).

The tools used in the proof of this theorem also enable us to show the following. To prove the MLC conjecture, it suffices to show:

($\star$) *The intersection of every sequence of nested copies is a point.*

Similarly, to prove the hyperbolicity conjecture it suffices to show:

($\star\star$) *This intersection has empty interior.*

It is becoming evident that the MLC conjecture implies the hyperbolicity conjecture. We note in conclusion that properties ($\star$) and ($\star\star$) are certainly necessary.

2. Hyperbolic Aspects

The goal of this section is to present some ideas that lead to Jakobson's theorem. This theorem (the precise statement of which is given later) is the point of departure for a new direction in the study of dynamical systems. In particular, it is a necessary step for understanding the work of Benedicks-Carleson on Hénon maps.

We start by pinning down the region of M left unexplored by the analysis in the last section; this region is the subject of Jakobson's theorem.

In Section 1 we saw that P_c is hyperbolic for every c in either $\mathbb{C} \setminus M$ or a hyperbolic component. On the other hand, P_c is not hyperbolic if c is in the boundary of a hyperbolic component.

Now consider a parameter c in M such that

(1) c is not in the closure of a hyperbolic component (i.e., all the periodic orbits are repelling);
(2) c is not in a copy of M (i.e., P_c is not renormalizable).

By Theorems 1.3 and 1.5, $K_c = J_c$ has empty interior and measure zero and is locally connected; the dynamics of P_c on K_c is controlled by the *recurrence* of the orbit of the critical point.

The following theorem, due to Jakobson [**11**], shows that if this recurrence is not too strong, then the dynamics is *(non-uniformly) hyperbolic*. For real c, this occurs with positive probability (in c).

2.1. Jakobson's theorem for real c. The statement of this theorem requires some preliminary definitions and notation.

2.1.1. *Preliminaries.* We begin by listing several known properties of polynomials P_c with c real and in M:

- For $c \leq 1/4$, the fixed points $\beta = (1+\sqrt{1-4c})/2$ and $\alpha = (1-\sqrt{1-4c})/2$ of P_c are real.
  The interval $A = [\alpha, -\alpha]$ will serve throughout the proof of the theorem as a test interval for the returns of a point in a neighborhood near 0.
- For $-2 \leq c \leq 1/4$, K_c is connected and $K_c \cap \mathbb{R} = [-\beta, \beta]$.
- For $c = 1/4$, $\beta = \alpha = 1/2$ is a parabolic point: $P_c'(\alpha) = 1$.
- For $1/4 > c > -3/4$, α is an attracting point.
- Let $c^{(2)}$ be the real root of the equation $P_c^2(0) = -\alpha$, i.e.

  $$c^3 + 2c^2 + 2c + 2 = 0.$$

  For $-3/4 \geq c \geq c^{(2)} = -1.54369\ldots$, we have

  $$\alpha \leq P_c^2(0) \leq -\alpha.$$

  The map $P^2_{c|A} : A \to A$ is quadratic and 0 is its only critical point, so P_c is 2-renormalizable.
- For $-2 \leq c < c^{(2)}$, $P_c^2(0) > -\alpha$ and P_c is not 2-renormalizable.

M. Jakobson published the following theorem in 1981 [**11**]:

THEOREM 2.1. *The following properties are satisfied for a set of parameters $c \in [-2, c^{(2)})$ of positive measure:*

(a) *There exists $\lambda = \lambda_c > 0$ such that*

$$\lim_{n \to +\infty} \frac{1}{n} \log |D_x P_c^n| = \lambda$$

for Lebesgue-almost every $x \in [-\beta, \beta]$.

(b) *There exists a probability measure that is invariant under P_c and absolutely continuous with respect to Lebesgue measure on $[-\beta, \beta]$.*

Before tackling the proof of the theorem, we give some more definitions and notation that will be needed in what follows. They are illustrated in Figure 7.

For $c < 0$, we denote by $\alpha^{(1)}$ the negative preimage of $-\alpha$ and by $\alpha^{(i)}$, $i > 1$, the negative preimage of $-\alpha^{(i-1)}$.

For $\alpha^{(i-1)} \geq c$, $\widetilde{\alpha}^{(i)}$ denotes the negative preimage of $\alpha^{(i-1)}$. The sequence $(\alpha^{(i)})$ is decreasing and has limit $-\beta$; the sequence $(\widetilde{\alpha}^{(i)})$ is increasing.

Furthermore,

$$-\beta < c < \alpha^{(1)}$$

for $c \in (-2, c^{(2)})$, and there exists $n > 1$ such that $c \in [\alpha^{(n)}, \alpha^{(n-1)}]$.

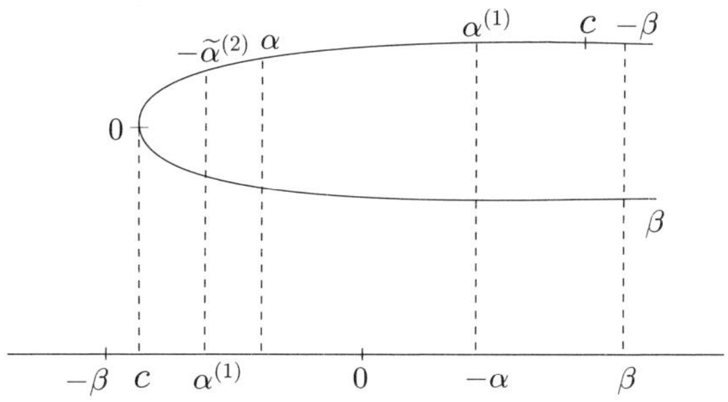

FIGURE 7

2.1.2. *The case* $c = -2$. This is a limiting special case: $\beta = 2 = -c$; 0 is preperiodic. The pair (λ, μ) can easily be computed explicitly in this case. It is nonetheless a good illustration of the approach followed for proving the theorem in general.

We have $\alpha = -1$, $P_{-2}(2\cos\theta) = 2\cos 2\theta$. To simplify notation we set $P = P_{-2}$. Let

$$h(x) = 1/\sqrt{4 - x^2};$$

then $|D_x P| = 2h(x)/h(P(x))$ and hence, for all $n > 0$,

$$|D_x P^n| = 2^n \frac{h(x)}{h(P^n(x))}.$$

Consider $\mu = \frac{1}{\pi} h(x)dx$ for $x \in [-2, 2]$. This is a probability measure that is invariant under P and absolutely continuous with respect to Lebesgue measure.

The sequence $(\widetilde{\alpha}^{(i)})_{i>0}$ is defined for $c = -2$, and the map T defined by the graph below is a first return map in $A = [-1, 1]$.

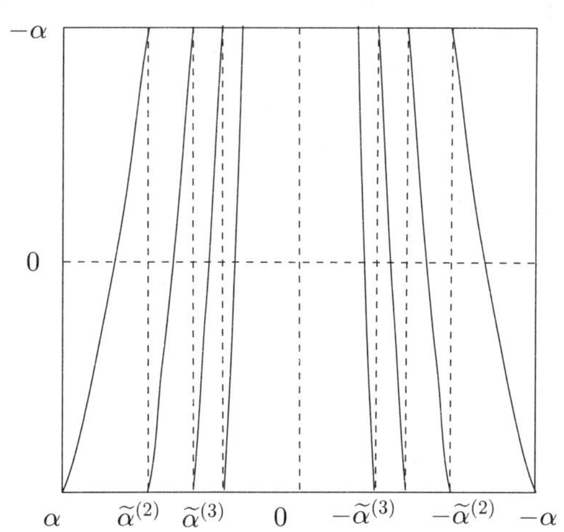

FIGURE 8

For $x \in A$, we set $N(x) = \inf\{n > 0 : P^n(x) \in A\}$ and $T(x) = P^{N(x)}(x)$. We also define $N_k(x) = \sum_{0 \leq i < k} N(T^i(x))$. Then $T^k(x) = P^{N_k(x)}(x)$. The normalized restriction $\mu_T = \mu_{|A}/\mu(A)$ is invariant under T.

For $k \geq 2$, we have

$$\mu_T(\{x : N(x) = k\}) = 2^{1-k},$$

and the random variables $N(T^i(x))$ are independent. Hence, for almost every $x \in A$,

$$\lim_{k \to +\infty} \frac{N_k(x)}{k} = \sum_{k \geq 2} k 2^{1-k} = 3.$$

For every $n > 0$, there exists k such that $N_k(x) \leq n < N_{k+1}(x)$. Hence

$$\log |D_x P^{N_{k+1}}| + (n - N_{k+1}) \log 4 < \log |D_x P^n| \leq \log |D_x P^{N_k}| + (n - N_k) \log 4.$$

It follows that, for almost every $x \in A$,

$$\lim_{n \to +\infty} \frac{1}{n} \log |D_x P^n| = \lim_{k \to +\infty} \frac{1}{N_k} \log |D_x P^{N_k}| = \log 2.$$

This property still holds for almost every $x \in (-2, 2)$ because such a point enters A after iteration by P. Thus the pair $(\lambda = \log 2, \mu)$ satisfies conditions (a) and (b) of Jakobson's theorem.

2.2. Regular parameters. If we want to generalize the strategy deployed for the case $c = -2$, we are led to consider the return map T from A to A. For $c \neq -2$, the domain of T is a union of intervals called regular (see the definition below).

More precisely, we first define the interval $\widehat{A} = [\alpha^{(1)}, -\alpha^{(1)}]$, which contains A and will allow us to control the distortion on A.

DEFINITION. An interval J is *regular* if there exist an integer $n > 0$ and a neighborhood $\widehat{J}$ of J such that the restriction $P^n_{|\widehat{J}}$ of P^n to $\widehat{J}$ is a *diffeomorphism from $\widehat{J}$ onto $\widehat{A}$* that maps J onto A.

If J is regular, then the neighborhood $\widehat{J}$ and the integer n are uniquely determined; J is said to be of order n, and n is called the order of J.

Let $M \geq 3$ be such that $\alpha^{(M-1)} \leq c < \alpha^{(M-2)}$. Then $P^M(0) = P^{M-1}(c) \in A$, and M is the *return time* of 0 to A. The intervals $[\alpha^{(i)}, \alpha^{(i-1)}]$, for $i > 0$, and $[\widetilde{\alpha}^{(i-1)}, \widetilde{\alpha}^{(i)}]$, for $1 < i < M - 1$, are regular of order i.

DEFINITION. A parameter $c \in [-2, c^{(2)})$ is *regular* if there exist $C > 0$, $\theta > 0$ such that, for all $n > 0$,

$$\text{meas}(\{x \in A : x \text{ is not in a regular interval of order} \leq n\}) \leq C 2^{-\theta n}.$$

EXAMPLES.

- For $c = -2$ and $n > 0$, the intervals $I_n = [\widetilde{\alpha}^{(n-1)}, \widetilde{\alpha}^{(n)}]$ and $-I_n$ are regular of order n and cover $A \setminus \{0\}$. A quick calculation shows that $\widetilde{\alpha}^{(n)} = \sin \pi/3 \cdot 2^n$, so $|I_n| \leq C\, 2^{-n}$ and $c = -2$ is a regular parameter.
- More generally, the parameters c for which 0 is preperiodic are regular; however, they form a set of measure zero.

We assume from now on that $c \in [-2, c^{(2)})$; the constants denoted C that appear are independent of M. Jakobson's theorem is a consequence of the following two theorems.

THEOREM A. *Conclusions (a) and (b) of Theorem 2.1 are satisfied if c is regular.*

THEOREM B. *The set of regular parameters in* $[-2, c^{(2)})$ *has positive measure.*

Indeed, the proportion of regular parameters in $[-2, -2+\varepsilon)$ approaches 1 as $\varepsilon \to 0$.

2.3. Sketch of the proof of Theorem A (classical). The proof sketched here is classical and consists of eight steps. In this subsection and the next, where the parameter c is fixed, we write $P_c = P$. If J is a regular interval of order n, we denote by $\widehat{J}$ the interval such that $P^n_{|\widehat{J}} : \widehat{J} \xrightarrow{P^n} \widehat{A}$ is a diffeomorphism, and we set $g_J = (P^n_{|\widehat{J}})^{-1}$.

2.3.1. *Bounded distortion.* Let J be a regular interval of order n. The distortion of g_J on A is bounded. The argument is as follows. The Schwarzian derivative

$$S(g_J) = D^2 \log|Dg_J| - \frac{1}{2}(D \log|Dg_J|)^2$$

is positive on $\widehat{A}$, so

$$|D \log|Dg_J(x)|| < C_0,$$

where $C_0 = \dfrac{2}{|\alpha - \alpha^{(1)}|}$. Let $C_1 = \exp(C_0|A|)$.

Then

$$\frac{1}{C_1}\frac{|J|}{|A|} \le |Dg_J| \le C_1 \frac{|J|}{|A|}$$

and, for every measurable subset B of A,

$$\frac{1}{C_1}\frac{\operatorname{meas}(B)}{|A|} \le \frac{\operatorname{meas}(g_J(B))}{|J|} \le C_1 \frac{\operatorname{meas}(B)}{|A|}.$$

2.3.2. *The return map T.* It follows from the definition that the endpoints of a regular interval of order n are two consecutive points in $P^{-n}(\{\alpha, -\alpha\})$. Hence two regular intervals either are nested or have disjoint interiors.

Let $\mathcal{J}$ be the set of intervals J that are regular, contained in but not equal to A (i.e., of order > 0), and maximal with these properties. We set $U = \bigsqcup_{J \in \mathcal{J}} \operatorname{int} J$.

For $J \in \mathcal{J}$ of order n and $x \in J$, we set $N(x) = n$ and $T(x) = P^n(x)$, thus defining maps $N : U \to \mathbb{N}$ and $T : U \to A$ such that

$$N_{|\operatorname{int} J} = n_J, \quad T_{|\operatorname{int} J} = P^{n_J},$$

where $n_J = \operatorname{ord}(J)$. The g_J, $J \in \mathcal{J}$, are the *inverse branches* of T.

For $x \in \bigcap_{j=0}^{k-1} T^{-j}(U)$, we set

$$N_k(x) = \sum_{j=0}^{k-1} N(T^j(x)),$$

so $T^k(x) = P^{N_k(x)}(x)$.

2.3.3. If $J_1, \dots, J_k$ is a finite sequence of elements of $\mathcal{J}$, we set

$$\underline{J} = g_{J_1} \circ \cdots \circ g_{J_k}(A);$$

this is a regular interval such that $g_{\underline{J}} = g_{J_1} \circ \cdots \circ g_{J_k}$. Conversely, every regular interval contained in A can be obtained in a unique way by this procedure. The set of such intervals can thus be identified with $\bigsqcup_{k \ge 0} \mathcal{J}^k$.

We set

$$C_2 = 1 - \frac{1}{C_1}\left(1 - \frac{1}{|A|}\max_{J \in \mathcal{J}} |J|\right);$$

then

$$0 < C_2 < 1.$$

For $\underline{J} \in \mathcal{J}^k$ and $x \in \underline{J}$, applying the estimates obtained in 2.3.1 gives

$$|\underline{J}| < C_2^k |A|; \quad |D_x T^k| \geq \frac{1}{C_1 C_2^k}.$$

We assume from now on that c is *regular*. In this case, $\text{meas}(A \setminus U) = 0$ and there exist $C > 0$, $\theta > 0$ such that, for every $n \geq 0$, $\text{meas}(\{x : N(x) \geq n\}) \leq C\, 2^{-\theta n}$.

2.3.4. Let $\mu = h(x)dx$ be a finite positive measure on A, absolutely continuous with respect to Lebesgue measure, and with density $h \in L^1(A)$. The image $T_\star \mu$ is also absolutely continuous with respect to Lebesgue measure and has density $\mathcal{L}h$, where

$$\mathcal{L}h(x) = \sum_{J \in \mathcal{J}} h \circ g_J(x) |Dg_J(x)|.$$

For $k > 0$, $\mathcal{L}^k h(x) = \sum_{\underline{J} \in \mathcal{J}^k} h \circ g_{\underline{J}}(x) |Dg_{\underline{J}}(x)|$.

2.3.5. Let $\mathbb{C}_{\widehat{A}}$ be the complex plane minus $(-\infty, \alpha^{(1)})$ and $(-\alpha^{(1)}, +\infty)$. For a regular interval J of order $n > 0$, the inverse branch g_J extends to a univalent map from $\mathbb{C}_{\widehat{A}}$ to $\mathbb{C}$ because the critical values of P^n are real. As n and J vary, these maps form a normal family (their images avoid $(-\infty, -\beta)$ and $(\beta, +\infty)$).

Let $\varepsilon_{\underline{J}}$ be the sign of $Dg_{\underline{J}}$ on A. The family

$$\mathcal{L}^k 1 = h_k = \sum_{\underline{J} \in \mathcal{J}^k} \varepsilon_{\underline{J}} Dg_{\underline{J}}$$

of holomorphic functions defined on $\mathbb{C}_{\widehat{A}}$ is also normal. Let h_T be a limit point of the sequence

$$\frac{1}{k} \sum_0^{k-1} h_j.$$

It is a holomorphic function on $\mathbb{C}_{\widehat{A}}$ that is positive on A and satisfies $\mathcal{L}h_T = h_T$.

2.3.6. The T-invariant measure $\mu_T = h_T(x)dx$ is ergodic. To see this, let E be a T-invariant subset of A of measure > 0, and let x_0 be a point of density of E. Since E is T-invariant, $x_0 \in \cap_k T^{-k}(U)$. Let $J(k) \in \mathcal{J}$ be the interval containing x_0. Then

$$\frac{\text{meas}(A \setminus E)}{|A|} \leq C_1 \frac{\text{meas}(J(k) \setminus (E \cap J(k)))}{|J(k)|}$$

by 2.3.1. Since $|J(k)| \to 0$ as $k \to \infty$, $\text{meas}(A \setminus E) = 0$. The measure μ_T is the unique T-invariant, absolutely continuous measure. In particular, the sequence $\frac{1}{k} \sum_0^{k-1} h_j$ converges to h_T (in fact, the sequence h_k itself converges to h_T).

2.3.7. *The exponent* λ. Since $\text{meas}(\{x : N(x) \geq n\}) \leq C\, 2^{-\theta n}$ for every $n \geq 0$, the map N is μ_T-integrable. Furthermore, $\log 1/C_1 < \log |DT| \leq (\log 4) N$; hence $\log |DT|$ is also μ_T-integrable. By Birkhoff's theorem,

$$\frac{1}{k} \log |D_x T^k| \xrightarrow{\text{a.e.}} \int \log |DT| d\mu_T; \quad \frac{1}{k} N_k(x) \xrightarrow{\text{a.e.}} \int N\, d\mu_T.$$

By the last inequality of 2.3.3,

$$\int \log |DT| d\mu_T = \frac{1}{k} \int \log |DT^k| d\mu_T > \log \frac{1}{C_2} > 0.$$

Finally,

$$\frac{\log|D_x P^{N_k(x)}|}{N_k(x)} \xrightarrow{\text{a.e.}} \lambda = \frac{\int \log|DT|\, d\mu_T}{\int N\, d\mu_T} > 0.$$

To pass from the subsequence $(N_k)_{k\geq 0}$ to the sequence $(n)_{n\geq 0}$, we proceed as in the case $c = -2$.

Condition (a) of Jakobson's theorem is satisfied at the point c.

2.3.8. *The invariant measure* μ_P. Finally, still for regular c, define S by

$$S : \mathcal{C}([-\beta, \beta]) \to L^1(A, d\mu_T)$$

and μ_P by

$$S\varphi(x) = \sum_{i=0}^{N(x)-1} \varphi \circ P^i(x), \qquad \int \varphi\, d\mu_P = \int S\varphi\, d\mu_T.$$

The measure μ_P is finite and positive on $[-\beta, \beta]$, invariant under P, absolutely continuous with respect to Lebesgue measure (equivalent to Lebesgue measure on $[c, P(c)]$), and ergodic. The measure $\dfrac{1}{\mu_P([-\beta,\beta])}\mu_P$ satisfies condition (b) of the theorem at c.

REMARKS.

- For the case $c = -2$, we recover $\mu_P = h_P(x)dx$, with $h_P(x) = \frac{1}{\pi}h(x)dx$, and $\mu_T = h_T(x)dx$, with $h_T = h_{P|A}$.
- μ_P is less regular than μ_T; h_P is not in $L^2([c, P(c)])$.

2.4. Some steps toward Theorem B. What has to be done now is to estimate the measure of the set of regular parameters. We sketch the strategy followed. First we construct a set R of parameters for which the successive returns of 0 to A will be good but not too close to 0. Next, we must show both that the elements of R are regular and that the conditions placed on these returns enable us to evaluate the size of R.

2.4.1. *Construction of* R. For every $m > 1$, there exists a unique solution $c^{(m)}$ in $[-2, c^{(2)})$ of the equation $\alpha^{(m-1)}(c) = c$. The sequence $(c^{(m)})$ is decreasing and has limit -2.

Given a very large integer M, we consider, in the rest of this subsection, the parameters $c \in (c^{(M)}, c^{(M-1)})$ for which

$$\alpha^{(M-1)}(c) < c < \alpha^{(M-2)}(c).$$

The integer M is the return time of 0 in A. Although $0 \notin U$, we agree to set $T(0) = P^M(0)$.

For $i \in [2, M-2]$, the intervals $[\widetilde{\alpha}^{(i-1)}, \widetilde{\alpha}^{(i)}]$, $[-\widetilde{\alpha}^{(i)}, -\widetilde{\alpha}^{(i-1)}]$ are maximal regular intervals of order i. We call them *simple intervals*. Every other element of $\mathcal{J}$ is of order $> M$.

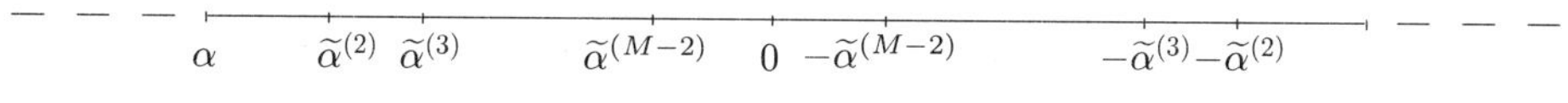

FIGURE 9

Now consider the following conditions for an integer $k \geq 0$:

$$(1_k) \qquad T(0) \in \bigcap_{\ell=0}^{\ell=k-1} T^{-\ell}(U).$$

Under this condition, the iterates $T^\ell(0)$ are defined for $0 < \ell \le k+1$. For $0 < \ell \le k$, $T^\ell(0)$ is in an interval $J(\ell) \in \mathcal{J}$ of order $n_\ell = N(T^\ell(0))$.

$$(2_k) \qquad \sum_{0<j\le\ell,\ n_j>M} n_j \le 2^{-\sqrt{M}}\ell \quad \text{for } 0<\ell\le k.$$

In particular, $J(\ell)$ is a simple interval for $\ell \le M\,2^{\sqrt{M}}$.

Observe that conditions (1_0), (2_0) are automatically satisfied and that conditions (1_{k+1}), (2_{k+1}) are more restrictive than conditions (1_k), (2_k).

Let R_k denote the set of parameters $c \in (c^{(M)}, c^{(M-1)})$ that satisfy conditions (1_k), (2_k), and let $R = \cap_{k\ge0} R_k$. Theorem B, and hence Jakobson's theorem, is a consequence of the following two assertions:

- (B') *Every parameter in R is regular.*
- (B'') *The relative measure of R in $(c^{(M)}, c^{(M-1)})$ approaches 1 as M approaches infinity.*

2.4.2. *Expansion along the critical orbit.* Suppose conditions (1_k) and (2_k) are satisfied. For $0 < \ell \le k+1$, we define a decreasing sequence of intervals $B(\ell)$ containing $T(0)$:

$$B(1) = A, \quad B(\ell+1) = g_{B(\ell)}(g_{J(\ell)}(A)).$$

We set

$$N_\ell = \sum_0^{\ell-1} N(T^j(0)) \quad \text{and} \quad g_{B(\ell+1)} = g_{B(\ell)} \circ g_{J(\ell)} = g_{J(1)} \circ \cdots \circ g_{J(\ell)}.$$

Then the intervals $B(\ell)$ are regular of order

$$\sum_1^{\ell-1} n_j = N_\ell - M.$$

When $J(\ell)$ is not simple, we have the trivial estimate

$$4^{-n_\ell} \le |Dg_{J(\ell)}| \le 1.$$

But we have a precise estimate for $|Dg_{J(\ell)}|$ in the simple case:

$$\left| |Dg_{J(\ell)}(x)|\, 2^{n_\ell} \frac{h(g_{J(\ell)}(x))}{h(x)} - 1 \right| \le C\,4^{n_\ell - M},$$

where $h(x) = 1/\sqrt{\beta^2 - x^2}$, $x \in A$. Combining these yields the following estimate for $B(\ell)$ (and this is where we use (2_k)):

$$2^{-N_\ell + M - CM^{-1}N_\ell} \le |Dg_{B(\ell)}(x)| \le 2^{-N_\ell + M + CM^{-1}N_\ell},$$

for every $0 < \ell \le k+1$ and every $x \in A$.

2.4.3. *The crucial measure estimate (assertion (B')).* This is the most delicate point in the proof. We continue to assume that conditions (1_k), (2_k) are satisfied. Let U_n denote the union of the intervals $J \in \mathcal{J}$ of order $\le n$, and $\mathcal{E}(n)$ the set of connected components of $A \setminus \overline{U_n}$.

We will show that for $0 < \theta < 1/2$, there exists $C > 0$ such that for every $n \le N_{k+1}$,

$$\sum_{E\in\mathcal{E}(n)} |E| \le C\,2^{-\theta n}.$$

We will consider only the special case where the following two properties are satisfied for all $0 < \ell \leq k$ (see Figure 10):

($\star$) The left endpoints of $B(\ell)$ and $B(\ell+1)$ are distinct.

($\star\star$) The leftmost connected component of $\widehat{B(\ell+1)} \setminus B(\ell+1)$ is a regular interval.

The general case is not much more complicated. We set

$$A(0) = [\widetilde{\alpha}^{(M-2)}, -\widetilde{\alpha}^{(M-2)}]$$

and, for $0 < \ell \leq k+1$, denote by $A(\ell)$ (resp. $\widehat{A(\ell)}$) the component of 0 in $P^{-M}(B(\ell))$ (resp. $P^{-M}(\widehat{B(\ell)})$). By ($\star$), $\widehat{A(\ell+1)} \subset A(\ell)$. By ($\star$) and ($\star\star$), the two components of $\widehat{A(\ell)} \setminus A(\ell)$ are regular intervals of order $N_{\ell+1}$. Note that $P^M(0)$ is the maximum of $P^M_{|\widehat{A(\ell)}}$.

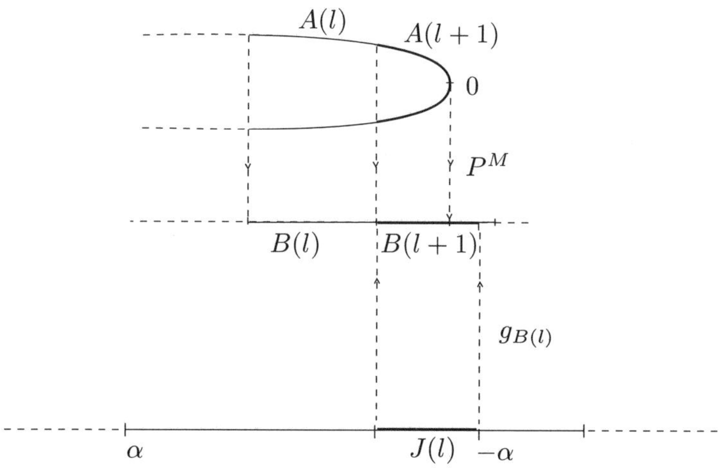

FIGURE 10

Let n be a (fixed) integer such that $N_k < n \leq N_{k+1}$. Then we can write

$$\mathcal{E}(n) = \bigsqcup_{0 \leq \ell < k} \mathcal{E}(n, \ell) \bigsqcup \mathcal{E}'(n),$$

where $\mathcal{E}(n,\ell)$ denotes the set of components that are contained in $A(\ell)$ but not in $\widehat{A(\ell+1)}$, and $\mathcal{E}'(n)$ is the set of components that are contained in $A(k)$. One then proves the following proposition:

PROPOSITION 2.2. *Let ℓ, n be such that $0 < \ell < k$ and $N_k < n \leq N_{k+1}$. Let $E \in \mathcal{E}(n,\ell)$.*

(i) *If $J(\ell)$ is simple and $g_{B(\ell)}(0)$ is to the left of $B(\ell+1)$, then $P^{N_\ell}(E)$ is a component of $\mathcal{E}(n - N_\ell)$.*

(ii) *If $J(\ell)$ is simple and $g_{B(\ell)}(0)$ is to the right of $B(\ell+1)$, then $P^{N_{\ell+1}}(E)$ is a component of $\mathcal{E}(n - N_{\ell+1} - 1)$.*

(iii) *In general, either $P^{N_\ell}(E)$ is a component of $\mathcal{E}(n - N_\ell)$ or there exists a regular interval J of order n_J, $0 < n_J < n_\ell$, such that $P^{N_\ell + n_J}(E)$ is a component of $\mathcal{E}(n - N_\ell - n_J)$. Moreover, the orders of such intervals J are distinct.*

From this proposition it is not hard to obtain the estimate

$$\sum_{E\in\mathcal{E}(n)} |E| \leq C\, 2^{-\theta n},$$

with a constant $\theta < 1/2$ that is close to $1/2$ when M is large. Indeed, this can be checked immediately for $n < M-1$. We have

$$\mathcal{E}(n) = \{[\widetilde{\alpha}^{(n)}, -\widetilde{\alpha}^{(n)}]\}.$$

By Proposition 2.2, one can now argue by induction on n, using the expansion given by the last inequality of 2.4.2.

2.4.4. *Structure in the parameter space.* The sequence (R_k) is decreasing, with $R_0 = (c^{(M)}, c^{(M-1)})$. Let us explain how to find the components of R_{k+1} contained in a given connected component of R_k.

Let $k \geq 0$, $c_0 \in R_k$. Then the intervals $B(\ell)$ and the integers N_ℓ are defined for P_{c_0} and $0 < \ell \leq k+1$.

Let I_k denote the component of c_0 in R_k; this is also the component of c_0 in $\{c : P_c^{N_{k+1}}(0) \notin \{\alpha, -\alpha\}\}$. Moreover, every preimage of α or $-\alpha$ of order $\leq N_{k+1}$ depends analytically on c in I_k. In particular, the endpoints of $J(\ell)$ (for $0 < \ell \leq k$) and of $B(\ell)$ (for $0 < \ell \leq k+1$) depend analytically on c in I_k. In contrast, for c in I_k, the integers n_ℓ (for $0 < \ell \leq k$) and N_ℓ (for $0 < \ell \leq k+1$) are independent of c.

If J is a regular interval for P_{c_0} of order $< N_{k+1}$, then the interval obtained by analytic continuation of the endpoints of J stays regular for every c in I_k. We set

$$n_{k+1}^{\star} = \max(M, 2^{-\sqrt{M}}(k+1))$$

and write

$$g = g_{[\alpha^{(M-1)}, \alpha^{(M-2)}]}.$$

One shows that if γ is an endpoint of an interval $J \in \mathcal{J}$ of order $< n_{k+1}^{\star}$ (in particular, if $\gamma = \alpha$ or $\gamma = -\alpha$), then

$$\left| \frac{\partial}{\partial c} g \circ g_{B(k+1)}(\gamma) - \frac{1}{3} \right| \leq C\, 2^{-M}$$

for every c in I_k. But it is also true that

$$\frac{\partial}{\partial c} g(P_c^M(0)) = \frac{\partial}{\partial c} P_c(0) = 1;$$

hence $c = P_c(0)$ ranges over $g(B(k+1))$ with a relative speed that is nonzero and essentially constant. To each $J \in \mathcal{J}$ of order $\leq n_{k+1}^{\star}$ there corresponds the interval $I_{k+1}(J) \subset I_k$, for which $P_c^M(0) \in g_{B(k+1)}(J)$. If c is in $I_{k+1}(J)$, then $T^{k+1}(0) \in J$ and hence $J = J(k+1)$.

The connected components of R_{k+1} contained in I_k are exactly those intervals $I_{k+1}(J)$ for which either $\operatorname{ord}(J) < M$ or

$$\Big(\sum_{0<\ell\leq k,\ n_\ell > M} n_\ell \Big) + \operatorname{ord}(J) \leq 2^{-\sqrt{M}}(k+1).$$

Finally, the measure estimate carries over to the parameter space.

For $0 < n \leq n_{k+1}^{\star}$, we have

$$\operatorname{meas}\Big(I_k \setminus \bigsqcup_{J\in\mathcal{J},\ 0<\operatorname{ord} J\leq n} I_{k+1}(J) \Big) \leq C'\, 2^{-\theta n} |I_k|.$$

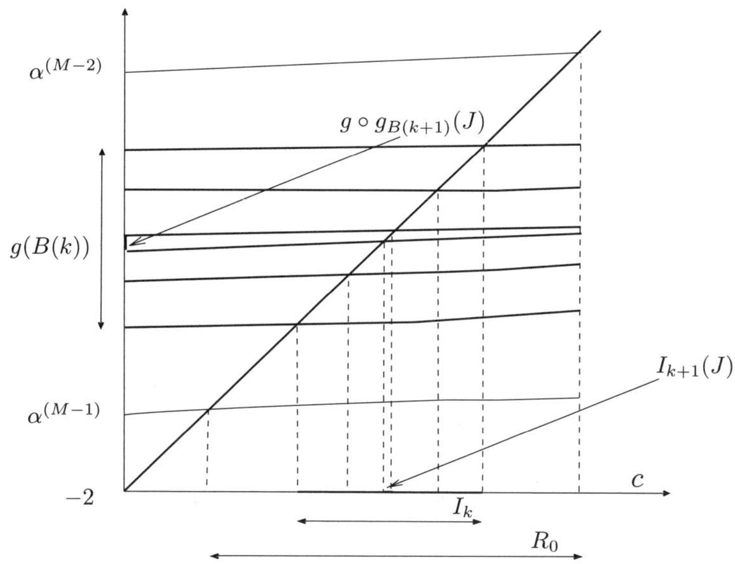

FIGURE 11

2.4.5. *The large deviation argument.* For $c \in R_0 = (c^{(M)}, c^{(M-1)})$, we define a sequence of *random variables* $X_k(c)$, $k > 0$, as follows. For $k \geq 0$:

- If $c \notin R_k$, then $X_{k+1}(c) = 0$;
- If $c \in R_k$, let I_k be its component in R_k.
  - If c is not in any of the intervals $I_{k+1}(J)$ (with $0 < \operatorname{ord}(J) \leq n^\star_{k+1}$) as above, we set $X_{k+1}(c) = n^\star_{k+1} + 1$.
  - If $c \in I_{k+1}(J)$, we set

$$X_{k+1}(c) = \begin{cases} 0 & \text{if } 0 < \operatorname{ord}(J) \leq M, \\ \operatorname{ord}(J) & \text{if } \operatorname{ord}(J) > M. \end{cases}$$

It is easy to see that R consists of exactly those c such that, for every $k > 0$,

$$\sum_1^k X_\ell(c) \leq 2^{-\sqrt{M}} k.$$

Moreover, the inequality

$$\operatorname{Prob}(X_{k+1} \geq n \,|X_1, \dots, X_k) \leq C'' 2^{-\theta n}$$

follows from 2.4.4. The left-hand side of this inequality indicates the probability that $X_{k+1} \geq n$, knowing that $X_1 = x_1, \dots, X_k = x_k$. Thus, if we set $S_k = \sum_1^k X_\ell$ and let Exp denote the expectation, we have

$$\operatorname{Exp}\Big(2^{\theta X_{k+1}/2} | X_1, \dots, X_k\Big) \leq 1 + C''' 2^{-\theta M/2}.$$

Hence

$$\operatorname{Exp}\Big(2^{\theta S_k/2}\Big) \leq \Big(1 + C''' 2^{-\theta M/2}\Big)^k$$

and

$$p_k = \operatorname{Prob}\Big(S_k \geq 2^{-\sqrt{M}} k\Big) \leq \frac{\mu^k - 1}{\lambda^k - 1},$$

where $\mu = 1 + C'''2^{-\theta M/2}$ and $\lambda = 2^{\theta 2^{-\sqrt{M}}/2}$.

Finally, for large M, we obtain

$$\sum_{k>0} p_k \leq 2^{-\theta M/3}$$

and

$$\operatorname{meas}\{c \in (c^{(M)}, c^{(M-1)}) : c \notin R\} \leq 2^{-\theta M/3} \left| c^{(M)} - c^{(M-1)} \right|.$$

3. Quasiperiodic Aspects

In this last section we present without proof some results on quasiperiodic dynamics, which apply in particular to quadratic polynomials. The first of these results shows that the only interesting univalent holomorphic dynamics in a neighborhood of a nontrivial (i.e. nonempty and not a single point) invariant connected compact set is the quasiperiodic dynamics.

3.1. Connected compact sets that are invariant under a univalent map. Let K be a nontrivial connected compact subset of $\mathbb{C}$, and let $\widetilde{F}$ be a map that is univalent (i.e. holomorphic and injective) in an open neighborhood V of K and satisfies $\widetilde{F}(K) = K$. We set $F = \widetilde{F}_{|K}$.

Let U be a connected component of $\mathbb{C} \setminus K$. Shrinking V if necessary, we may assume that $V \cap U$ is connected and $\widetilde{F}(V \cap U)$ is contained in a component U' of $\mathbb{C} \setminus K$. If U is bounded (resp. unbounded), let $h : \mathbb{D} = \{z : |z| < 1\} \to U$ (resp. $h : \mathbb{C} \setminus \overline{\mathbb{D}} \to U$) be a conformal representation.

When $U' = U$, the map $g = h^{-1} \circ \widetilde{F} \circ h$ extends by Schwarz reflection to an analytic circle diffeomorphism of $\mathcal{S}^1$, with rotation number $\rho(U, F)$.

The following theorem gives a classification of the different types of dynamical behavior of $(K, F = \widetilde{F}_{|K})$.

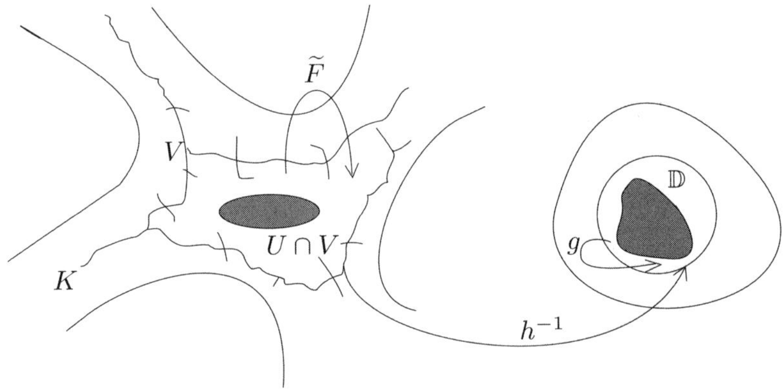

FIGURE 12

THEOREM 3.1. *With the hypotheses and notation above, exactly one of the following cases holds:*

(1) *F has finite order.*
(2) *F is of Morse-Smale type; i.e., F has finitely many periodic orbits, and the ω-limit and α-limit sets of each orbit are periodic orbits.*

(3) *K is full or annular, with irrational rotation number. The iterates of F accumulate at the identity (in the topology of uniform convergence on K).*

We give some comments on and supplementary results to this theorem, which is due to E. Risler [**17**] and also to R. Pérez Marco [**16**] for the full case.

(1) Recall that K is full if $\mathbb{C}\setminus K$ is connected, and annular if $\mathbb{C}\setminus K$ has two connected components.

(a) If K is full and $\rho = \rho(\mathbb{C}\setminus K, F)$ is irrational, then F has a fixed point with multiplier $\exp 2i\pi\rho$, and this point is the unique periodic point.

If the interior $\operatorname{int} K$ of K is nonempty, then $\operatorname{int} K$ is a Siegel disk, i.e. a simply connected domain on which F is conjugate to the rotation

$$z \longmapsto (\exp 2i\pi\rho)z.$$

(b) If K is annular and we are in case (3) of the theorem, then $\widetilde{F}$ preserves the two connected components U_{int}, U_{ext} (where U_{int} denotes the bounded component). In this situation,

$$\rho(U_{\text{int}}) = \rho(U_{\text{ext}}) \in \mathbb{R}\setminus\mathbb{Q},$$

and ρ is called the rotation number of (K, F). The map F has no periodic orbit.

If $\operatorname{int}(K) \neq \emptyset$, then $\operatorname{int} K$ is an annulus on which F is conjugate to the map $z \mapsto (\exp 2i\pi\rho)z$.

(c) Pérez Marco [**16**] has shown that in both cases above, the dynamics is quasiperiodic: (F^{q_n}) converges uniformly to the identity, where p_n/q_n is the nth convergent to ρ in its continued-fraction expansion.

When $\operatorname{int}(K) = \emptyset$, the topology of K becomes very complex; hence such K are called *hedgehogs* (cf. 3.4).

(2) Suppose that F is not of finite order and we are not in any of the preceding situations. In this case, F is of Morse-Smale type (cf. point (2) of Theorem 3.1). If a component U of $\mathbb{C}\setminus K$ is preserved by some iterate F^k of F, then $\rho(U, F^k)$ is rational. Furthermore, $\rho(U, F^k)$ is zero if K is not full or annular.

The periodic points are attracting, repelling, or parabolic. Possibly after conjugating $\widetilde{F}$ by a quasiconformal homeomorphism and enlarging K, we have

$$K = \{\text{periodic points of } F\} \cup \left(\bigcup_{i=1,\dots,r} \overline{P_i} \right),$$

where P_i is a petal that is invariant under an iterate F^k. That is, P_i is a connected open set with the following properties: $F^k(P_i) = P_i$; $F^k_{|P_i}$ has no fixed point; and there exist two points x_α and x_ω such that $\lim_{r\to+\infty} F^{rk}(x) = x_\omega$ and $\lim_{r\to-\infty} F^{rk}(x) = x_\alpha$ for every $x \in P_i$.

The petals P are of three distinct types:

- those for which $x_\omega \neq x_\alpha$,
- those for which $x_\omega = x_\alpha$ and $\overline{P}$ is full,
- those for which $x_\omega = x_\alpha$ and $\overline{P}$ is annular.

Figure 13 represents a dynamics in which each type of petal apppears.

Furthermore, there is a "Mather-Ecalle-Voronin"-type analytic classification of the pairs $(K, \widetilde{F})$.

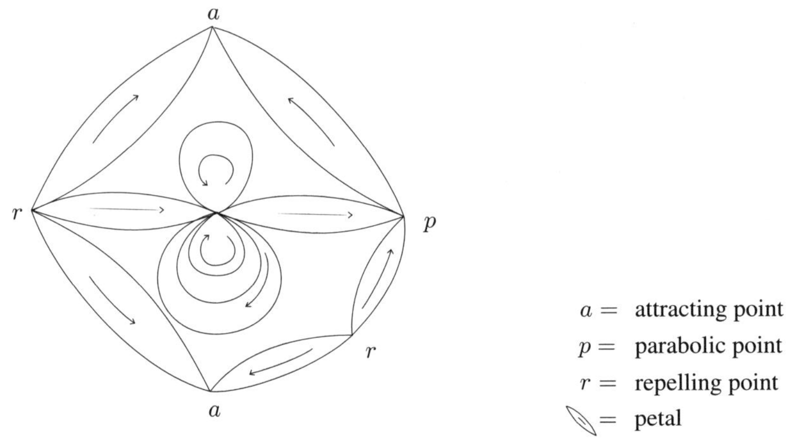

FIGURE 13

We will say nothing further about Morse-Smale type. The following subsections deal with case (3), separating the linearizable case (Subsections 3.2, 3.3) and the nonlinearizable case (hedgehogs, Subsection 3.4).

3.2. The Brjuno condition and the Brjuno function. Let $\alpha \in \mathbb{R}\backslash\mathbb{Q}$ have continued-fraction expansion

$$\alpha = a_0 + \cfrac{1}{a_1 + \cfrac{1}{a_2 + \cfrac{1}{\ddots}}}.$$

We set $\alpha_0 = \alpha$, $\beta_{-1} = 1$, $a_0 = p_0$, $q_0 = 1$, and, for $n > 0$, define the natural numbers α_n, p_n, q_n, β_n by

$$\alpha = a_0 + \cfrac{1}{a_1 + \cfrac{1}{\ddots + \cfrac{1}{a_n + \alpha_n}}} = \frac{p_n + p_{n-1}\alpha_n}{q_n + q_{n-1}\alpha_n}$$

$$\beta_n = |q_n\alpha - p_n| = \prod_{i=0}^{n} \alpha_i.$$

We also set $\beta_0 = \alpha_0$. The sequence (β_n) measures how well α is approximated by p_n/q_n.

The Brjuno numbers were defined in §1.3.2 by the condition

$$\phi(\alpha) = \sum_{n>0} q_n^{-1} \log q_{n+1} < \infty.$$

But rather than considering the function ϕ, we prefer to work with the Brjuno function, which maps $\mathbb{R} \setminus \mathbb{Q}$ to $\mathbb{R} \cup \{\infty\}$ and is defined by

$$B(\alpha) = \sum_{0}^{\infty} \beta_{i-1} \log \alpha_i^{-1}.$$

This function is periodic with period 1: $B(\alpha+1) = B(\alpha)$. For $0 < \alpha < 1$, it satisfies the functional equation

$$B(\alpha) = \log \alpha^{-1} + \alpha B(1/\alpha).$$

In fact, there exists a constant $C > 0$ such that

$$|B(\alpha) - \phi(\alpha)| < C,$$

so the set of Brjuno numbers is

$$\mathcal{B} = \{\alpha \in [0,1) : \mathcal{B}(\alpha) < +\infty\}.$$

Let $f(z) = (\exp 2i\pi\alpha)z + O(z^2)$ be a univalent function on $\mathbb{D}$. We denote by Δ_f the Siegel disk of f, i.e. the (possibly empty) maximal domain on which f is analytically conjugate to the rotation $z \mapsto (\exp 2i\pi\alpha)z$.

The Brjuno function provides an estimate of the size of the Siegel disk Δ_f as a function of α. More precisely, Brjuno [**2, 3**], Siegel [**19**], and Yoccoz [**22**] prove the following theorem.

THEOREM 3.2. *There exist $0 < C_0 < C_1$ (universal constants) such that the following conditions hold for every $\alpha \in \mathbb{R} \setminus \mathbb{Q}$:*

(1) *For $\alpha \in \mathcal{B}$,*
  (1.1) *if $f(z) = (\exp 2i\pi\alpha)z + O(z^2)$ is univalent in $\mathbb{D}$, then f is linearizable and $\Delta_f \supset D(0, C_0 \exp(-B(\alpha)))$;*
  (1.2) *there exists $f(z) = (\exp 2i\pi\alpha)z + O(z^2)$, univalent in $\mathbb{D}$ and quadratic-like, such that $\Delta_f \not\supset D(0, C_1 \exp(-B(\alpha)))$.*

(2) *For $\alpha \notin \mathcal{B}$, the polynomial $P : z \mapsto (\exp 2i\pi\alpha)z + z^2$ is not linearizable and $\Delta_P = \emptyset$.*

Analysis of the Brjuno function. This work was carried out jointly by Marmi, Moussa, and Yoccoz [**14**]. The starting point is the functional equation for B:

$$B(x) = -\log x + xB\left(\frac{1}{x}\right).$$

One is thus led to consider the set

$$\Sigma = \{f : \mathbb{R} \to \mathbb{R} : \ f \text{ is measurable and } f(x) = f(x+1)\}$$

and the operator $T : \Sigma \to \Sigma$ defined by $T(f)(x) = xf(1/x)$. The functional equation above can also be written as

$$(\mathrm{Id} - T)(B) = t,$$

where $t(x) = -\log x$ for $0 < x < 1$. For every $p \geq 1$, T acts on $L^p(\mathbb{T}) \cap \Sigma$, where $\mathbb{T} = \mathbb{R}/\mathbb{Z}$; its spectral radius is < 1, so $\mathrm{Id} - T$ is invertible, and we have

$$B = \sum_{n \geq 0} T^n(t).$$

Since the function t is not in $L^\infty(\mathbb{T})$, we cannot expect B to be in $L^\infty(\mathbb{T})$. On the other hand, we have the following inclusions:

$$L^\infty(\mathbb{T}) \subset \mathrm{BMO}(\mathbb{T}) \subset \bigcap_{p \geq 1} L^p(\mathbb{T}).$$

Recall that

$$\mathrm{BMO}(\mathbb{T}) = \left\{\varphi \in L^1(\mathbb{T}) : \sup_I \frac{1}{|I|} \int_I |\varphi - \varphi_I| dx < +\infty\right\},$$

where I ranges over the intervals in $\mathbb{T}$ and φ_I is the average of φ on I. Since the function t is in $\mathrm{BMO}(\mathbb{T})$, we immediately obtain assertion (a) of the following theorem:

THEOREM 3.3. *The function B satisfies the following conditions.*
(a) *B is in* $\mathrm{BMO}(\mathbb{T})$.
(b) *The harmonic conjugate of B is in $L^\infty(\mathbb{T})$.*

We sketch the proof of assertion (b). Consider the space

$$E = \{h : \overline{\mathbb{C}} \setminus [0,1] \to \overline{\mathbb{C}},\ \text{holomorphic},\ h(\infty) = 0\}.$$

Let $\mathcal{M}$ denote the monoid in $\mathrm{GL}(2,\mathbb{Z})$ consisting of the identity and matrices of the form $\begin{pmatrix} a & b \\ c & d \end{pmatrix}$ satisfying $0 \le a \le b \le d$ and $a \le c \le d$. For $g \in \mathcal{M}$, $h \in E$, the function

$$L_g(h) = (a - cz)\left[h\left(\frac{dz-b}{a-cz}\right) - h\left(-\frac{d}{c}\right)\right] - \frac{ad-bc}{c}h'\left(-\frac{d}{c}\right)$$

is in E. This defines an action $L : \mathcal{M} \times E \to E$

$$L(g,h) = L_g(h).$$

We now set $t_1(z) = -\frac{1}{\pi}\mathrm{Li}_2(1/z)$, where Li_2 is the dilogarithm:

$$\mathrm{Li}_2(z) = \sum_{n>0} \frac{z^n}{n^2}.$$

The function t_1 is in E. It can be shown that the series

$$B_1 = \sum_{g\in\mathcal{M}} L_g(t_1)$$

defines an element of E satisfying
- $B_1(\overline{z}) = \overline{B_1(z)}$;
- for $0 < x < 1$, $B(x) = \lim_{y\to 0+} \operatorname{Im} B_1(x+iy)$;
- $B_2(z) = \operatorname{Re} \sum_{n\in\mathbb{Z}} B_1(z+n)$ is bounded on $\{z : \operatorname{Im}(z) > 0\}$.

The case of the quadratic polynomial $Q_\lambda(z) = \lambda(z - z^2)$. The results obtained for Q_λ that we mention here are due to Yoccoz [**22**].

For $\lambda \in \mathbb{C}^*$, not a root of unity, let H_λ be the unique formal series of the form $H_\lambda(z) = z + a_2 z^2 + \cdots$ that satisfies

$$H_\lambda(\lambda z) = Q_\lambda(H_\lambda(z)).$$

Let $R(\lambda)$ be the radius of convergence of H_λ; H_λ extends continuously and injectively to $\{z : |z| \le R(\lambda)\}$.

If $|\lambda| < 1$, then 0 is an attracting fixed point of Q_λ and $R(\lambda)$ is nonzero. Set

$$U(\lambda) = \lim_{n\to\infty} \lambda^{-n} Q_\lambda^n(1/2).$$

The function $\lambda \to U(\lambda)$ is well defined, is holomorphic on $\mathbb{D}$, and satisfies
- $|U(\lambda)| = R(\lambda)$, $H_\lambda(U(\lambda)) = 1/2$;
- $U(\lambda)$ is the only singularity of H_λ in $\{z : |z| \le R(\lambda)\}$.

The case where λ has modulus 1 but is not a root of unity is clearly the most interesting but also the hardest. We point out that in this case

– $|U|$ has nontangential limit $R(\lambda)$ at λ.

3.2.1. *Some questions about the functions B and U.*

– For $\alpha \in \mathcal{B}$, does there exist $C' > 0$ such that

$$R(\exp 2i\pi\alpha) \leq C' \exp(-B(\alpha)) \ ?$$

In other words, is the function $\alpha \to \log|U(\exp 2i\pi\alpha)| + B(\alpha)$ in $L^\infty(\mathbb{T})$? Recall that $R(\exp 2i\pi\alpha) > C\exp(-B(\alpha))$. Moreover,

$$R(\exp 2i\pi\alpha) \leq C(\varepsilon)\exp(-(1-\varepsilon)B(\alpha))$$

for every $\varepsilon > 0$.

– For which α does $\operatorname{Arg} U$ have a limit at $\lambda = \exp(2i\pi\alpha)$? When this limit exists, $U(\lambda)$ is defined. Is it still true that $H_\lambda(U(\lambda)) = 1/2$? We expect a positive answer to these questions. By a famous theorem of Fatou [**4**], it is true almost everywhere. This would imply in particular that the critical point $c = 1/2$ is on the boundary of the Siegel disk.

– Is $\operatorname{Arg} U$ bounded in $\mathbb{D}$?

The function $\log U = \log|U| + i\operatorname{Arg} U$ is a dynamical analogue of the function B_2. In this dictionary, the counterpart of the boundedness of B_2 should be the boundedness of $\operatorname{Arg} U$. On the other hand, $\operatorname{Re} U$ could be positive, so that $\log|U|$ would be in $\operatorname{BMO}(\mathbb{T})$.

3.3. The boundary of the Siegel disk. For $\alpha \in \mathbb{R}\backslash\mathbb{Q}$, consider the sequence

$$\Delta_0(\alpha) = 0, \ldots, \Delta_{n+1}(\alpha) = \delta_{\log \alpha_n^{-1}}(\Delta_n(\alpha)),$$

where $\delta_u(t)$ is the function defined by

$$\delta_u(t) = \begin{cases} e^t & \text{if } t \leq u, \\ e^u(t-u+1) & \text{if } t \geq u. \end{cases}$$

Set

$$\mathcal{H}_0 = \{\alpha : \exists\, n_0(\alpha) : \Delta_n(\alpha) \geq \log \alpha_n^{-1} \quad \forall n \geq n_0(\alpha)\}.$$

The elements of $\mathcal{H}_0$ are those for which the growth of $\log \alpha_n^{-1}$ is controlled by $\Delta_n(\alpha)$, so that α_n is not too small; these are also the α that are not very well approximated by rationals. Set

$$\mathcal{H} = \{\alpha : \alpha_n \in \mathcal{H}_0 \text{ for every } n \geq 0\}.$$

In fact,

$$\mathcal{H} \subset \mathcal{H}_0 \subset \mathcal{B},$$

and these sets are all distinct.

Consider the action of $\operatorname{PGL}_2(\mathbb{Z})$ on $\mathbb{R}\setminus\mathbb{Q}$ via the generators $\alpha \mapsto \alpha + 1$ and $\alpha \mapsto 1/\alpha$, which produce the continued-fraction expansion of α. We see that $\mathcal{B}$ is the set of α whose orbits intersect $\mathcal{H}_0$ and that $\mathcal{H}$ is the set of α whose orbits are contained in $\mathcal{H}_0$.

The following theorem of Yoccoz [**22**] characterizes the elements of $\mathcal{H}$.

THEOREM 3.4. *$\alpha \in \mathcal{H}$ if and only if every $f \in \operatorname{Diff}^\omega_+(\mathbb{T})$ with rotation number α is analytically linearizable.*

We return to quadratic polynomials. The following theorem of Herman [**8**] gives, in the linearizable case, a (partial) description of the boundary of the Siegel disk.

THEOREM 3.5. *Suppose $\alpha \in \mathcal{B}$. Let S be the boundary of the Siegel disk of $Q_\alpha(z) = (\exp 2i\pi\alpha)(z - z^2)$, and let $c = 1/2$ be its critical point.*

(1) *If $\alpha \in \mathcal{H}$, then $c \in S$.*
(2) *There exists $\alpha \in \mathcal{B} \setminus \mathcal{H}$ such that S is a quasicircle not containing c.*
(3) *α is of constant type (i.e., there exists $C > 0$ such that $\alpha_i \leq C$ for every i) if and only if S is a quasicircle containing c.*
(4) *The set of α for which S is a Jordan curve containing c is of full measure.*

REMARKS.

(a) Assertion (4) was announced by Herman and Yoccoz.
(b) The set of α of constant type is a proper subset of $\mathcal{H}_0$.
(c) The proof of (3) uses the results of Swiatek [**21**].

The following are natural questions if one wants to be able to complete this description of S:

QUESTIONS

- Is the converse of (1) true, so that $\alpha \in \mathcal{H}$ if and only if $c \in S$?
- Is S always a Jordan curve?

3.4. Hedgehogs. These objects were studied by R. Pérez Marco [**15**], and the results cited in this subsection are due to him.

We return to the study of case (3) of the classification of compact sets that are invariant under a univalent map, given in Theorem 3.1.

First we consider the case of a full compact set. Let $\alpha \in \mathbb{R} \setminus \mathbb{Q}$, and let f be univalent in a neighborhood of $\overline{\mathbb{D}} = \{z : |z| \leq 1\}$ and of the form

$$f(z) = (\exp 2i\pi\alpha)z + O(z^2).$$

For $0 < r \leq 1$, we set $\overline{\mathbb{D}}_r = \{z : |z| \leq r\}$ and denote by K_r the connected component containing 0 of the maximal invariant compact set $\cap_{n\in\mathbb{Z}} f^{-n}(\overline{\mathbb{D}}_r)$. This is a full compact set that intersects the boundary of $\overline{\mathbb{D}}_r$.

The rotation number $\rho(\mathbb{C} \setminus K_r, f)$ is independent of r; indeed, it is the rotation number of the circle diffeomorphism derived from f by a conformal representation of $\mathbb{C} \setminus K_r$ (cf. 3.1), which also equals $\rho(\mathbb{C} \setminus K_1, f)$ by point (1.b) following Theorem 3.1. Since $\rho(\mathbb{C} \setminus K_r, f)$ approaches α as r approaches 0, it equals α for all $r \in [0, 1]$.

R. Pérez Marco proves the following results:

- If $K \subset \overline{\mathbb{D}}$ is a nonempty, full, invariant compact set, then there exists a number r, $0 \leq r \leq 1$, such that $K = K_r$ (we set $K_0 = \{0\}$).
- If $\operatorname{int} K_1 = \emptyset$, then $f_{|K_1}$ is uniquely ergodic; the unique invariant measure is the Dirac measure at 0.
- Let p_n/q_n be the nth convergent to α. Then the sequence $f^{q_n}_{|K_1}$ converges uniformly to $\mathrm{Id}_{|K_1}$; the subgroup of iterates of $f_{|K_1}$ in the group of homeomorphisms of K_1 is not discrete; and the centralizer of $f_{|K_1}$ is not countable.

We now briefly consider the annular case. Let K_0 be an annular compact set surrounding 0, and let h_{int} (resp. h_{ext}) be the conformal representation of $\overline{\mathbb{D}}$ (resp. $\mathbb{C} \setminus \overline{\mathbb{D}}$) onto the bounded (resp. unbounded) component of $\mathbb{C} \setminus K_0$ that fixes 0 (resp. ∞).

For $R_0 \leq 1 \leq R_1$, we set

$$\overline{A}_{R_0,R_1} = \mathbb{C} \setminus [\, h_{\text{int}}(\{z : |z| < R_0\}) \cup h_{\text{ext}}(\{z : |z| > R_1\})\,] .$$

Let f be a univalent map in a neighborhood of an annulus $\overline{A}_{R_0,R_1}$ that preserves K_0 and each component of $\mathbb{C}\setminus K_0$. Suppose the rotation number α of $f_{|K_0}$ is irrational.

For $R_0 \leq r_0 \leq 1 \leq r_1 \leq R_1$, we denote by K_{r_0,r_1} the connected component of K_0 in the maximal invariant compact set $\bigcap_{n\in\mathbb{Z}} f^{-n}(\overline{A}_{r_0,r_1})$. This is an annular compact set that intersects the two boundary components of $\overline{A}_{r_0,r_1}$. As before, the rotation number of K_{r_0,r_1} equals α. Every invariant annular connected compact set that contains K_0 and is contained in $\overline{A}_{R_0,R_1}$ is one of the K_{r_0,r_1}. The sequence of iterates $f^{q_n}_{|K_{r_0,r_1}}$ converges uniformly to $\mathrm{Id}_{|K_{r_0,r_1}}$.

Various possibilities for the compact sets K_r and K_{r_0,r_1} are represented in Figure 14:

- If $\alpha \in \mathcal{H}$, then K_r is a disk whose boundary is an analytic curve; K_{r_0,r_1} is an annulus whose two boundary components are analytic curves.
- If $\alpha \in \mathcal{B}$, then the interior of K_r contains 0 when $r > 0$. In the annular case, the interior of K_{r_0,r_1} contains an annular open set surrounding 0 if f is sufficiently close to a rotation.

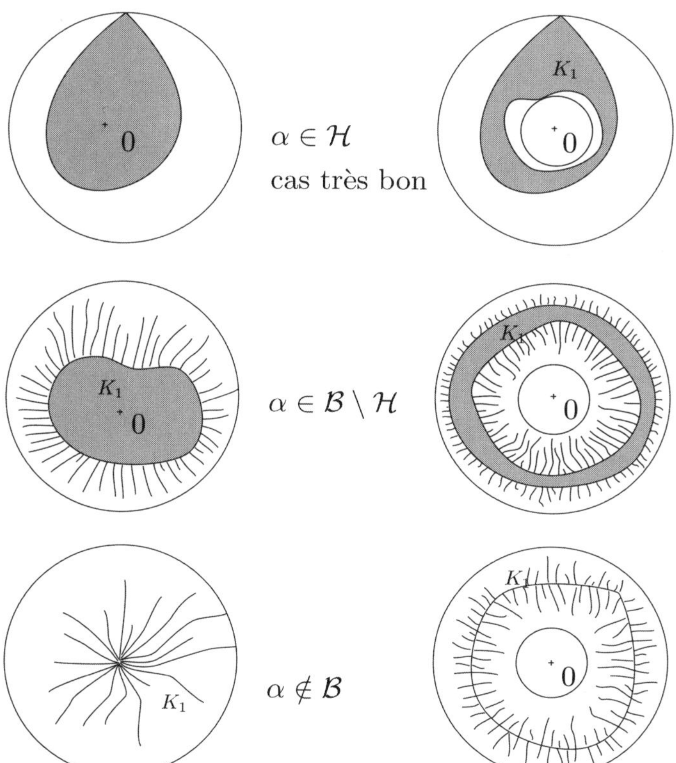

FIGURE 14

3.5. Herman rings. Let $\mathbf{Rat}_d$ be the set of rational maps of degree d. If V is the subvariety of reducible rational maps, then $\mathbf{Rat}_d = \mathbb{CP}^{2d+1} \setminus V$. For fixed $\alpha \in \mathbb{R}\setminus\mathbb{Q}$, let $W(\alpha)$ be the subset of $\mathbf{Rat}_d$ consisting of those rational maps R that have a Herman ring of period 1 and rotation number α.

The following theorem is due in part to M. Herman [**9**] and to E. Risler [**18**].

THEOREM 3.6. *If $\alpha \in \mathcal{B}$, then $W(\alpha)$ is a complex analytic hypersurface.*

Little is actually known about $W(\alpha)$; it is not an algebraic hypersurface. We mention some questions to which we would like to have answers, and for which there are partial answers.

QUESTIONS

(1) Is $W(\alpha)$ nonsingular?
It is easy to see that $W(\alpha)$ is nonsingular at a point that admits a unique Herman ring with rotation number α, when the modulus of this ring is large.
(2) What is the dependence of $W(\alpha)$ on α as α ranges over $\mathcal{B}$?
When α ranges over certain compact subsets of $\mathcal{B}$ defined by uniform arithmetic conditions on α, the $W(\alpha)$ depend in a $\mathcal{C}^{\infty}$ way on α and hence form a partial foliation.
(3) What can be said about the closure $\overline{W(\alpha)}$ in $\mathbb{CP}^{2d+1}$?
(4) What happens when α leaves $\mathcal{B}$?

Bibliography

[1] L. Ahlfors and L. Bers, *Riemann's mapping theorem for variable metrics*, Ann. of Math. (2) **72** (1960), 385–404.

[2] A. Brjuno, *Analytical form of differential equations*, Trans. Moscow Math. Soc. **25** (1971), 131–288.

[3] ———, *Analytical form of differential equations*, Trans. Moscow Math. Soc. **26** (1972), 199–239.

[4] L. Carleson and T. W. Gamelin, *Complex dynamics*, Springer-Verlag, New York, 1993.

[5] A. Douady and J. H. Hubbard, *Étude dynamique des polynômes complexes*, Publications Mathématiques d'Orsay, 1984.

[6] ———, *On the dynamics of polynomial-like mappings*, Ann. Sci. École Norm. Sup. (4) **18** (1985), 287–343.

[7] J. Graczyk and G. Światek, *Generic hyperbolicity in the logistic family*, Ann. of Math. (2) **146** (1997), 1–52.

[8] M.-R. Herman, *Are there critical points on the boundaries of singular domains?*, Comm. Math. Phys. **99** (1985), 593–612.

[9] ———, *Simple proofs of local conjugacy theorems for diffeomorphisms of the circle with almost every rotation number*, Bol. Soc. Brasil. Mat. **16** (1985), 45–83.

[10] J. H. Hubbard, *Local connnectivity of Julia sets and bifurcation loci: three theorems of J.-C. Yoccoz*, Topological methods in modern mathematics (Stony Brook, NY, 1991), Publish or Perish, Houston, TX, 1993, pp. 467–511.

[11] M. V. Jakobson, *Absolutely continuous invariant measures for one-parameter families of one-dimensional maps*, Comm. Math. Phys. **81** (1981), 39–88.

[12] M. Lyubich, manuscript, 1993.

[13] R. Mañé, P. Sad, and D. Sullivan, *On the dynamics of rational maps*, Ann. Sci. École Norm. Sup. (4) **16** (1983), 193–217.

[14] S. Marmi, P. Moussa, and J.-C. Yoccoz, *The Brjuno functions and their regularity properties*, Comm. Math. Phys. **186** (1997), 265–293.

[15] R. Pérez Marco, *Topology of Julia sets and hedgehogs*, Publications Mathématiques d'Orsay, vol. 94–48 (1994).

[16] ———, *Classification dynamique des continua pleins invariants par un difféomorphisme holomorphe*, preprint, 1996.

[17] E. Risler, *Linéarisation des perturbations holomorphes des rotations*, Thesis, École polytechnique, Palaiseau, 1996.

[18] ———, *Linéarisation des perturbations holomorphes des rotations et applications*, Mémoires, vol. 77, Société mathématique de France, 1999.

[19] C. L. Siegel, *Iteration of analytic functions*, Ann. of Math. (2) **43** (1942), 607–612.

[20] D. Sullivan, *Quasiconformal homeomorphisms and dynamics. I. Solution of the Fatou-Julia problem on wandering domains*, Ann. of Math. (2) **122** (1985), 401–418.

[21] G. Świątek, *Rational rotation numbers for maps of the circle*, Comm. Math. Phys. **119** (1988), 109–128.

[22] J.-C. Yoccoz, *Petits diviseurs en dimension* 1, Astérisque, vol. 231, Société mathématique de France, 1995.

[23] ———, *Jakobson's theorem*, notes from a course given at ETH (Zurich), 1996–1997.

Titles in This Series